Intermediate GNVQ Engineering

Intermediate GNVQ Engineering

Colin Chapman
Alan Darbyshire
Roger Timings

LONGMAN

Addison Wesley Longman Limited
Edinburgh Gate, Harlow
Essex CM20 2JE, England
and Associated Companies throughout the world

First published 1997

British Library Cataloguing-in-Publication Data
A catalogue entry for this title is available from the British Library.

ISBN 0-582-29089-9

Set by 32 in 9½/11 Sabon
Printed in Great Britain by Henry Ling Ltd, at the Dorset Press, Dorchester, Dorset

Contents

Preface **vii**
Acknowledgements **viii**

Unit 1 Engineering Materials and Processes 1
Alan Darbyshire
1.1 Selection of materials and components for engineered products **2**
1.2 Selection of processes to make engineered products **16**
1.3 Production of an engineered product to specification **40**
1.4 Unit test **46**

Unit 2 Graphical Communication in Engineering 49
Colin Chapman
2.1 Sketching, drawing and CAD for conceptualising and communicating ideas **50**
2.2 Diagrams and charts for communicating information **55**
2.3 Schematic and circuit diagrams **61**
2.4 Engineering drawing **67**
2.5 Unit test **81**

Unit 3 Science and Mathematics for Engineering 83
Alan Darbyshire
3.1 Physical quantities and their units **84**
3.2 Mathematical techniques for manipulating and evaluating data **86**
3.3 Static systems **91**
3.4 Dynamic systems **98**
3.5 Thermal systems **104**
3.6 Electrical systems **107**
3.7 Measurement of physical quantities **113**
3.8 Unit test **122**

Unit 4 Engineering in the Society and the Environment 125
Roger Timings
4.1 The application of engineering technology in society **126**
4.2 Career options and paths in engineering **143**
4.3 Unit test **154**

Answers to numerical self-assessment tasks **157**
Answers to unit tests **158**
Index **159**

Preface

Qualifications for technician occupations associated with engineering are undergoing development by the National Council for Vocational Qualifications. This has resulted in a qualification known as the General National Vocational Qualification: Engineering. The General National Vocational Qualification (GNVQ) replaces the former BTEC First Diploma and National Diploma awards, and is available at Foundation, Intermediate and Advanced levels.

The GNVQ qualifications are awarded by the City & Guilds of London Institute, the Royal Society of Arts (RSA) and EdExcel (formerly the Business and Technology Education Council). GNVQs consist of mandatory, key skills, optional and additional units; the syllabi for the mandatory and key skill units are common to all three awarding bodies. This textbook has been designed to meet the revised requirements of the three bodies for the four mandatory units of the Intermediate award.

While the text has been primarily designed to satisfy the requirements of the Intermediate GNVQ, it will also be useful reference for the Foundation and Advanced awards, plus the relevant optional and additional units, and National Vocational Qualifications (NVQs).

The book has been written in units which follow the order and titles of the four mandatory units of the Intermediate award. These are:

- Engineering Materials and Processes
- Graphical Communication in Engineering
- Science and Mathematics for Engineering
- Engineering in Society and the Environment

Each **unit** is divided into sections which generally reflect the various **elements** of the units. These are then further sub-divided into more specific topics which address the requirements of the **performance criteria** and their associated **range statements**. Key words and phrases related to these are highlighted throughout the book.

Spread throughout each unit are **self-assessment tasks** designed to encourage the reader to reinforce their learning, some of which integrate the key skills of Communication, Application of Number and Information Technology. At the end of each unit are sample **multiple-choice questions** which provide examples of the questions which the student might meet in the **external mandatory tests** for the units.

The presentation of the information in the individual units has not always slavishly followed the exact order of the unit specifications. This has been done to allow the text to be presented within a more logical structure and allow relevant and complementing issues to be placed together for the reader's benefit. It is hoped, however, that the careful organization and sign-posting of information, together with the comprehensive index, will allow students to easily satisfy the requirements of each unit.

The book is not meant to be an exemplar of engineering and associated information, but rather to indicate the basic approaches which may be adopted to fulfil the various requirements. Where dimensions are stated, they are intended to give an idea of scale rather than be prescriptive in meeting particular requirements.

The authors, in writing this book, have been conscious of the need to reflect the philosophy of the GNVQ awards and realise that there will be important omissions apparent to the informed reader. These omissions have been made in order to reduce any possible confusion in the student due to the inclusion of material not required by the various syllabi. The authors and publishers, however, would be pleased to receive any constructive comments or suggestions that may be incorporated into future revisions.

Acknowledgements

The authors and publishers are grateful to the following individuals and organisations who have provided help and materials during the preparation of this text, and have given permission to reproduce copyright material:

GEC Alsthom plc; RJB Mining (UK) Ltd; Severn Trent; Tom Morgan (De Aston School, Market Rasen); West Midlands Regional Waste Disposal Authority.

Dave Eastbury for our Fig. 2.9; Crocodile Clips Limited for our Fig. 2.31; TecQuipment Limited for our Figs 2.33–2.35 & 2.58–2.61; Butterworth Heinmann for our Fig. 4.3; Rio Tinto plc for our Figs 4.4 & 4.5; PowerGen plc for our Fig. 4.6 (from *Environmental Performance Report 1995*); British Gas plc for our Fig. 4.8 (from *Continuing to care for the environment*); ICI plc for our Figs 4.9 & 4.10 (from *Environmental Performance 1990/1995*); British Nuclear Fuels Limited for our Fig. 4.11; Biffa Waste Services Limited for our Figs 4.12 & 4.13 (from *Waste: Somebody Else's Problem? Our Choices and Responsibilities*); West Midlands Industrial Club for our Fig. 4.15; EnTra Publications Limited for our Fig. 4.16; City of Birmingham Careers Service for our Figs 4.17–22.

Extracts from British Standards Institution publications (our Figs 2.29, 2.37, 2.43, 2.45–2.47, 2.49, 2.50 & 2.53–2.57 from PP7307:1989 *Graphical Symbols for use in Schools and Colleges* and PP7308:1986 *Engineering Drawing Practice for Schools and Colleges*) are reproduced with the permission of BSI. Complete copies can be obtained by post from BSI Customer Services, 389 Chiswick High Road, London W4 4AL; Telephone: 0181 996 7000.

Engineering Materials and Processes

Alan Darbyshire

Engineered products are important to the nation's economy. They include motor vehicles, machine tools, aircraft, ships, agricultural equipment, electrical generation and supply equipment, domestic appliances and electronic goods.

A successful engineered product must be fit for its intended purpose. This means that it must do the job for which it was designed with a minimum of maintenance and for a reasonable period of time. To be successful, an engineered product must also have a selling price which is competitive. To help achieve these goals, design and production engineers must select the most appropriate materials, components and production processes. For a product to be successful, a sound knowledge of these areas is thus essential.

The key areas covered by this unit are:

- Selection of materials and components for engineered products.
- Selection of processes needed to make electro-mechanical engineered products.
- Production of electro-mechanical products to specification.

After reading this unit the student will be able to:

- Identify possible materials and components for an engineered product and select those which are most suitable.
- Identify and select the most suitable processes and process sequence required to make an engineered product.
- Identify the techniques, tools, equipment, safety procedures and safety equipment relevant to the selected processes.
- Carry out the processes to produce an engineered product to specification using the relevant safety procedures and equipment.
- Maintain tools, equipment and the working area in good order during and after processing.

1.1 Selection of materials and components for engineered products

Some engineered products are made in one piece from a single material. Examples include nuts, bolts, pipes, tubes, castings and moulded plastic products. These in turn may be assembled into larger and more complex products for which they are the component parts. Motor cars, washing machines, hi-fi equipment and computers are complex engineered products which are assembled from components.

The materials and components selected for engineered products must be fit for their intended purpose. This means that they must do their job reliably and be suitable for their service conditions. As far as possible they should be chosen from readily available materials and components. This is to ensure a reliable supply and to keep their cost as low as possible.

Topics covered in this section are:

- Identification of possible materials for a mechanical product.
- Selection of suitable materials for a mechanical product.
- Selection of suitable mechanical components for a mechanical product.
- Identification of possible electrical and electronic components for an electrical product.
- Selection of suitable electrical and electronic components for an electrical product.
- Selection of suitable materials and components for an electromechanical product.

Identification of possible materials for a mechanical product

The main types of material used in mechanical products are:

- Metals
- Polymer materials
- Ceramics
- Composite materials

Metals

Metals are widely used in engineered products. Sometimes they are used in an almost pure form but more often they are mixed with other metals and non-metallic elements to improve their properties. Such a mixture is called an alloy. Some of the metals used in engineering have been known since ancient times. In addition to the precious metals gold and silver, the base metals such as copper, tin, lead and iron have a long history of use. Metals may be subdivided into:

- Ferrous metals
- Non-ferrous metals

Ferrous metals

These include iron and alloys in which iron is the chief constituent. The name is derived from the Latin word for iron which is 'ferrum'. Iron is a soft grey metal whose softness makes it unsuitable for use as a structural material. It does not machine easily to a good surface finish and, when molten, it tends to be pasty, making it difficult to cast. As a result pure iron is seldom used except in laboratory experiments.

If small amounts of the non-metallic element carbon are added to iron its properties are greatly improved. When the resulting alloy contains up to 1.4% of carbon it is called 'Plain Carbon Steel' and when it contains between 3.2% and 3.5% carbon it is known as 'Cast Iron'. The different compositions of the alloy have their own special properties and uses. They are classified as:

- Dead mild steel (0.1–0.15% carbon content) – This is stronger than pure iron but remains quite ductile and malleable. It is used for pressed components such as motor body panels. It is drawn out to make steel wire, rod and tube and it is also used for rivets.
- Mild steel (0.15–0.3% carbon content) – This is the most common structural material in use. It is supplied in the form of bar, sheet, plate, tube and girders. It is widely used as a general workshop and building material because it is easy to machine and has a fairly high tensile strength. When hot, it is easy to form to shape by forging, rolling and drawing. It may also have its surface hardness increased by a heat treatment process known as case hardening.
- Medium carbon steel (0.3–0.8% carbon content) – This is stronger and tougher than mild steel. It is widely used for forged components which need to have good impact resistance. Its hardness and toughness may be further improved by heat treatment.
- High carbon steel (0.8–1.4% carbon content) – This is the hardest and most wear resistant of the plain carbon steels and, like medium carbon steel, its properties may be further improved by heat treatment. It is used for cutting tools such as wood chisels, files, screw cutting taps and dies and for other sharp-edged hand tools such as trimming knives.
- Grey cast iron (3.2–3.5% carbon content) – When carbon is added to iron in amounts of more than about 1.7% content, it is found that, after the metal has cooled down, flakes of carbon in the form of graphite can be seen when it is viewed under a microscope. The alloy is then called 'grey cast iron' and the high carbon content has several effects on its properties. Compared to the plain carbon steels, it is weak in tension but remains strong in compression. When molten, it is very fluid compared to pure iron and steel. This makes it easy to cast into intricately shaped components. A further benefit of the free graphite flakes is that they make the metal self-lubricating. This allows it to be easily machined without a cutting fluid. The free graphite flakes also enable the metal to absorb energy from vibrations. This makes grey cast iron an ideal material for machine beds and frames. Indeed grey cast iron is used for a wide range of engineering components, which are intricately shaped but not loaded too heavily in tension.

Plain carbon steel is an alloy of iron and carbon. When other elements are added to improve the properties of the metal, the resulting mixture is known as an alloy steel. Two of the most common alloy steels are:

- Stainless steel – In addition to iron and carbon, this contains chromium and nickel which are added to give improved corrosion resistance. Stainless steel is widely used in the food and drink industries and for medical and surgical equipment.
- High-speed steel – This alloy steel contains tungsten, chromium and vanadium which enable the material to maintain its hardness and sharp cutting edge at high cutting speeds. High-speed steel is widely used for drills, lathe tools and milling cutters.

Table 1.1 Ferrous metals

Ferrous metal	Tensile strength (N/mm^2)	Melting point (°C)	Density (kg/m^3)
Mild steel	500	1500	7800
Medium carbon steel	750	1450	7800
High carbon steel	900	1400	7800
Grey cast iron	200	1200	7200
Stainless steel	600	1500	7900
High-speed steel	—	1400	7900

Non-ferrous metals

These are metals which contain little or no iron. Iron is sometimes added as an alloying element to improve the properties of these metals but only in very small amounts. The most common non-ferrous base metals used in engineering are:

- Copper – This is one of the oldest non-ferrous metals to be used by man. Copper is a malleable and ductile material, it is corrosion resistant and is an excellent conductor of heat and electricity. Copper has fairly good tensile strength and is easily joined by soldering and brazing.
- Zinc – This is a soft but rather brittle material and, like most non-ferrous metals, it is corrosion resistant. Its main applications are as a protective coating for mild steel products and for alloying with copper to form the range of non-ferrous alloys known as brasses.
- Tin – This is a soft metal which is very malleable and highly corrosion resistant. It is used as a protective coating for the mild steel sheet used in food canning and for alloying with copper to form the range of non-ferrous alloys known as tin-bronzes. Tin is also mixed with lead to form the range of alloys known as soft solders.
- Aluminium – This is the lightest of the non-ferrous metals in common use. It is very malleable and ductile and a good conductor of both heat and electricity. It is also corrosion resistant but in its pure form it has a low tensile strength. It is widely used in the overhead transmission lines of the electricity supply grid. These have a core of high tensile steel cable, to carry the weight of the spans, surrounded by aluminium conductors to carry the electricity. Aluminium is alloyed with copper to give the range of non-ferrous alloys known as aluminium bronzes and also with smaller quantities of copper, silicon, iron, manganese, magnesium and other elements to give a specialist range of aluminium alloys.

Table 1.2 Non-ferrous metals

Non-ferrous metals	Tensile strength (N/mm^2)	Melting point (°C)	Density (kg/m^3)
Copper	232	1083	8900
Tin	15	230	7300
Zinc	200	420	7100
Aluminium	93	660	2700

There are a great many non-ferrous alloys used in engineering. Some of the most common types are:

- Brasses – Copper and zinc together with small amounts of tin and lead are mixed in varying proportions to give the range of alloys called brasses. As a general rule brasses with higher copper content are the more ductile. They are used for cold forming operations such as pressing or drawing. The brasses with the higher zinc content are less ductile and are used for hot-forming operations such as extrusion, hot stamping and casting.
- Tin-bronzes – Copper and tin, together with small amounts of phosphorus, zinc and lead, are mixed together in varying proportions to give the range of non-ferrous alloys called tin-bronzes. As a general rule, the bronzes with the higher copper content are the most ductile and are used for cold-forming operations. Those with the higher tin content are the least ductile and are used for casting.
- Aluminium-bronzes – Copper is mixed with small amounts of aluminium, nickel, iron and manganese to give the range of non-ferrous alloys called aluminium-bronzes. Although they are more expensive than tin-bronzes they have better corrosion resistance at high temperatures.
- Cupro-nickel alloys – Copper is mixed with nickel and small quantities of iron and manganese to give the range of non-ferrous alloys known as cupro-nickels. These are very strong and have good corrosion resistance. They have many engineering uses and they are also used for 'silver' coinage.
- Aluminium alloys – A wide range of aluminium alloys has been developed for applications where good strength, low weight and corrosion resistance are required. Some have been designed for casting and other for cold-forming processes. In some cases the properties of the alloy can be further improved by heat treatment. One of the best-known aluminium alloys is 'Duralumin' which is widely used in aircraft production.

Self-assessment tasks

1. Why is pure iron not used as a structural material?
2. What are the constituents of plain carbon steel?
3. What grade of plain carbon steel is most widely used for the girders in buildings and for general workshop use?
4. What grade of plain carbon steel is most widely used for wood chisels and other sharp-edged hand tools?
5. Grey cast iron contains carbon in the form of graphite flakes. Name two ways in which this benefits the material.
6. How do the tensile and compressive strengths of cast iron compare?
7. What kind of alloy steel is used for twist drills and milling cutters?
8. What are the properties of copper which make it suitable for use in plumbing and central heating systems?

9. Name two applications of zinc as an engineering material?
10. What are the properties of aluminium which make it suitable for use in overhead electricity transmission lines?
11. What are the main constituents of brass?
12. Which metal is alloyed with tin to form tin-bronzes?
13. Name a common use for cupro-nickel alloy.
14. What kind of an alloy is 'Duralumin', and where is it widely used?

Polymer materials

These include the different types of plastic and rubber. Their structure consists of long intertwined chains of molecules known as polymers. The name, plastics, is a little misleading, since at normal temperatures most plastics are quite elastic. It refers to their condition when they are being moulded into shape. At this stage the raw material may be in the form of a resin or it may be in the form of a powder or granules which are heated into liquid form before moulding.

The different types of plastic and rubber have widely different properties but they are all good electrical insulators, they have low weight compared to metals and they are resistant to many of the chemicals which cause metals to corrode. Polymer materials may be classified into:

- Thermoplastics
- Thermo-setting plastics
- Rubbers

Thermoplastics

These are less rigid than thermo-setting plastics and can be softened and remoulded by heating. Unfortunately some degradation occurs if the material is reheated too often or is overheated. Scrap components which are recycled should be remoulded into low specification products.

There are many thermoplastics in everyday use as engineering materials. Some of the most common types are:

- Polyalkenes – These include polythene which is used for wrapping and packaging, polypropylene which has similar uses but greater strength, polystyrene which is easily moulded into a variety of products and polyvinyl chloride (PVC) which is tough, easily moulded and widely used to insulate electric wiring and cable.
- Acrylics – These include Perspex which is transparent, tough and can easily be moulded into shape.
- Fluorine plastics – These include polytetrafluoroethylene (PTFE). This has an extremely low coefficient of friction. It is used as a coating for bearing surfaces and also as a coating for non-stick cooking utensils.
- Polyamides – These include nylon which is tough and has a low coefficient of friction. It can be drawn out into strong fibres or moulded into components such as gears and bearings.
- Polyesters – These include terylene which is strong and has good electrical insulation properties. They can be drawn out into fibres and they are also available as film. Polyester fabric is used as a reinforcement for rubber in drive belts, hoses and tyres.

Thermo-setting plastics

These differ from thermoplastics in that when they are moulded, a chemical reaction occurs. Cross-links are formed between the polymer chains which cannot be broken by reheating. The cross-polymerisation process can result from heat and pressure being applied to the plastic powder or from mixing the plastic resin with a chemical called a hardener.

Filler materials are used to modify the properties of thermo-setting plastics. Glass fibre and carbon fibre are probably the best known fillers, but wood flour, calcium, carbonate, mica granules, cloth and paper are also used. Filler materials modify the properties of thermo-setting plastics by making them tougher and reducing shrinkage during moulding. Their use also economises on the amount of plastic material used. Some of the more common thermo-setting plastics used as engineering materials are:

- Phenol-methanal – This is better known as bakelite and was one of the earliest plastics to be produced. It is still used for handles, knobs, and laminates. Wood flour is a common filler used to give it strength and its products are mainly black or brown in colour. It is strong, scratch resistant and a good thermal and electrical insulator.
- Urea-methanal – This is better known as formica. It has similar properties to bakelite but can be produced in a variety of colours. It is used to make switches, plugs, buttons and also as a bonding adhesive for plywood.
- Melamine-formaldehyde – This is highly water resistant, tough, tasteless, odourless, and an excellent electrical insulator. It can be produced in all colours and is used for electrical fittings and table wear.
- Epoxy resins – These are mixed with a suitable hardener and are reinforced with glass, fibre, carbon fibre or cloth. They are water resistant and have good adhesion to metals. They are used to mould motor vehicle panels and boat hulls. When reinforced with cloth or paper they are used to make 'Tufnol' rod and sheet. This is used for making light duty gears, handles and printed circuit board.
- Polyester resins – These are also mixed with a hardener and used to make glass fibre and carbon fibre reinforced composites. They are used for motor vehicle panels, boat hulls, safety helmets, fishing rods and archery bows. Polyester resins are also used in paints and enamels.

Rubbers

As with plastics, rubbers are polymer materials. They are known as 'elastomers', which have the property of returning to their original size and shape after being subjected to considerable amounts of deformation. Rubbers have many applications where flexibility is of importance. They are often reinforced with fabric or wire mesh as in motor vehicle tyres and conveyor belts. Other applications include hoses, drive belts, seals, gaskets, cable sheathing, and flexible mountings and couplings. Some of the more common types of rubber are:

- Natural rubber – The tree from which natural rubber is obtained is native to South America, although it is also grown widely in the Far East. Natural rubber is readily attacked by oil, petrol and ozone, which cause it to perish. It is not used much by engineers except when mixed with some of the following synthetic rubbers.
- Styrene rubber – This is also known as GR-S rubber and was developed in America to combat the shortage of

natural rubber during the second world war. Since then it has been widely used for motor vehicle tyres, footwear and many other applications. It is resistant to oils and petrol and is sometimes blended with natural rubber.

- Neoprene – This is a synthetic rubber which is closely related to natural rubber. It is resistant to vegetable and mineral oils and can withstand high temperatures. It is used for oil seals, gaskets, hoses and many other engineering applications.
- Butyl rubber – This is a synthetic rubber which contains small amounts of natural rubber. It has good resistance to heat and chemicals and is impermeable to air and other gases. It is used for moulded diaphragms, tank linings, tyre inner tubes and many other engineering applications.
- Silicone rubber – This is a synthetic rubber which is resistant to extreme temperatures. Most synthetic rubbers become brittle at −20 °C, but silicone rubber can maintain its flexibility to −80 °C. It can also withstand temperatures up to 300 °C and is widely used in aircraft and other applications where a wide variation in temperature can occur.

Ceramics

The term 'ceramics' is derived from the Greek word for potter's clay. It covers a wide group of products such as building bricks and tiles, fire bricks and cement for furnace linings, glass, porcelain, and the abrasive grits used for grinding wheels and cutting tools. The chief ingredients of ceramics are clays and mixtures containing silicon, aluminium and magnesium oxides.

The main properties of ceramic materials are that they are good electrical insulators, they are hard, wear resistant and strong in compression. They do, however, tend to be brittle and weak in tension. Many of them can resist very high temperatures which make them suitable for use in furnaces and for cutting tools. The main types of ceramic material are:

- Amorphous ceramics – These are non-crystalline and include the various grades of glass used for windows, containers, lenses and the glass fibre matting used to reinforce thermo-setting plastics. They have a melt at around 400 °C which is lower than most other ceramics.
- Crystalline ceramics – These include the crystals of aluminium oxide, better known as 'emery' which is used as an abrasive in lapping paste and on emery paper. They also include magnesium oxide which is used to insulate the conductors in mineral-insulated electrical cables.
- Bonded ceramics – These are moulded from clay mixtures and fired in kilns. Vitrification occurs giving a structure which is part amorphous and part crystalline. Bonded ceramic products include pottery, porcelain insulators and grinding wheels.
- Cements – These include Portland cement which is used to make mortar in the building industry, fire clays used in furnaces and special cements used to make concrete.
- Semiconductors – These include silicon, germanium and certain metal oxides. They are poor conductors of electricity at low temperatures but their conductivity increases as the temperature rises. Semiconductor materials and, in particular, silicon are used in a wide range of electrical and electronic components such as thermistors, diodes, transistors and integrated circuits. Small amounts of other elements such as aluminium, boron and phosphorus are added to give the required electrical properties for these applications. The process is known as 'doping'.

Composite materials

A composite is made up of two or more materials which are combined together in a way which improves their mechanical properties. Concrete, reinforced with steel, is an example of a composite material which has been in use for many years in building and engineering. Two other main groups of composites in common use are:

- Fibre-reinforced plastics – The most common of these are glass fibre-reinforced plastics and carbon fibre-reinforced plastics. In both cases the fibres give toughness and durability to thermo-setting plastics which tend to be brittle when used alone.

  Both types of reinforced plastic are used for car body panels and small boat hulls. There are many other uses for glass-reinforced plastic mouldings while carbon fibre-reinforced plastics are also used for fishing rods, dingy spars and archery bows.
- Wood composites – The most common of these are plywood, blockboard and chipboard which are bonded together with adhesives. Plywood consists of thin sheets which are bonded together with their grain directions alternately perpendicular to each other. Blockboard consists of thin wooden laths bonded together and sandwiched between thin sheets. Chipboard consists of bonded sawdust or wood chips.

  Wood composites are often given a veneer of a more expensive wood for decorative purposes an outer layer of plastic or plastic composite. They are widely used as building materials and in furniture and boats. Chipboard is also produced with a layer of bitumen felt bonded to one side for use as roofing board.

Self-assessment tasks

1. What is the main difference between thermoplastic and thermo-setting plastic materials?
2. What type of thermoplastic material is polythene?
3. Name a common use for polyvinyl chloride (PVC) and state the properties which make it suitable for this application.
4. What type of thermoplastic is Perspex?
5. Which thermoplastic material is used as a non-stick coating for cooking utensils?
6. What is phenol-methanol better known as, and what are some of its uses?
7. Why is natural rubber not used much by itself as an engineering material?
8. Name two typical uses of the synthetic rubber, neoprene.
9. Which synthetic rubber has a working temperature range from −80 °C to 300 °C ?
10. What are the main ingredients of ceramic materials?
11. Name a common amorphous ceramic material and the form in which it is supplied when used to reinforce thermo-setting plastics.
12. What is 'emery' and what is it used for?
13. How does the electrical conductivity of silicon and germanium change with temperature?
14. Give two examples of a composite material.
15. Which thermo-setting plastic resins are reinforced with glass fibre and carbon fibres to make composite materials?

Selection of suitable materials for a mechanical product

The selection criteria for materials which are to be used in mechanical products are their:

- Mechanical properties
- Thermal properties
- Electrical properties
- Resistance to chemical attack
- Cost
- Availability

Mechanical properties

The mechanical properties of a material describe its behaviour when loaded in different ways. They include:

- Tensile strength – The Ultimate Tensile Strength (UTS) gives a measure of how strong a material is when being pulled in tension. It is the maximum load in newtons which each square millimetre of cross-sectional area can carry before breaking. These units are often written as N/mm^2 . The tensile strength of a material is obtained by stretching a specimen until it breaks and recording the maximum load carried. It is then calculated using the following formula:

$$\text{Ultimate tensile strength} = \frac{\text{Maximum load carried}}{\text{Original cross-sectional area}}$$

- Hardness – This is a measure of a material's resistance to wear and abrasion. It is measured by pressing a hard steel ball (Brinell test) or a pyramid-shaped diamond (Vickers pyramid test) into the surface of the material for a given time and using a given load. The dimensions of the indentation can then be used to calculate the Hardness Number of the material, although many modern hardness testing machines automatically display the value.
- Elasticity – This is a measure of a material's ability to return to its original shape when loaded and unloaded. Springs must be made of an elastic material. The opposite of elastic is plastic. When a plastic material is deformed it does not return to its original shape. Plastic materials may have one or both of the next two properties.
- Ductility – This is a measure of the amount by which a material can be stretched before breaking when being pulled in tension. There are different ways of measuring ductility, one of the most common is to calculate the Percentage Elongation of a specimen which has been stretched until it fractures. It is calculated using the formula

$$\text{Percentage elongation} = \frac{\text{Increase in length}}{\text{Original length}} \times 100$$

  For standard specimens the value can be read off directly by placing the fractured pieces on a special gauge. Materials which are formed to shape by pulling them through a die to make rod wire or tube must be ductile. Materials with a very low ductility are said to be brittle and so brittleness is the exact opposite of ductility.
- Malleability – This is sometimes confused with ductility but the two are not quite the same. Malleability is a measure of a material's ability to be deformed by compressive forces. Any material which is pressed or forged into shape must be malleable.
- Toughness – This is a measure of a material's resistance to shock loading or impact. It is measured by applying a sudden force to a standard specimen and recording the amount of energy, measured in joules, which it has absorbed when it breaks. Materials used for hammers and forging tools must be tough.

Thermal properties

These describe the behaviour of a material when its temperature changes. They include:

- Thermal conductivity – This is a measure of a material's ability to transmit heat energy. Metals are mainly good conductors of heat energy whereas plastic and ceramic materials tend to be bad conductors.
- Thermal expansivity – This is a measure of the effect which temperature change has on the dimensions of a material. Metals tend to have the higher values of thermal expansivity than other solid materials and some metals expand or contract much more than others for the same temperature change. This property is put to use in bi-metallic switches and bimetallic thermostat temperature-sensing elements.

Electrical properties

These describe the behaviour of a material under the effects of electrical potential difference. They include:

- Electrical conductivity – This is a measure of a material's ability to allow electric current to pass through it. Metals tend to be good conductors of electricity while plastics and ceramics tend to be bad conductors. Between these extremes are the semiconductor materials which are poor conductors of electricity at low temperatures, but whose conductivity improves with temperature rise.
- Electrical resistivity – This term is more often used to describe and measure the ability of a material to pass electric current. A material which is a good conductor of electricity is said to have a low resistivity (low electrical resistance) while a bad conductor is said to have a high resistivity (high electrical resistance).
- Temperature coefficient of resistance – This is a measure of the change of resistance of an electrical conductor when its temperature changes. The resistance of metals tends to increase with temperature rise and they are said to have a positive temperature coefficient of resistance. The resistance of most semiconductor materials, however, decreases with temperature rise and they are said to have a negative temperature coefficient of resistance.

Resistance to chemical attack

This is the ability of a material to withstand degradation due to corrosion or solvent attack:

- Corrosion – All metals corrode with time although some are more resistant to corrosion than others. The rate of corrosion depends on service conditions. Metals need to be carefully chosen for use in hostile chemical environments. High temperatures, atmospheric pollution and the presence of moisture also tend to cause corrosion.

The most common corrosion process is oxidation where a metal reacts chemically with oxygen. Non-ferrous metals are generally resistant to oxidation. Copper, tin, zinc, lead and their alloys all react with oxygen but the oxide film which forms on their surface is very dense and protects them from further attack. Ferrous metals however, such as plain carbon steels and cast iron, form oxide films which are loose and porous and as a result their corrosion is progressive.

At high temperatures the oxide formed on steel is a flaky black scale known as 'millscale'. The process is called dry corrosion. In the presence of moisture, plain carbon steels and cast iron produce oxide in the form of the loose brownish-red scale known as rust. This process is called electrolytic corrosion.

Electrolytic corrosion occurs where two dissimilar metals are in contact such as at fixing points where screwed fasteners or rivets may be of a different composition to the material being joined. It also occurs where there are dissimilar crystals or 'grains' within a metal such as occurs in steels and cast iron.

If the materials become covered with moisture, this acts as an electrolyte, i.e. a liquid through which an electric current can pass. An electric cell is set up and a small current flows through the materials and the electrolyte. This causes one of the materials to corrode. Although corrosion cannot be completely eliminated, it can be slowed down by the use of protective coatings. Paints, plastic coatings and coating with a more corrosion-resistant metal are common forms of surface protection.

- Solvent attack – Solvents are liquids which attack engineering materials such as plastics and rubbers. They can cause the material to dissolve or they can degrade its mechanical properties. Care should be taken when selecting plastics and rubbers which will come into contact with chemical substances. The same applies to materials which will come into contact with petrol, fuel oil and lubricants as these can also act as solvents.

Cost

The cost of a material depends on a number of factors. Some of these are:

- The scarcity of its natural resource.
- The amount of processing which is required to convert it from a raw material to a useable form.
- The form in which it is supplied.
- The quantity in which it is purchased.

Wherever possible, materials should be selected which are available in standard forms of supply and require as little processing as possible. Savings can also be made by purchasing materials in large quantities or by placing regular repeat orders with a supplier.

Availability

Engineering materials are supplied in many different forms. These include:

- Ingots
- Granules
- Liquids
- Castings and mouldings
- Forgings and pressings
- Barstock
- Sheet and plate
- Pipe and tube

Most of these forms are supplied in standard sizes and quantities. It is in these that they are most readily available. Castings, mouldings, forgings and pressings initially require the production of special patterns and dies. This is a highly skilled operation and can take a considerable time. Once the patterns and dies have been made, however, repeat orders can be obtained more quickly.

Self-assessment tasks

1. A steel wire of cross-sectional area 2.5 mm^2 is loaded in tension and breaks when the load is 1200 N. What is the tensile strength of the steel?
2. How is the hardness of an engineering material commonly measured?
3. What is elasticity and which engineering components need to have this property?
4. Which property is required in materials which are drawn through a die to make wire or tube?
5. What is the difference between the properties of ductility and malleability in engineering materials?
6. What kind of loads do hammers and forging tools experience and what property must they have to withstand them?
7. What is the property which measures the ability of a material to allow heat energy to pass through it?
8. What does the thermal expansivity of a material measure?
9. How do the electrical resistivities of metals and plastics compare?
10. What generally happens to the electrical resistance of metals when their temperature rises?
11. Why are non-ferrous metals such as copper and tin more resistant to corrosion than ferrous metals such as mild steel and cast iron?
12. What kind of corrosion is likely to occur where two different metals are joined together and moisture is present around the joint?

Selection of suitable mechanical components for a mechanical product

Engineered products such as motor vehicles, aircraft and domestic appliances are assembled using a variety of mechanical fixing components. If access is required for maintenance, and if components will need to be removed or replaced, screwed fastenings are generally used.

The ISO metric screw thread is now used by most countries for general engineering purposes and will eventually replace other systems. Alongside the ISO metric, the British Association Thread (BA) is used internationally for the small sized screwed fastenings in instruments and electrical equipment. It provides a specialist range of small diameter fine threads not catered for by the ISO metric system.

Other screw thread systems are to be found on older products and they are still used in the manufacture of replacement parts and units. They include the British Standard Whitworth Thread (BSW), the British Standard Fine Thread (BSF), the Unified Course Thread (UNC) and the Unified Fine Thread (UNF) which were all produced in imperial sizes.

If components are intended to be assembled permanently they may be riveted in position. Rivets are available in different sizes and materials and with different shaped heads.

The most common fixing components used on mechanical products are:

- Nuts and bolts
- Setscrews
- Studs and nuts
- Self-tapping screws
- Rivets

Nuts and bolts

Metric threads are specified on drawings and technical documents in a particular way. For example, a metric bolt with a hexagonal head and its hexagonal nut (Fig. 1.1) might be specified as:

Steel, Hex Hd Bolt – M8 × 1.25 × 50
or
Steel, Hex Nut – M8 × 1

where M indicates the ISO metric system
8 indicates 8 mm diameter
1.25 indicates 1.25 mm pitch
50 indicates 50 mm length of the bolt.

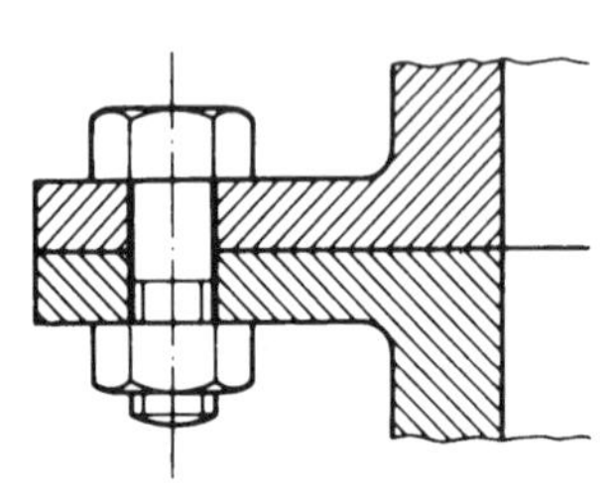

Figure 1.1 Hexagonal nut and bolt: section through a bolted joint

The diameter of bolts selected for a particular purpose depends on the load to be carried. Bolts are so designed that, in tension, there is an equal chance of the bolt failing by:

- The head pulling off.
- The thread stripping.
- Failure across the threaded part of the shank.

The plain part of the shank should extend through the joint face so that the threaded part is not carrying any shearing load. Although it is not shown in Fig. 1.1, it is advisable to fit a washer under the nut and sometimes under the bolt head to spread the load and to avoid damaging the surfaces of the components.

There is a tendency for nuts to become loose where vibration is present. To prevent this a range of special nuts and locking devices has been developed. They can be divided into two classes:

- Friction locking devices – These include lock nuts, various shapes of spring washer and Simmonds-type friction nuts (Fig. 1.2). It is good practice to replace spring washers and friction nuts with new ones when they have been removed for maintenance purposes.

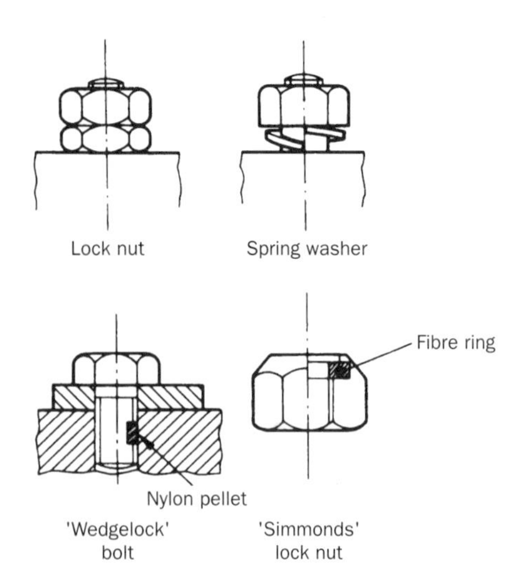

Figure 1.2 Friction locking devices

- Positive locking devices – Positive locking devices are used where the presence of a loose or detached nut or bolt could cause a serious accident (Fig. 1.3).

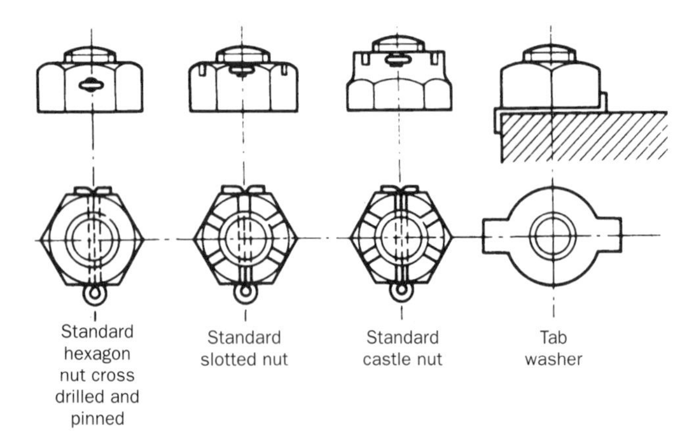

Figure 1.3 Positive locking devices

It is essential to replace split pins and tab washers with new ones after they have been removed for maintenance purposes.

Setscrews

Setscrews are used to fasten parts together where one of the parts is too large for a nut and bolt to be used. The larger part is drilled and tapped with a suitably sized screw thread. The smaller part is often spot-faced to allow the screw head to seat properly or it may be counterbored or countersunk in applications where a flat external surface is required. Setscrews have a variety of head shapes to suit different applications (Fig. 1.4).

Setscrews need not be threaded over the compete length of the shank, except when fixing thin plate components. They should be selected with the plain part of the shank slightly shorter than the thickness of the component through which it passes.

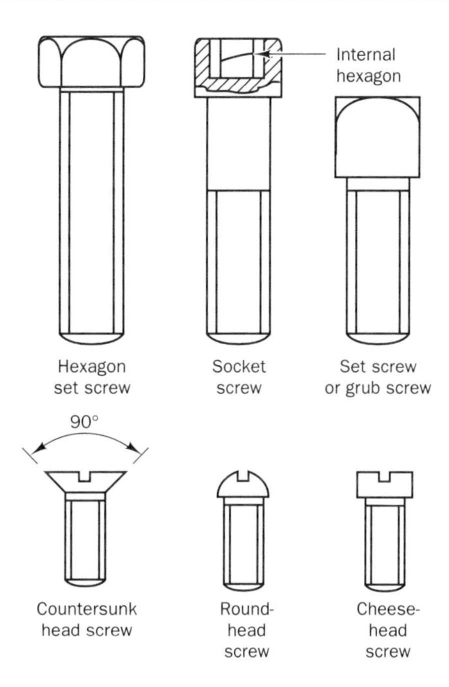

Figure 1.4 Setscrews

Studs and nuts

For applications such as securing inspection covers which may have to be removed regularly, it is possible that screwed fastenings may become damaged over a period of time. Here, studs are often used instead of setscrews or machine screws. A stud is a length of bar which is threaded at each end, as in Fig. 1.5. The end with the shorter thread is screwed tightly into the body of the main component and the end with the longer thread holds the nut.

The advantage of using studs and nuts is that any wear will occur on the outer thread of the stud and not in the expensive body of the main component. If the wear is excessive, the stud can be replaced quite cheaply. As with other screwed fastenings, it is advisable to employ a locking device where vibration is present.

Self-tapping screws

These are often used with plastic and sheet metal components. They may be of the thread-forming or thread-cutting types depending on the materials being joined. For each type of self-tapping screw a pilot hole must be made whose diameter is equal to the root diameter of the thread, i.e. the diameter measured across the bottom of the threads.

The first of the screws in Fig. 1.6 forms a thread as it is screwed in by displacing the softer material. The second type cuts material away to form a thread and can be used on harder substances. The third type is designed to be hammered home for applications where removal is not intended. It will be noted that it has a multi-start thread, i.e. more than one spiral groove, which requires fewer turns to drive it home.

Rivets

Rivets are used where it is intended to make a permanent joint. When designing a riveted joint the size of the rivets is dictated by the thickness of the plates or sheets being joined. Riveted joints should not be used to withstand tensile forces as their strength lies in resisting the shearing forces which act across the shanks of the rivets.

Rivets are manufactured from different materials and with different shaped heads to suit different applications (Fig. 1.7). A common requirement is that the rivet material must be fairly malleable to allow the head to be formed when making the joint.

Snap and pan head rivets are used where maximum strength is required. Flat and mushroom headed rivets are used where the rivet head is not required to protrude very far

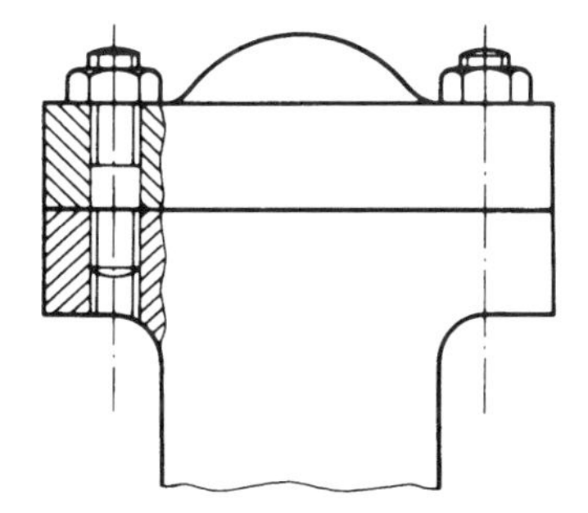

Figure 1.5 Use of studs: fixing for an inspection cover

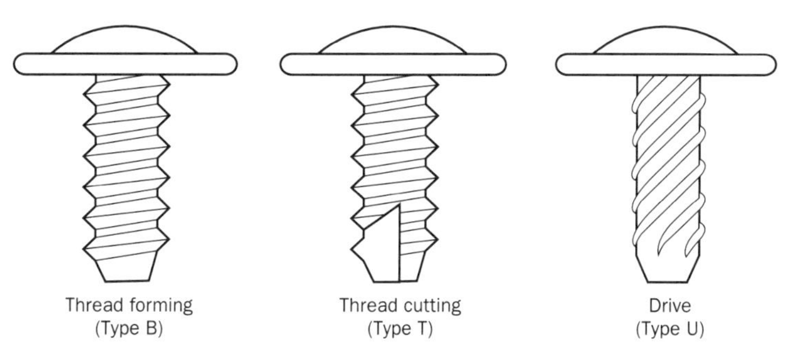

Figure 1.6 Self-tapping screws

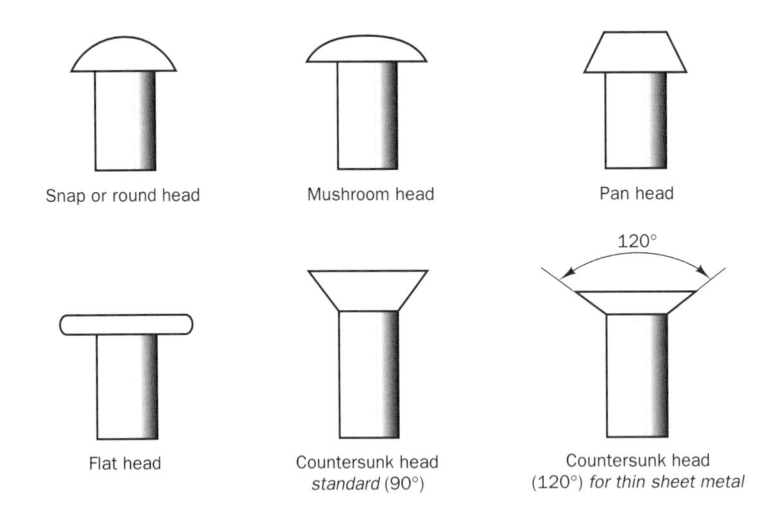

Figure 1.7 Standard types of rivet head

above the joint surface. Countersunk heads are used where a flush surface is required.

Wherever possible the rivets should be made from the same material as those being formed to prevent electrolytic corrosion which may occur if moisture is present.

Where access can only be gained to one side of a joint it is possible to join thin plate materials using pop-rivets. When a sealed joint is required, the solid type is used which does not leave a central hole.

Self-assessment tasks

1. A steel hexagonal headed bolt is specified as $M10 \times 1.5 \times 100$. What do these numbers indicate?
2. What kind of engineering applications is the British Association (BA) screw thread used for?
3. Why is it advisable to fit a plain washer under a nut?
4. Describe two different friction locking device which can prevent a screwed fastening from becoming loose due to vibration.
5. Where should a positive locking device be used in preference to a friction type?
6. Where are setscrews used to join components together rather than nuts and bolts?
7. Why is it often best to use studs and nuts to secure inspection covers which have to be removed regularly?
8. What are the two basic types of self-tapping screw and what kinds of material would they be used with?
9. What kind of rivet head is used where the joint is required to have a flush surface?
10. For what kind of application are pop-rivets used?

Identification of possible electrical and electronic components for an electrical product

Electrical products such as domestic appliances, radio, television and information technology equipment are continually being improved and up-dated. Some of them are very complex engineering systems but they can usually be broken down into a number of less-complicated subsystems which are easier to understand. Many of these are simple circuits containing basic electrical and electronic components. Some of the most commonly used materials and components are:

- Cable and connecting wire
- Insulators
- Resistors
- Capacitors
- Inductors
- Diodes
- Bipolar transistors

Cable and connecting wire

The connecting wire and cable used in electric and electronic circuitry usually consists of a conducting copper core covered by an insulating material. Some of the most common types are:

- Insulated wire – The copper core may be single or multi-strand and it may be tin plated to aid soldering. For normal temperature applications the insulation is usually PVC, but silicone rubber is also used for environments where ozone, radiation and solvents are present. For higher temperature applications, up to 200 °C, PTFE is used and the copper may be silver plated
- Multi-cored cable – This can contain up to fifty separate insulated wires which are colour coded and enclosed in an outer insulating sheath. The cores may also be surrounded by braided metal or metal foil screening within the outer sheath. Screened cable is used for signal transmission applications. The screening protects the conductors from the effects of stray magnetic fields which may interfere with the signals being carried.
- Ribbon cable – This can consist of up to sixty-four parallel copper conductors which may be separately insulated and colour coded before being laminated between layers of PVC film to form a flat ribbon. Ribbon cables are widely used on computers, peripherals, interface units and audio and digital equipment.
- Mineral-insulated cable – With this kind of cable the conductors are insulated from each other by compressed mineral powder which is contained in an outer sheath of copper. This gives more protection than PVC for industrial and commercial applications.
- Armoured cable – This contains conductors which are insulated with PVC and surrounded by galvanised steel wire to protect them in heavy duty applications. The outer cover is made from tough PVC.

Insulators

Insulating materials are as essential as conducting materials to the functioning of electrical and electronic equipment. Insulation is necessary to prevent human contact with live conductors and components. It is also necessary to prevent unwanted contact between conductors and also between conductors and their surrounding metal casings or framework as this can cause a short circuit.

The types of insulation material in use are many and varied. They include:

- Thermoplastics – Sleeving made from PVC and PTFE is available in a range of colours for insulating bare conductors.
- Thermo-setting plastics – A variety of control knobs, terminals, insulating caps, covers and spacers are moulded in thermo-setting plastic materials.
- Rubbers – A variety of moulded caps, covers and shrouds are available in rubber for use in environments which may be hostile to plastic materials.
- Ceramics – Ceramic materials are used to make insulators which range from small distance pieces and spacers to the large insulators used on overhead power lines.

Many of the above components are produced in standard designs and sizes which are readily available and listed in suppliers' catalogues.

Resistors

Resistors are used to control the flow of current in electric and electronic circuits. The different types of resistor include carbon compound, carbon film, metal film and metal oxide. Wire wound resistors are also used for certain applications.

Resistors are produced with different power ratings ranging upwards from 1/4 watt. They are colour coded or marked with a number and letter code which enables their value in ohms and their tolerance to be read off. The tolerance is given as a percentage of the indicated value i.e. a 10% tolerance means that the value is within 10% of the indicated value.

Capacitors

Capacitors are devices which can store electrical charge. Their capacitance, measured in microfarads or picofarads, is a measure of their storing capacity. In their simplest form they consist of two metal plates separated by an insulating (dielectric) material. This may be air, waxed paper, plastic or mica. They are sometimes colour coded in the same way as resistors except that their value is read off in picofarads.

Capacitors can be used as a smoothing device to absorb surges of current which can occur when a circuit is switched on. For this purpose they are connected across the switch contacts to prevent arcing. Capacitors are also used in power supply units to smooth out the pulsations in a rectified direct current. Another of their properties is that while they block the flow of direct current, they will allow an alternating current to pass. They can thus be used as a filtering device to remove the direct current component where an alternating and direct current are superimposed on each other.

Electrolytic capacitors are a more complicated type in which the dielectric is a metal oxide coating on one of the plates. They can be made with a much higher capacitance than simple plate capacitors, but they are more expensive and must always be connected with the correct polarity. Failure to do this results in damage to the capacitor.

Inductors

In their simplest form these are wire coils wound on a core which may be made from an insulating material or a ferromagnetic substance.

An inductor will oppose a change in the current flowing through it. When a current is switched on it will slow down its growth and when it is switched off it will slow down its decay. They are available in a variety of sizes and their value is quoted in millihenrys or microhenrys.

Like capacitors, inductors can be used as a device to smooth out the pulsations in a direct current. They offer increased opposition to alternating currents as the frequency increases, and if the frequency is high enough they will almost completely block an alternating current. This is the opposite effect to that of a capacitor.

Diodes

A diode is a device which will allow electric current to flow through it in one direction only. Some of the most common types are:

- Junction diodes
- Photodiodes
- Light-emitting diodes
- Bipolar transistors

Junction diodes

A junction diode consists of two pieces of p-type and n-type semiconductor material in contact (Fig. 1.8).

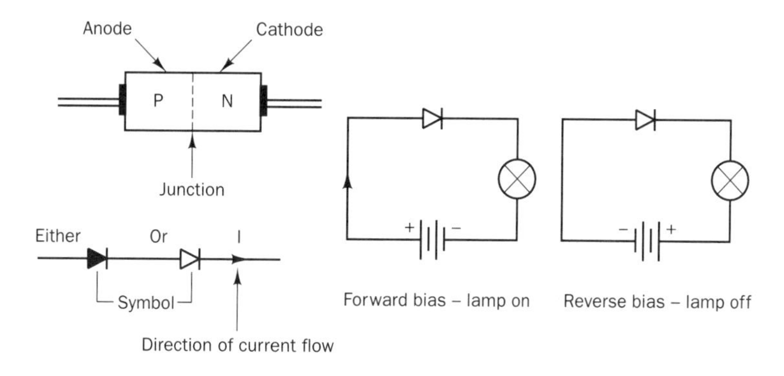

Figure 1.8 Junction diode

The term 'forward bias' means that a voltage is applied in the direction of current flow. The term 'reverse bias' means that a voltage is applied in the opposite direction to that in which current can flow.

Junctions diodes are widely used as rectifiers to change an alternating current into a pulsating direct current by permitting only the forward half-cycle of the current to pass.

Photodiodes

These are of similar construction to the junction diode except that the case has a transparent window through which light can enter.

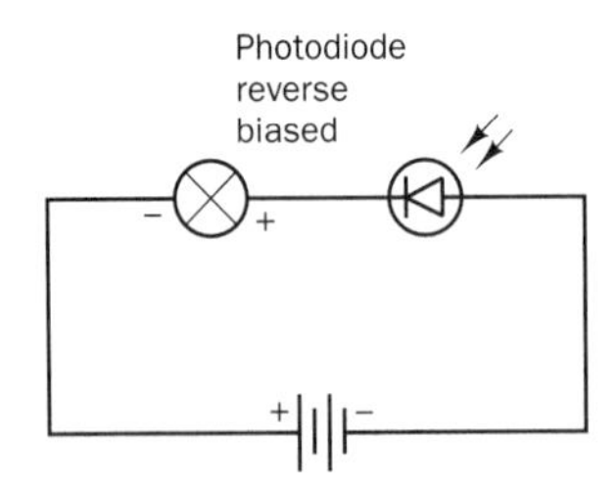

Like all diodes, photodiodes (light sensitive junction diodes) will not normally conduct when reverse biased. However they become conductive when bright light falls on them.

Figure 1.9 Photodiode

A potential is applied in the reverse direction and when no light is entering no reverse current will flow. When light falls on the semiconductor material it releases current carriers and a reverse current flows. The photodiode can thus be used as a light sensitive switch (Fig. 1.9).

Light-emitting diodes

These are junction diodes in which the materials used are gallium arsenide and gallium phosphide mounted in a transparent case. This may be moulded to form a lens at one end.

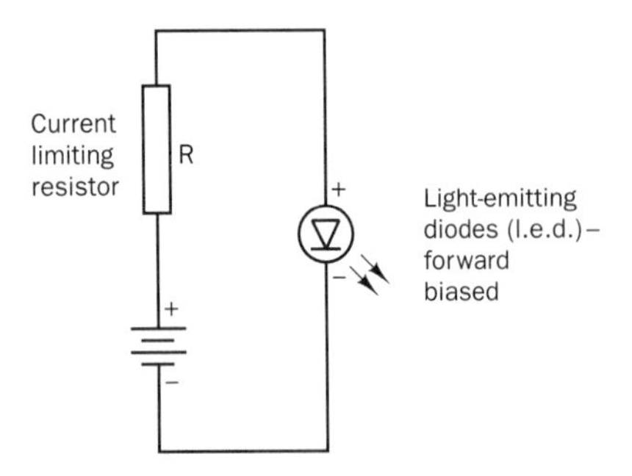

Figure 1.10 Light-emitting diode (l.e.d.)

When biased in the forward direction, a current passes and light is emitted (Fig. 1.10). This may be red, yellow or green depending on the composition of the materials used. Light-emitting diodes can be used as visual current indicators. They are also used in seven-segment light-emitting-diode elements in which different diode combinations can display the digits 0 to 9 (Fig. 1.11).

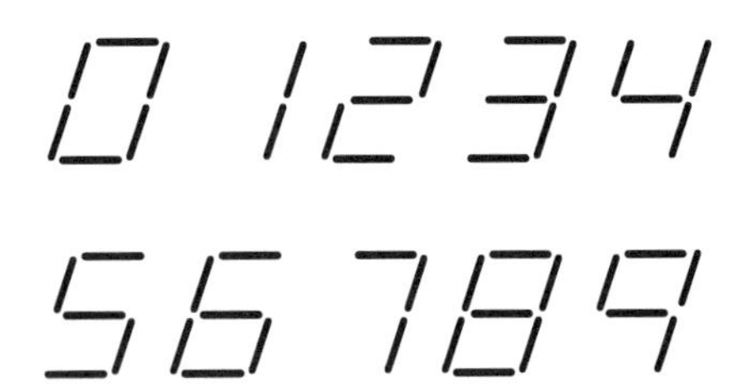

Figure 1.11 The digits 0–9 on a seven-segment display

Bipolar transistors

Transistors are used as high-speed switches and amplifiers. They are made in a variety of sizes and shapes.

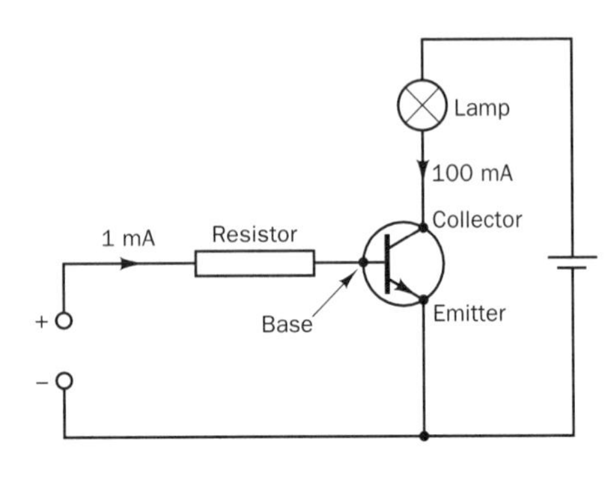

Figure 1.12 Transistor circuit

Bipolar transistors are made from p-type and n-type semiconductor materials which are sandwiched together in a particular way. They have three connections known as the base, the collector and the emitter.

In the circuit shown in Fig. 1.12 a transistor is being used as a switch to light the lamp. The lamp will not light until the current flowing in the circuit which connects the base and the emitter has reached a particular value. This enables a weak current in the base–emitter circuit, say 1 mA, to switch on a much larger current, say 100 mA, in the collector–emitter circuit which contains the lamp.

When a transistor is used as signal amplifier, the input signal produces small variations in the base–emitter current. These produce much larger variations in the collector–emitter current which are an amplified copy of the input signal.

Self-assessment tasks

1. Which are the main materials used to make the insulated connecting wire used in electric and electronic circuits?
2. Why are cables used for signal transmission sometimes surrounded by a braided metal, or metal foil sheath?
3. Where do you find ribbon cables being widely used?
4. What kind of cable would you use for a heavy duty application where there is a possibility of damage from nearby work activities?
5. Which thermoplastic insulating material can be used as an alternative to PVC for temperatures which might rise to 200 °C?
6. Name three kinds of insulating component which might be moulded from a thermo-setting plastic material.
7. Name three kinds of resistor used in electric and electronic circuits.
8. The value of resistors and capacitors might be quoted as having a 10% tolerance. What does this mean?

9. Capacitors are often used in power supply units where alternating current is changed to direct current. What is their purpose in this kind of application?
10. What effect does an inductor have when the current flowing through it is switched on or switched off?
11. What controlling effect does a junction diode have on current flow?
12. Describe one possible use of a photodiode?
13. What is the difference between a photodiode and a light emitting diode?
14. Name two uses for a bipolar transistor?
15. What are the names of the three connections to a bipolar transistor?

Selection of suitable electrical and electronic components for an electrical product

Electrical and electronic components are selected to perform a particular function under given operating conditions. These conditions include the maximum expected current, voltage and operating temperature. Where energy transformation is taking place they might also include the required power rating. The selection of a particular component may depend on its:

- Voltage rating
- Current rating
- Power rating
- Temperature rating
- Resistance or conductivity
- Tolerance

The common electrical and electronic components whose function have been described are:

- Cable and connecting wire
- Insulators
- Resistors
- Capacitors
- Inductors
- Diodes
- Transistors

Cable and connecting wire

This is selected on the basis of its current, voltage and temperature rating. In the case of power cable, there are regulations which state the cross-sectional area of the conductors which must be used for particular applications. The sizes specified for some domestic circuits are given in Table 1.3.

Table 1.3 Cables for domestic circuits

Lighting and bell circuits	1.5 mm^2
Ring main and water heater circuits	2.5 mm^2
Cooker circuits	10 mm^2

Insulators

Insulators are selected by their voltage rating. This is the voltage they can withstand before their high resistance is broken down.

They are also selected to suit different service conditions. Ceramics are resistant to solvents and can be used for high-temperature applications. Plastics and rubbers may deteriorate under these conditions but they are less rigid and brittle. They can thus be used for applications where vibration and flexing occur.

Resistors

Resistors are selected by their nominal value, tolerance and power rating. Their value in ohms and their tolerance may be marked on them or is displayed in colour code as shown in Fig. 1.13.

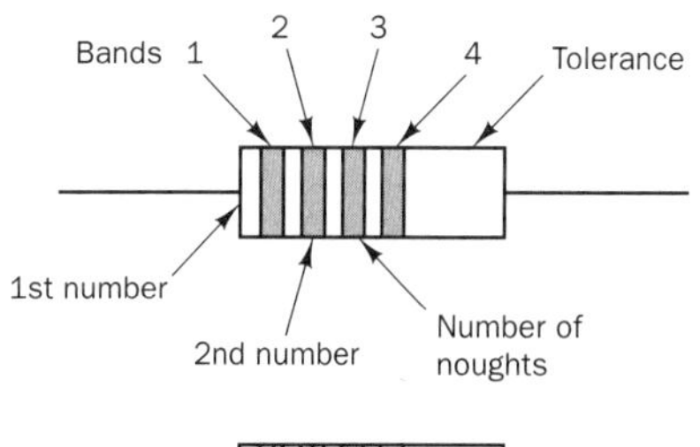

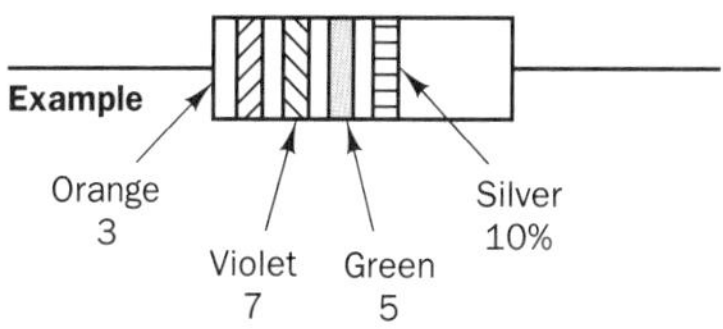

3 700 000 Ω tol. 10%
Resistor marking

Digit	*Colour*
0	Black
1	Brown
2	Red
3	Orange
4	Yellow
5	Green
6	Blue
7	Violet
8	Grey
9	White
Tolerance	*Colour*
5%	Gold
10%	Silver
20%	No colour band

Figure 1.13 Resistor colour code

The properties and uses of the different types of resistor are given in Table 1.4.

Table 1.4 Resistor types

Property	Type of resistor			
	Carbon comp.	Carbon film	Metal oxide film	Wire wound
Max. value	20 MΩ	10 MΩ	100 MΩ	270 Ω
Tolerance	±10%	±5%	±2%	±5%
Power rating	0.125–1 W	0.25–2 W	0.5 W	2.5 W
Uses	General	General	Accurate work	Low values

Capacitors

Capacitors are selected by their nominal value, tolerance, working voltage, leakage current and working temperature.

Table 1.5 Capacitor types

Property	Non-polarised capacitors			Electrolytic capacitors	
	Polyester	Mica	Ceramic	Aluminium	Tantalum
Values	0.01–10 μF	1 pF–0.01 μF	10 pF–1 μF	1–100 000 μF	0.1–100 μF
Tolerance	±20%	±1%	−25 to +50%	−10 to +50%	±20%
Leakage	Small	Small	Small	Large	Large
Uses	General	High frequency	Decoupling	Low frequency	Low voltage

Their value in microfarads or picofarads and their tolerance may be marked on them. Alternatively the information may be displayed in colour code. This is read in the same way as for resistors except that the value is in picofarads.

The properties and uses of the different types of capacitor are given in Table 1.5. The material quoted is the dielectric.

Inductors

Inductors are selected by their nominal value (in microhenrys and millihenrys) and their tolerance. Depending on their application, they are also selected by their direct current, resistance ratings and a.c. frequency rating.

Air-cored inductors have small inductance values measured in millihenrys. They are used in radio and communications equipment to stop high-frequency currents from entering certain parts of a circuit.

An iron core greatly increases the strength of the magnetic field produced around an inductor. As a result, iron-cored inductors have much higher values of inductance measured in henrys (e.g. 10 H). They are used as low-frequency smoothing devices in power supply units.

Iron-dust, and ferrite-cored inductors are used for applications such as in the tuning circuits of radio equipment. Tuning circuits contain both inductors and capacitors which enable a frequency of one particular frequency to be selected. While both of these core materials have magnetic properties, iron-dust has a very high resistance and ferrite is a non-conducting material. This property prevents stray currents, known as eddy currents, from flowing in the inductor core. Eddy currents can cause energy loss which affects the performance of the component.

Diodes

Diodes are selected by their voltage, current and temperature ratings. They include signal diodes, power diodes, light-emitting diodes and photodiodes.

Signal diodes are used in radio, communications and computer equipment where it is required to allow small currents, measured in milliamperes, to pass in one direction only.

Power diodes are designed for use in power supply equipment, typically to convert alternating current to direct current. They can be selected to handle voltages up to 1000 V and currents of over 100 A.

Light-emitting diodes are available for different voltage applications ranging from a few volts up to mains voltage. They are selected for the required light intensity of their output which is measured in millicandelas (mcd).

Photodiodes are available with different lens sizes to suit different applications. They are selected for the speed at which they can respond to a change of light intensity, the range of light wavelengths to which they are sensitive and for their operating temperature range.

Transistors

Transistors are classified in different groups ranging from low power, low frequency to high power, high frequency and switching applications. They have different current, voltage and power ratings given in the form of a code. In the continental system the first letter of the code gives the semiconductor material, e.g. A for germanium and B for silicon. The second letter gives the application, e.g. C for an audio frequency amplifier and F for a radio frequency amplifier. A transistor coded BC109 is thus a silicon transistor for audio frequency amplification.

The American system is different, however, and some manufacturers use their own systems of coding. Fortunately the main suppliers' catalogues contain all the relevant data for each different transistor together with its application.

Self-assessment tasks

1. What should be the cross-sectional area of the conductors in the cable to a domestic electric cooker?
2. To what kind of circuits should cable with conductors of 1.5 mm^2 cross-sectional area be limited?
3. State one advantage and one disadvantage of ceramic materials when used for insulation purposes.
4. If a 270 Ω resistor with a tolerance of +10% is required for a circuit, what will be its colour code?
5. A resistor of value 75 MΩ with a power rating of 0.5 W is required for accurate work. Which type of resistor would be best for this application?
6. A non-polarised capacitor of value 100 pF ± 1% is required for a circuit which will carry high-frequency currents. What is the most suitable type?
7. A 10 H inductor is to be used as a smoothing device in a power supply unit. What kind of core material would you expect it to have?
8. State the typically different uses of signal diodes and power diodes.
9. What kind of diode would you select to indicate that current is flowing in part of a circuit?
10. What do the first two letters indicate in the continental system of transistor coding?

Selection of suitable materials and components for an electromechanical product

Having read the previous sections of this unit, the student should be able to select suitable materials and components for a range of engineered products.

Portfolio suggestions for mechanical products are:

- Hacksaw with an adjustable frame
- Woodworking plane
- Foot pump
- Fishing reel
- Top-pan spring balance

Portfolio suggestions for electrical products are:

- Extension lead
- Electrician's screwdriver
- Inspection lamp
- Continuity tester
- Electric soldering iron

The selection and reporting activities will need to be planned as a number of tasks. A suggested plan is as follows:

- Task 1 – Choose one appropriate mechanical and one appropriate electrical product for your report.
- Task 2 – Obtain or produce drawings of the chosen products. These may be sectioned drawings, exploded drawings or circuit diagrams which clearly show the different parts and components of the products. Number the items which require material or component selection, if this has not already been done, and draw up a parts list for each product. These should give the item number, name of the item and the quantity required.
- Task 3 – Select suitable materials or components for the numbered items. Write a short note for each item outlining the reasons for your choice and explaining why alternative materials and components were not selected.

1.2 Selection of processes to make engineered products

When the design of an engineered product has been finalised and its materials and components have been selected, the next stage in its realisation is to plan its production. In fact the two stages usually overlap. Discussions often take place between the designers and production engineers during the design stage. This ensures that the final designs are practical and that the product can be made with the production facilities available.

Production processes must be selected which will produce the quantities required at the times when they are required. They must also be able to meet the standards of quality which are specified in the design.

The cost of production is equally important and the processes selected should be the most economic processes available. When it is complete, the product must sell at a competitive price which will cover its costs and return a profit to the engineering company.

Topics covered in this section are:

- Identification of suitable processes for making electro-mechanical products.
- Identification of specific techniques appropriate for each process.
- Selection of suitable processes for making a given electro-mechanical product.
- Selection of the most appropriate specific techniques necessary to perform each process.
- Identification of relevant safety procedures and equipment for the processes selected.

Identification of suitable processes for making electromechanical products

Engineered products may have to pass through a number of production processes before they are completed. These may include:

- Material removal
- Joining and assembly
- Heat treatment
- Chemical treatment
- Surface finishing

Material removal

Some engineering products can be cast or moulded to their finished shape and specified dimensions without much further work being done to them. Injection moulded and die cast components in particular, have very little excess material and that which does remain can be easily and quickly trimmed off.

Many components however, have dimensional and material specification which make it impossible to produce them in a single process. In the raw state their material is often supplied in the form of oversized castings, mouldings, forgings and barstock. These must then be machined to the specified dimensions using a variety of material removal processes. Three of the most common are:

- Drilling
- Turning
- Milling

Drilling

The most common are the portable power drill, the sensitive bench drill and the pillar drill. There are various types of drilling machine. Portable hand-held drills may be powered by electricity or compressed air. They are generally used for drilling small diameter holes in large assemblies or structures.

The sensitive bench drill (Fig. 1.14) is used for light work and the pillar drill is used for larger components. The larger

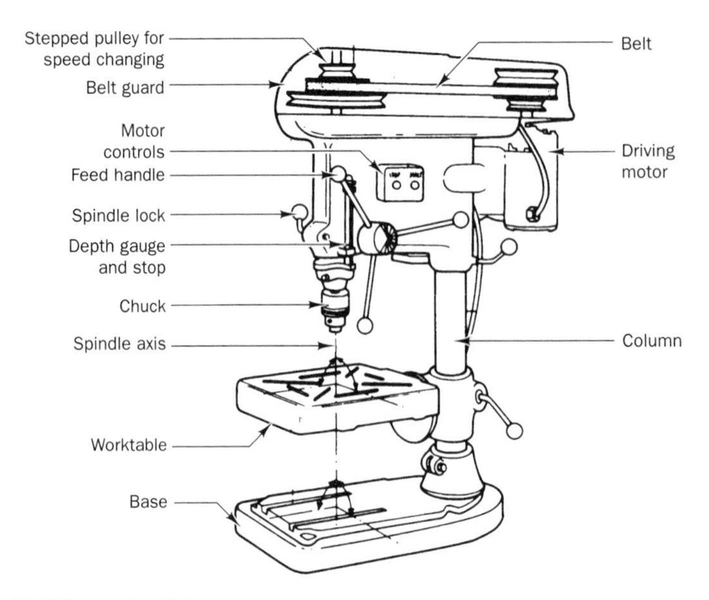

Figure 1.14 Sensitive bench drill (tilting worktable)

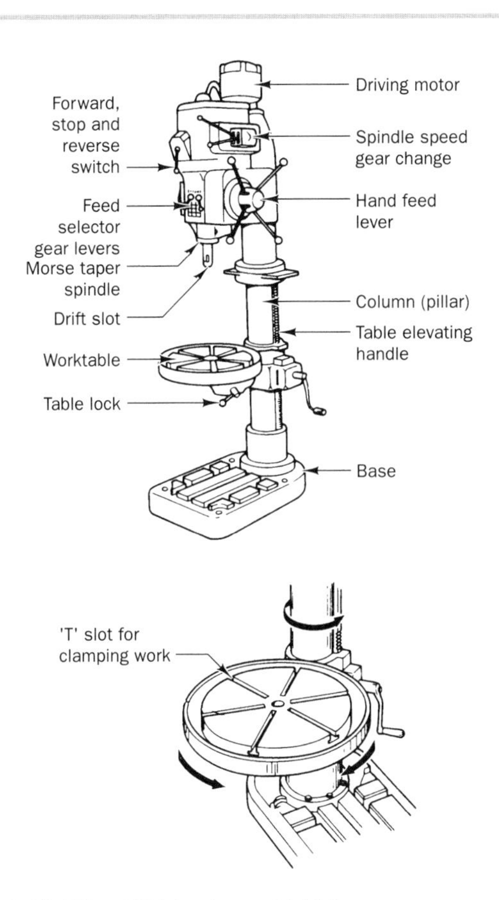

Figure 1.15 Pillar drill (circular worktable)

pillar drill (Fig. 1.15) is equipped with a gearbox to give a range of rotational cutting speeds. On the smaller pillar drills and sensitive bench drills the rotational speed is changed by moving the position of the drive belt on stepped pulleys.

The twist drills used for metal cutting (Fig. 1.16) are made from high-speed steel (HSS). They have two cutting edges, or lips, and two helical flutes cut along their length. Each flute has a thin raised land running along it to reduce friction between the body of the drill and the workpiece.

The twist drill should be regarding as a roughing out tool because the hole produced may be larger than the drill size and can also be slightly out-of-round. If a hole of accurate size and roundness with a good surface finish is required, it should be drilled undersize and finished off by reaming.

Machine reamers (Fig. 1.17) are multi-fluted tools which are used to finish off drilled holes to an accurate size and good surface finish. There are two types of reamer. One type which cuts on the ends of (and also along) the flutes is called a rose-action reamer. This gives good results with materials such as cast iron, bronzes and plastics which tend to close and grip the reamer. The other type cuts on the ends of the flutes only and gives good results with mild steel.

It will be noted that the helix on the flutes of a reamer is left handed, which is opposite to that on twist drills. This is to prevent the reamer from being pulled into the hole and also prevents the chips from being drawn back up the hole where they might damage the surface finish.

Turning

Turning is done on a lathe; the basic type of lathe is called a centre lathe (Fig. 1.18). In the basic turning process a single point cutting tool or a twist drill removes material from a rotating workpiece. Turning can be used to produce cylindrical and tapered components. It can also produce flat surfaces by machining across the end faces of a component.

The centre lathe spindle is driven through a gearbox which gives a selection of different cutting speeds and tool feed rates. The spindle can carry a chuck, a faceplate or a catchplate and centre for gripping or supporting the workpiece. The tailstock, with its centre, is used to support the free end of the workpiece. It can also be used to carry a Jacobs-type chuck for holding drills and screwcutting taps or a die holder for cutting external threads.

The lathe bed is made from cast iron and contains V-shaped slideways which are surface hardened. The tail stock and carriage assembly can be moved along the slideways and clamped in position if required. The carriage assembly, or saddle, contains the cross-slide, which is at right angles to the spindle axis, and the compound slide which can be set at different angles for turning tapers. Mounted on top of the compound slide is the tool post.

The front of the carriage assembly, or apron, contains the feed control hand wheel for moving the carriage along the bed to cut cylindrical surfaces. It also contains the feed control hand wheel for the cross-slide. This moves the cutting tool at right angles to the spindle axis, to increase the depth of cut and for facing operations. Most centre lathes have automatic carriage and cross-slide feed. They are engaged by means of levers which are also situated on the front of the carriage assembly.

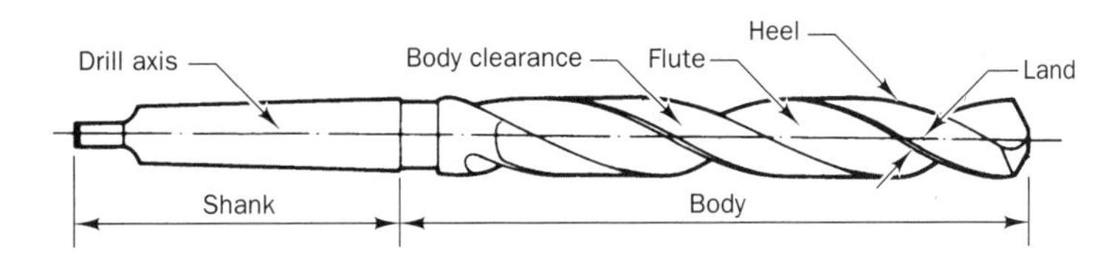

Figure 1.16 Twist drill

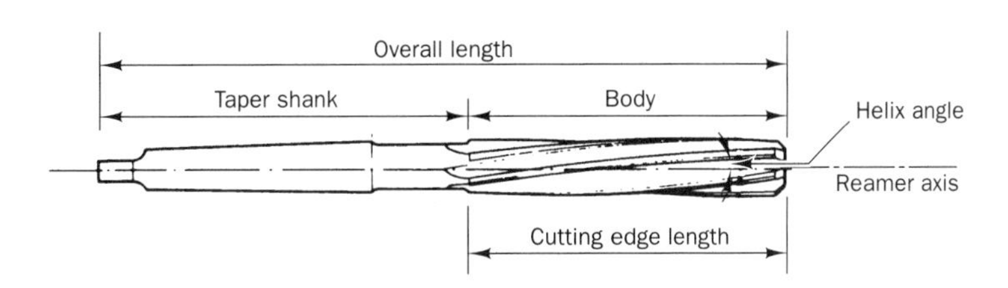

Figure 1.17 Machine reamer

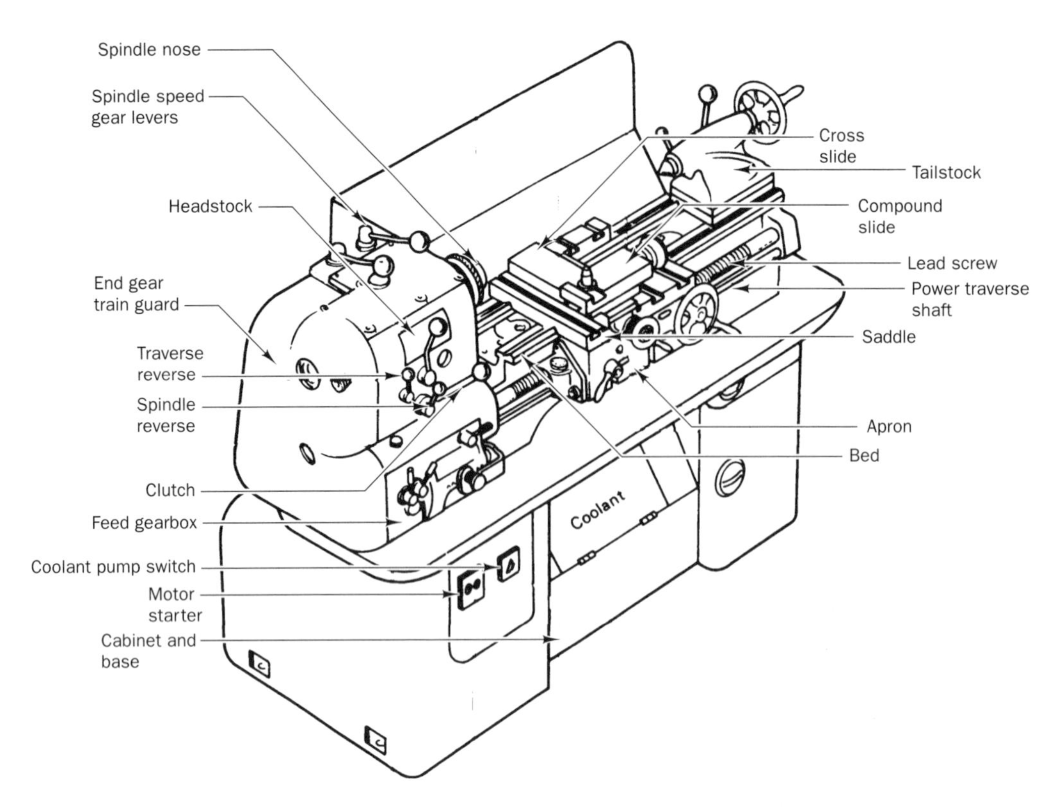

Figure 1.18 The centre lathe

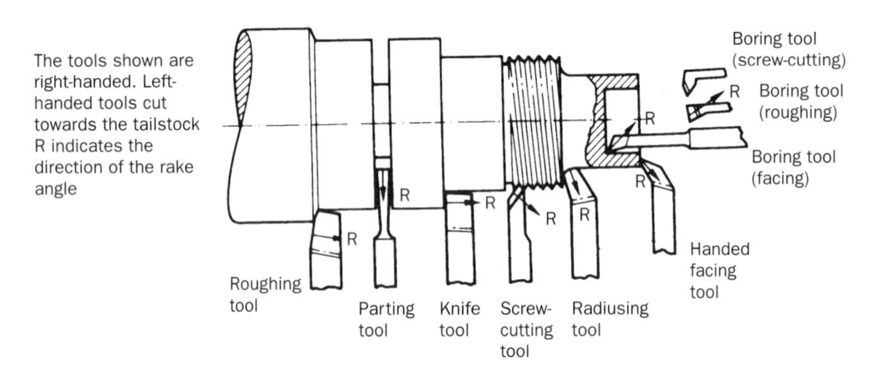

Figure 1.19 Turning tools

There are many different types of lathe cutting tool. Some of them are shown in Fig. 1.19.

All of the cutting tools are ground to a characteristic wedge shape with what are known as rake and clearance angles, as shown in Fig. 1.20.

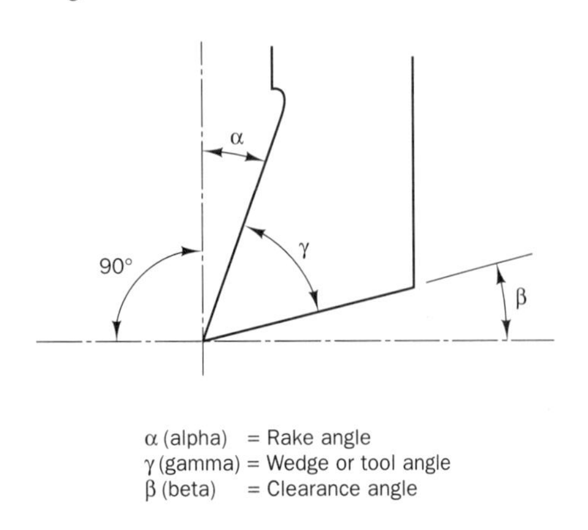

Figure 1.20 Tool-cutting angles

In the case of the knife tool and the parting-off tool, the cutting edge is at right angles to the direction of feed (Fig. 1.21). This is called 'orthogonal' cutting and the rake and clearance angles of the cutting edge are easy to see.

The cutting angles of the other tools shown in Fig. 1.19 are more complex and their cutting action is said to be oblique

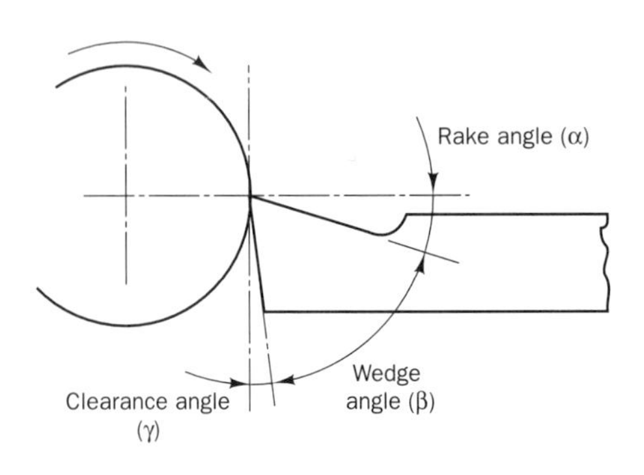

Figure 1.21 Orthogonal cutting

rather than orthogonal. Different cutting angles are recommended for cutting different materials and in all cases regrinding should be left to an expert.

Milling

In the milling process the workpiece is secured to the machine worktable and fed under a rotating multi-toothed cutter. There are two basic types of milling machine:

- Horizontal mill – This has its cutter axis in the horizontal plane.
- Vertical mill – This has its cutter axis in the vertical plane.

Both types of mill are used to machine flat surfaces and to cut steps and slots. The main parts of typical horizontal and vertical milling machines are shown in Figs 1.22 and 1.23.

In both types of milling machines the worktable can be raised and lowered and also moved horizontally in two perpendicular directions. The spindle which holds the cutter may be driven by a variable speed motor or through a gearbox, for selecting the required cutting speed. On most milling machines, the worktable can be set to traverse automatically under the cutter.

There is a wide variety of milling cutters. All of them have a series of wedge-shaped teeth, ground with rake and clearance angles.

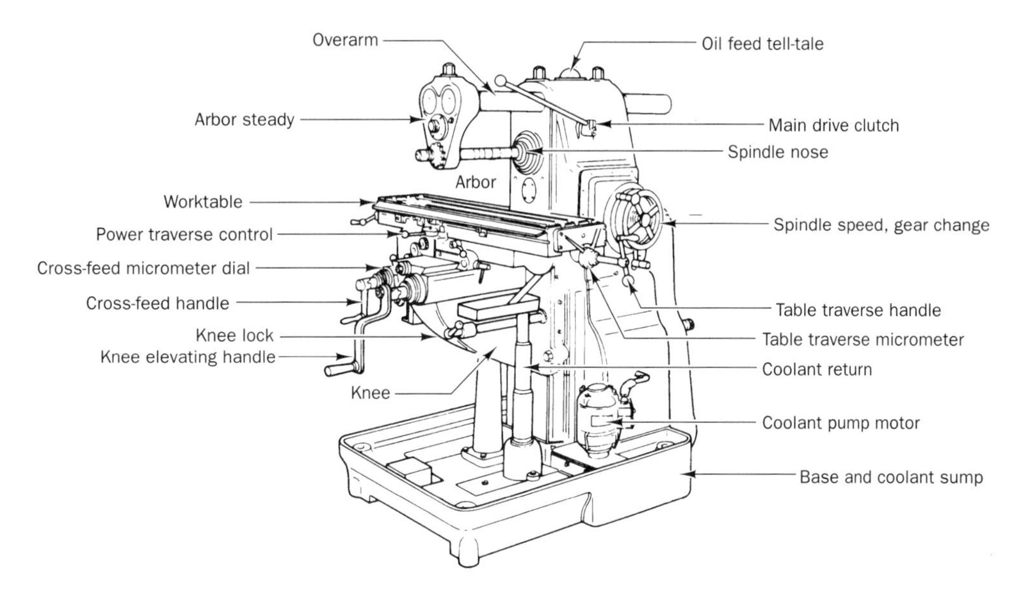

Figure 1.22 Horizontal milling machine

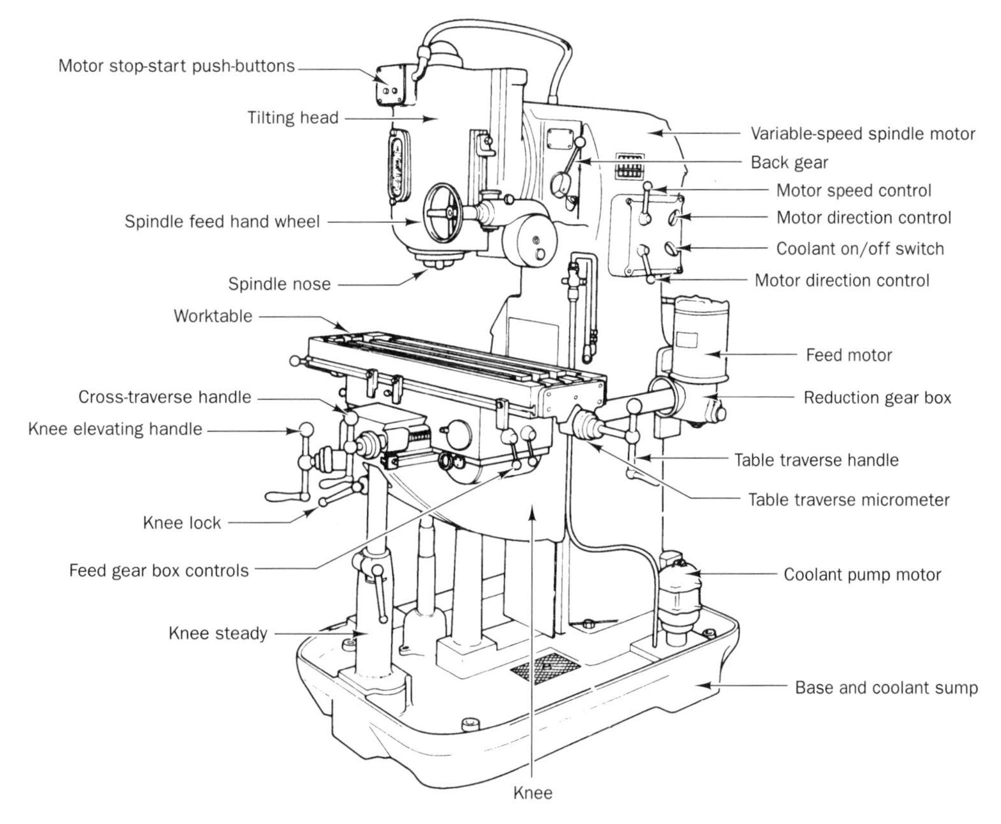

Figure 1.23 Vertical milling machine

Four of the most common cutters used on the horizontal milling machine (Fig. 1.24) are:

- Slab cutters – These are also called slab mills and roller mills. They are used to produce wide flat surfaces
- Side and face cutters – These have teeth on the periphery and the side faces. They are used for light facing operations and for cutting slots and steps.
- Slotting cutters – These are thinner than either of the above cutters and have teeth on the periphery only. They are used for cutting narrow slots and keyways.
- Slitting saws – These are the thinnest of the cutters. They are used to cut narrow slots and to cut material to size.

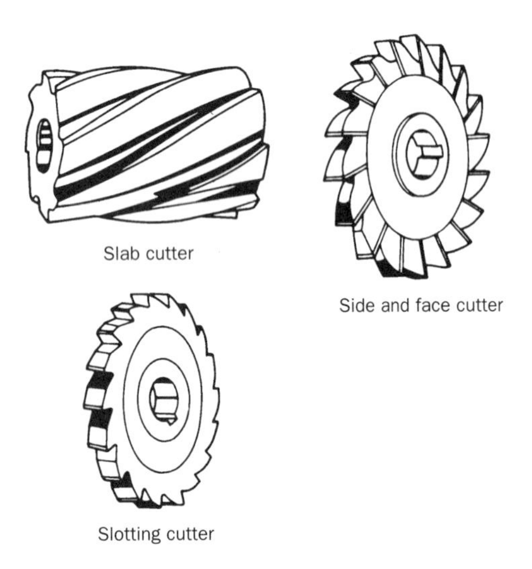

Figure 1.24 Cutters for the horizontal milling machine

Four of the most common cutters used on the vertical milling machine (Fig. 1.25) are:

- Face mills – These are used to produce wide flat surfaces and steps. They have teeth on the periphery and the end face. They produce flat surfaces more accurately than slab mills because they 'generate' the surface. That is to say that every part of every tooth on the end face passes over the whole surface.
- Shell end mills – These are also used for generating flat surfaces. They are smaller than face mills and mounted on a stub arbor.

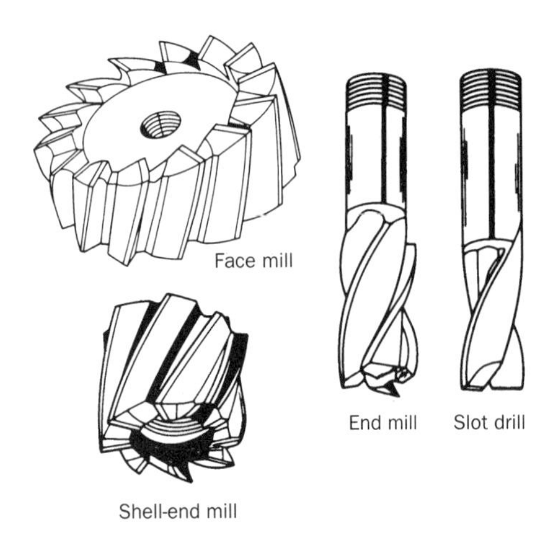

Figure 1.25 Cutters for the vertical milling machine

- End mills – End mills also have teeth on the periphery and end face. They are used for light facing operations, profiling and for milling slots.
- Slot mills – These are end mills with two cutting lips. Their prime use is for accurately milling slots and keyways.

Self-assessment tasks

1. Name two common types of drilling machine.
2. What material are the twist drills for metal working generally made from?
3. What are machine reamers used for?
4. What are the different surface shapes which can be machined on a centre lathe?
5. Name three different ways in which a workpiece can be supported in a centre lathe.
6. Which controls are to be found on the apron of a centre lathe carriage assembly?
7. Which kind of lathe cutting tool would you use for taking large initial cuts from a workpiece?
8. What are the two basic types of milling machine?
9. What kind of cutter would you use for machining a wide, flat surface on a horizontal milling machine?
10. What kind of cutter would you use for accurately machining narrow slots and keyways on a vertical milling machine?

Joining and assembly

The way in which the component parts of engineered products are joined together depends on the degree of permanence required. The choice of joining method also depends on the service loads which a product must carry and its service environment. Some of the most common methods are:

- Screwed fastening
- Riveting
- Soft soldering
- Hard soldering
- Welding
- Adhesive bonding

Screwed fastening

The different types of screwed fastening and locking device have been described previously. Screwed fastenings are used for joining those parts and components of a product which may need to be removed or replaced during service.

When parts have to be accurately located, fitted bolts are used. These have the diameters of their shanks machined to a fine tolerance and the holes in which they fit are drilled so that there is very little clearance. This ensures the precise positioning of mating parts and allows very little relative movement when loads are applied.

When it is desirable to have some slight movement or adjustment between mating parts, 'black bolts' are used. These are forged with a larger tolerance on the shank diameter and the holes through which they pass are drilled with a more generous clearance. This allows for quick assembly and adjustment of position between mating parts before the bolts are tightened.

In applications where strength is required, bolts and screws are made from forged high-tensile alloy steels. They are heat treated to give a high-tensile strength and their threads are formed by the process of thread rolling. This gives a stronger and more wear-resistant thread than those which have been cut using a die.

Nuts, bolts and screws are often given surface treatment to improve their corrosion resistance. Cadmium and zinc plating are the most common of these processes.

Where metric and non-metric screwed fastenings are in use they should be kept separate and clearly labelled. Only those of the specified thread form and size should be used.

Riveting

Riveting is a permanent method of joining materials together. Rivets are available made from steel, copper, brass, aluminium and aluminium alloys. They should be chosen with a composition as close as possible to that of the materials which are being joined. This reduces the risk of electrolytic corrosion around the rivets.The different shapes of head to suit different applications have been described previously.

An advantage of riveted joints (Fig. 1.26) is that they are not as rigid as those which are soldered, brazed or welded. They are able to flex a little under load which may be desirable in applications where shock loads occur. A disadvantage is that it is sometimes difficult to make riveted joints perfectly liquid- or gas-tight. They have been replaced by welded joints in many applications, but they are still widely used for joining materials such as aluminium alloys which are not so easy to weld.

The simplest type of riveted joint is the lap joint where overlapping sheets or plates are joined by one or more rows of rivets. As an alternative, the edges of the sheets or plates can be butted together and covered with a narrow plate, or 'strap', which is riveted to both of them. For maximum strength, straps can be positioned above and below the butted edges to give a double-strap butt joint.

Soft soldering

Soft solder is a low-melting point alloy of tin and lead which is used to form a bond between metal surfaces. Copper, brass and mild steel are metals to which molten solder readily adheres. Soft solder is used to make joints in plumbing, sheet metal fabrications and electrical and electronic circuits.

In the joining process a liquid or paste known as a flux is applied to the joint surfaces. This prepares the surfaces and enables the solder to adhere to them. Heat is applied to the joint and molten solder is drawn between the joint surfaces by capillary action. The solder is said to 'wet' the surfaces and forms a permanent film called an 'amalgam' which is difficult to remove.

Soft solder contains tin and lead, and a little antimony to improve its fluidity when molten. Some of the more common compositions are shown in Table 1.6.

Table 1.6 Soft solders

BS type	Composition (%)			Melting range (°C)
	Tin	Lead	Antimony	
A	65	34.4	0.6	183–185
K	60	39.5	0.5	183–188
F	50	49.5	0.5	183–212
G	40	59.6	0.4	183–234
J	30	69.7	0.3	183–255

Type A is used in electrical and electronic circuits. It is supplied in the form of wire with a core of resin flux. Types

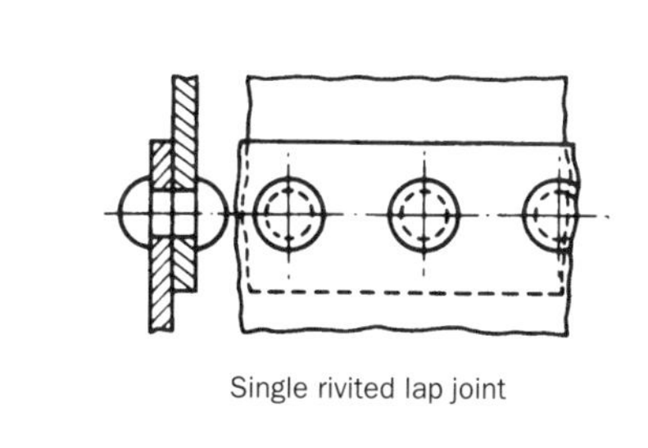
Single rivited lap joint

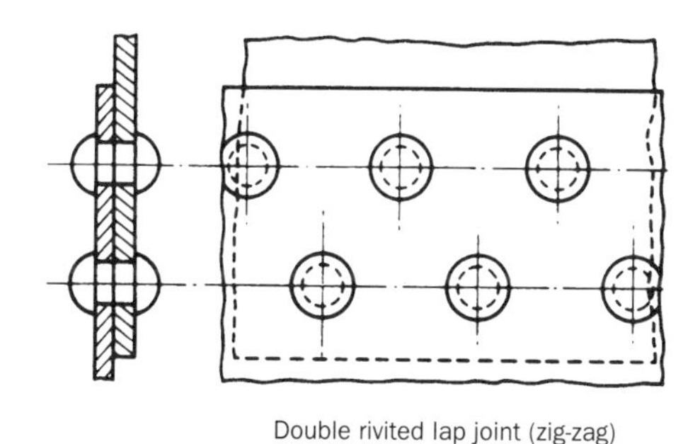
Double rivited lap joint (zig-zag)

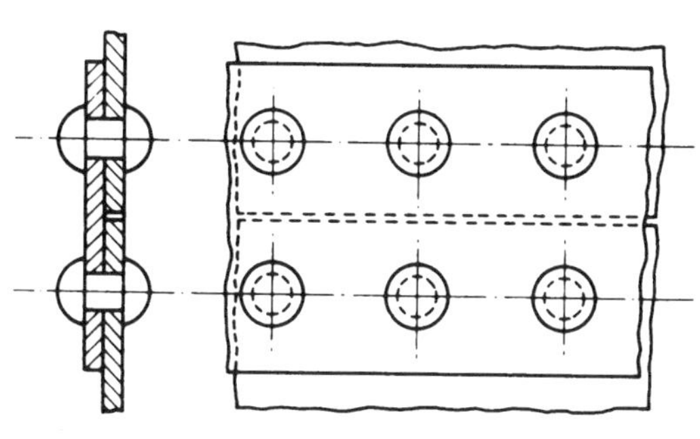
Single-strap butt joint (chain)

Double-strap butt joint (chain)

Figure 1.26 Types of riveted joint

K, F and G are used in sheet metal work, and type J is used in plumbing. They are supplied mainly in stick form but can also be obtained as wire or ribbon.

The fluxes used in soft soldering are usually supplied in the form of a paste or liquid which is applied to the surfaces being joined after they have been cleaned to remove dust, grease and metal oxide. There are two types of flux, active and passive.

Acidified zinc chloride solution, also known as Bakers Fluid, is an active flux. When the metal surfaces are heated, it cleans them of any remaining grease or metal oxide and protects them from further oxidation until solder has adhered to them. Its main disadvantage is that it is corrosive and any remaining flux should be washed off immediately after soldering.

Resin fluxes are supplied as a paste and are contained as a core in electrician's solder. They are passive fluxes which do not have a cleaning action. A resin flux merely protects surfaces which have been cleaned manually from any further oxidation when they are heated.

Hard soldering

Hard soldering includes brazing where brass is the bonding material, and silver soldering where the bonding material is an alloy of silver, copper, antimony, zinc and sometimes cadmium. Copper, brass, mild steel and cast iron are metals which can be hard soldered.

Hard solder has more strength and a much higher melting point than soft solder. Heat is applied from a blowtorch fed with a mixture of air and natural or propane gas. As with soft solder, the molten hard solder forms a bond with the parent metals and is drawn through a joint by capillary action.

The type of hard solder chosen for a particular application must have a melting point below that of the materials being joined. Brass, known as brazing spelter, is suitable for joining copper, mild steel and cast iron. Silver solder is more expensive, not quite so strong, and has a lower melting point than brazing spelter. It can be used to join the same metals and, because of its lower melting point, it is also suitable for joining with brass components. Table 1.7 shows the composition and melting points of two typical hard solders.

Table 1.7 Hard solders

BS type	Category	Composition (%)				Melting
		Silver	Copper	Zinc	Cadmium	range (°C)
3	Silver solder	50	15	15	20	620–640
10	Brazing spelter	–	60	40	–	885–890

As with soft soldering, a flux is used to clean away oxide film and protect the surfaces of a joint against further oxidation. Different fluxes are used with different hard solders, but most of them contain sodium borate, better known as borax. They are generally supplied in the form of a powder which is mixed with water to make a stiff paste. This is then applied to the joint faces before assembly.

Welding

In the welding process, the materials being joined are heated until they fuse together in the molten state. Additional material may be added from a filler rod which has the same composition as the material being joined. When formed, the joint should be equal in strength to the parent material. The two basic manual welding processes are:

- Oxy-acetylene welding – In this process, oxygen and acetylene gas burn to produce heat. The gases are stored under pressure in steel cylinders and released through a pressure regulator. Each regulator contains two pressure gauges which indicate the internal cylinder pressure and the line pressure to the welding torch.

  The gases are mixed together in the torch and burn to give temperatures which can exceed 3000 °C. This is well above the melting point of steel which is around 1500 °C. No flux is required when welding steel by this method, but the joint surfaces must be clean and properly prepared.
- Manual metal arc welding – Here the heat source is an electric arc which is struck between the filler rod and the material being joined. The arc is in fact a prolonged spark in which the temperature can reach 6000 °C. At this temperature, oxidation is a problem with steel. To overcome this the filler rod is coated with a flux which, as it burns, forms a gas shield around the arc to exclude the oxygen from the atmosphere.

  As with oxy-acetylene welding, the filler rod is of the same composition as the metal being joined. A mains transformer is used to supply the low voltage and high current which are required to maintain the arc.

Adhesives

Traditional glues and gums made from animal and vegetable matter are still used for some low-strength applications, particularly because of their non-toxic properties. In recent years, however, a wide range of high-strength synthetic adhesives have been developed. These have replaced many of the traditional substances and found many new applications in engineering. Some of the most common adhesives in use are:

- Thermoplastic adhesives – These may be heat-activated or solvent activated. With the heat-activated type, the adhesive is softened by heating and then spread evenly over the joint surfaces. These are then brought together and pressure is applied until the adhesive has cooled and set. With the solvent-activated type, the adhesive is supplied in liquid form. It is spread evenly on the joint surfaces which have been brought together and held in position until the solvent has evaporated. As this happens, the adhesive sets to form the joint.
- Impact adhesives – These are also solvent-based adhesives which are spread evenly on the joint surfaces. These are then left open to the atmosphere until the solvent has evaporated. The dry surfaces are then brought together and a bond is immediately formed.
- Thermo-setting adhesives – These are applied to the joint faces which are then brought together and held in position. The adhesive undergoes a chemical change when heat is applied and starts to solidifiy. As this proceeds, permanent cross-links are formed between the polymer chains in the adhesive. The heat can be applied externally or from an internal chemical reaction by mixing the adhesive with a chemical hardener before application. Once formed, the bond is permanent and cannot be softened by heat or solvents.
- Cyanoacrylate adhesives – These are more commonly known as 'super-glues' which cure in the presence of

moisture. They are very fluid and should be applied in a thin coating to one of the joint surfaces only. When the surfaces are brought into contact and light pressure is applied, bonding takes place almost immediately.

Heat treatment

In the case of metals and in particular alloys, a number of heat treatment processes have been developed to improve and enhance their properties. Some processes are designed to make a material more malleable and ductile while others seek to improve the hardness and toughness. In all processes, the temperature to which the material is raised and the rate of cooling are critical to the final properties of the material.

Chemical treatment

Processes which involves the use of chemicals are used to clean material, remove material and for surface finishing. The preparation processes include de-greasing in chemical solvents and pickling in acids to remove scale. They are used to prepare material and components for machining, for assembly or to receive some form of protective coating. Chemical treatment processes also include etching, which is used to remove unwanted material in the production of printed circuit boards.

The finishing processes include electroplating and anodising. With electroplating, metals are given a protective coating of another metal for protection or for decorative purposes. With anodising, a metal is given a thickened oxide film on its surface to increase its corrosion resistance.

Surface finishing

Finishing processes are chosen to meet the service conditions and aesthetic requirements of engineered products. The service conditions might dictate that some form of protective coating is required to guard against wear and corrosion. Painting, plating and coating with rubber or plastic material are common protective finishing processes.

Alternatively, a product may need to have a very smooth surface finish for a particular application. This may need a specialist machining process, such as surface grinding or cylindrical grinding, to produce it.

Self-assessment tasks

1. What are fitted bolts and what kind of application are they used for?
2. Why is it advisable to choose rivets which are made from the same material as the parts being joined?
3. Name three common metals or alloys which can be joined by soft soldering.
4. Which type of soft solder is used to make joints in electrical and electronic circuits?
5. What are the two basic types of hard solder?
6. How is heat applied to the joint in hard-soldering processes?
7. What are the two basic manual-welding processes?
8. Why is the filler rod used in manual metal arc welding coated with a flux?
9. Describe how two surfaces are joined together using an impact adhesive.
10. What are cyanoacrylate adhesives more commonly called?

Identification of the specific techniques appropriate for engineering processes

There are specific techniques and procedures associated with the processes described above. They have been developed over the years as the engineering processes have evolved. They should be followed to ensure that product quality is maintained and to provide a safe working environment for process operators and technicians. The specific techniques which will be described are:

- Material removal techniques
- Joining and assembly techniques
- Heat treatment techniques
- Chemical treatment techniques
- Surface-finishing techniques.

Material removal techniques

Material removal processes each have their own particular methods of tool holding and work holding. They also have their own specific operating procedures and techniques. Three of the most common material removal processes are:

- Drilling
- Turning
- Milling

Drilling

There are two basic methods of securing twist drills and reamers in drilling machine spindles.

- Morse taper – The spindles of drilling machines are hollow and contain an internal taper known as a 'morse taper'. The larger diameter drills and reamers have tapered shanks which locate in the internal taper. They are held in the spindle and driven through frictional contact alone.

  Jacobs chucks also have morse-tapered shanks and are held in the machine spindle in the same way. It is essential to see that tapered shanks are clean and undamaged before inserting them in the spindle. They should be removed using a drift which locates in an elongated hole at the top of the spindle taper. A sharp tap from a hammer will release the taper and a piece of wood positioned under the drill or chuck will prevent damage to the drill point or machine table.
- Jacobs chuck – The small sizes of drills and machine reamers have parallel shanks for gripping in a Jacobs-type chuck.

  When using a Jacobs-type chuck the drill should be securely gripped along as much of its shank as possible and securely tightened using the chuck key. The chuck key should then be removed and placed away from the workpiece.

There are different ways of holding a workpiece for drilling, but they all seek to restrain it from moving vertically and horizontally and from rotating with the drill. The most common methods of workholding are:

- Clamping to the machine table – Large or irregularly shaped components are often clamped directly to the machine worktable using bolts and clamps (Fig. 1.27).

They may also be clamped to an angle plate which is itself secured to the machine worktable. V-blocks should be used to support cylindrical components. The clamp bolts should be positioned as close to the workpiece as possible for the clamps to exert maximum force. At least two clamps should be used with large workpieces.

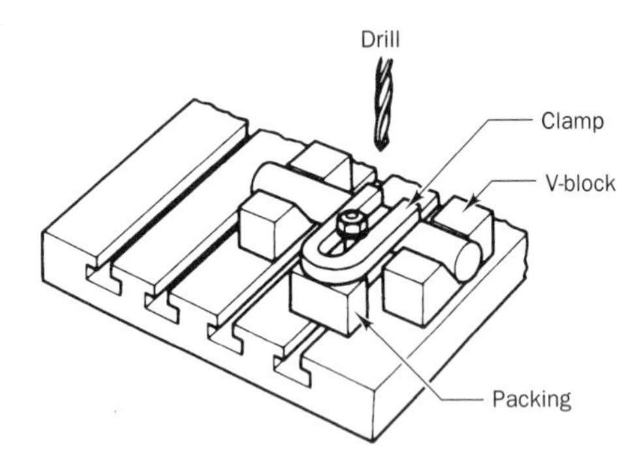

Figure 1.27 Methods of workholding

- Gripping in a machine vice – Smaller components may be held in a machine vice which can itself be bolted to the machine table. Parallel bars should be used to support the work to ensure that the drilled hole is perpendicular to the lower face of the work and to protect the drill and vice as the drill breaks through. A V-block is used to support cylindrical components (Fig. 1.28).

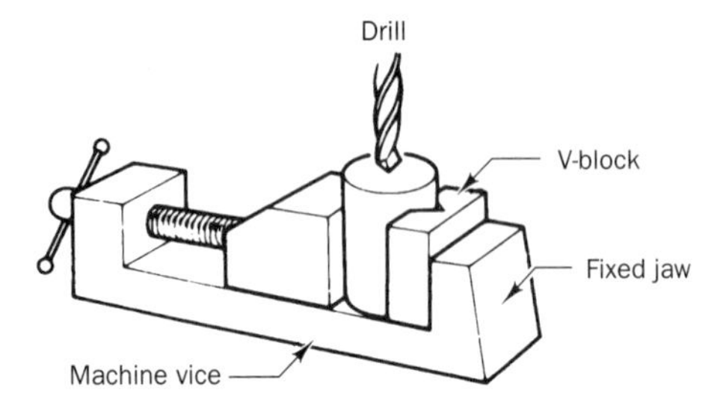

Figure 1.28 Use of machine vice

Having secured the cutting tool and workpiece in a drilling machine, it is essential to follow the correct procedures for:

- Setting the spindle speed
- Hand feeding the drill
- Removing swarf

Setting the spindle speed
As a general guide, the cutting speeds for drilling different materials with high-speed steel twist drills are given in Table 1.8.

Table 1.8 Cutting speeds for drilling

Material being drilled	Cutting speed (m/min)
Aluminium	70–100
Brass	35–50
Bronze (phosphor)	20–35
Cast iron (grey)	25–40
Copper	35–45
Steel (mild)	30–40
Steel (medium carbon)	20–30
Thermo-setting plastic	20–30

The required spindle speed for a given diameter of twist drill must be calculated because the bigger the diameter the smaller its rotational speed will need to be to give the recommended cutting speed. The following formula is used:

$$N = \frac{1000S}{\pi d}$$

where N = spindle speed in revolutions per minute
S = cutting speed in metres per minute
π = 3.142
d = drill diameter in millimetres

Having calculated the required spindle speed, the nearest value to it is selected from the range available. This is done using the speed change levers on machines fitted with a gearbox. On other machines the speed is selected by first isolating the machine, and then altering the position of the drive belt on its cone pulleys.

Hand feeding the drill
Care should be taken not to put too much pressure on a drill when feeding it by hand. This can cause small-diameter drills to break and the larger diameter drills to overheat. When deep holes are being drilled it may be advisable to use a coolant to prevent overheating.

Care should be taken as a drill is about to break through the material as at this point there is often a tendency for it to grab. The pressure should be eased off as breakthrough approaches. Most drills are fitted with a depth gauge which gives an indication of how far the drill has entered the material. It can also be set to limit the travel of the drill as it breaks through the material and when drilling blind holes.

Removing swarf
A drill which is correctly ground and used with the correct speed and feed will often produce continuous lengths of swarf from each of the cutting lips. This occurs especially when drilling mild steel and aluminium, but if the swarf is allowed to become too long, it can rotate with the drill and cause injury.

The drilling pressure should be eased off occasionally to break the swarf and prevent this from happening. Swarf should not be allowed to accumulate on drill worktables. After completing a drilling operation, the table and its tee- slots should be swept clean and the swarf deposited in bins provided.

All drills should be fitted with an appropriate safety guard. This should always be placed in position before drilling.

Turning

When using the centre lathe for material removal operations, the cutting tools are held in the tool post or the tailstock:

- The tool post – On centre lathes the tool post is mounted on top of the compound slide. It holds the single point cutting tools used for cylindrical turning, facing, boring, parting-off, etc.

  Small centre lathes may have a pillar-type tool post which holds a single tool. The larger machines often leave a four-way tool post which can be rotated to bring the different tools into operation (Fig. 1.29). The tools must be set with their cutting edges exactly at the height of the spindle axis. Packing pieces are placed under them to achieve this. The pillar type tool post has a curved boat-piece beneath the cutting tool which allows the height of the tool point to be easily adjusted.

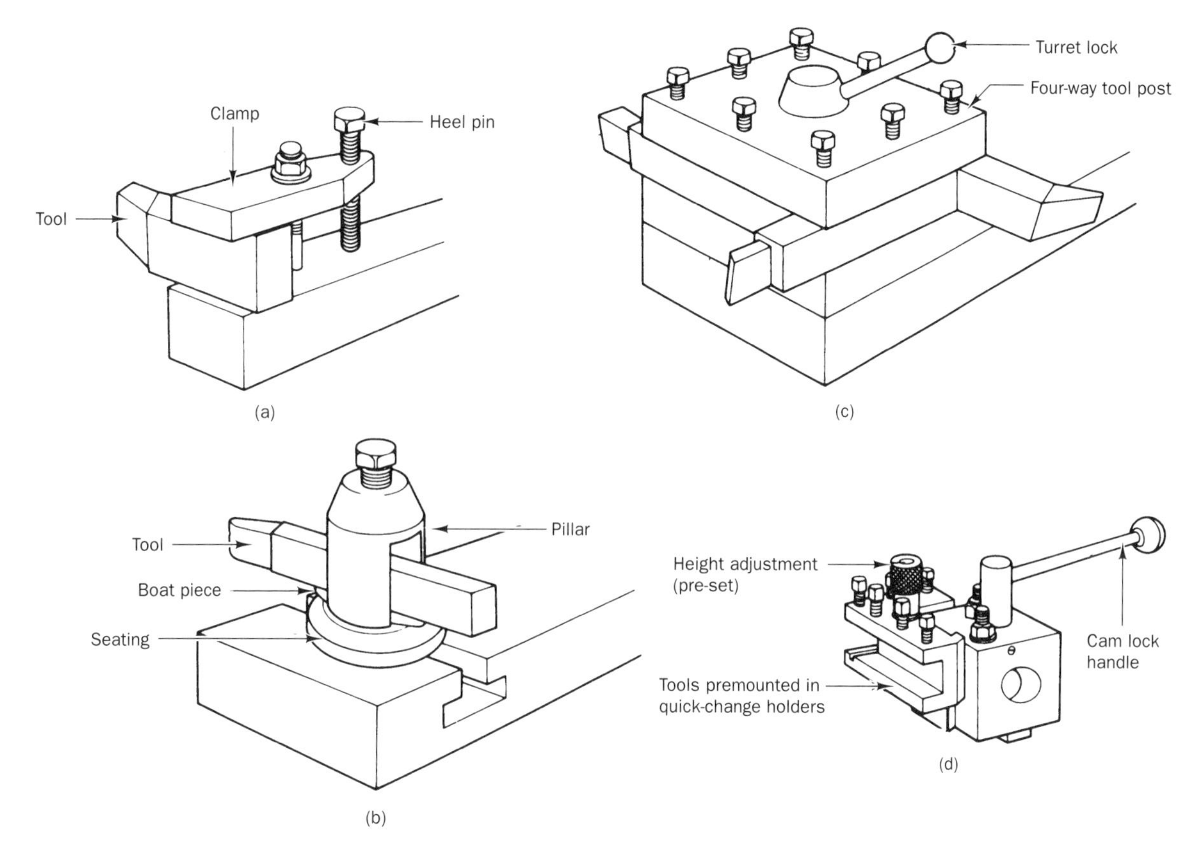

Figure 1.29 Tool posts: (a) English (clamp) type; (b) American (pillar) type; (c) turret (four-way) type; (d) quick-release type

If the centre height above the lathe bed is known, the tool height can be set using a steel rule. Alternatively, the tool cutting edge can be set at the height of the tailstock centre. As a check, the tool can be run up to the stationary workpiece against an upright steel rule. If the rule appears to stand vertically the tool will be close to centre height.

Some centre lathes are fitted with quick-release tool clamps. These have adjusting screws for setting the tool to the correct centre height. They can be preset and removed from the tool post while awaiting use at some later time.

- The tailstock – The tailstock is used when drilling in the centre lathe. Its spindle contains an internal morse taper which can hold a Jacobs chuck and taper-shank drills. The tailstock is clamped to the lathe bed during drilling operations and the drill is fed into the rotating workpiece using the spindle feed handwheel.

  The tailstock is also used when cutting screw threads with taps and dies. Screw cutting taps are held in the Jacobs chuck and the dies are held in a diestock fitted with a morse-tapered shank which locates in the tailstock spindle. The unclamped tailstock is fed by hand to the workpiece. At the same time, the workpiece is rotated backwards and forwards by hand with the lathe isolated.

Their are different ways of holding a workpiece in the lathe but they all seek to restrain it from any form of movement except rotation with the lathe spindle. The workpiece is most commonly held using the following:

- Three-jaw self-centring chuck
- Four-jaw independent chuck
- Faceplate
- Catchplate carrier and centres

Three-jaw self-centring chuck
These can be used to hold a wide range of cylindrical and hexagonal workpieces. The three jaws move inwards and outwards together to ensure that the centre line of the workpiece lies on the spindle axis. They are driven by the scroll plate which is rotated by the chuck key (Fig. 1.30).

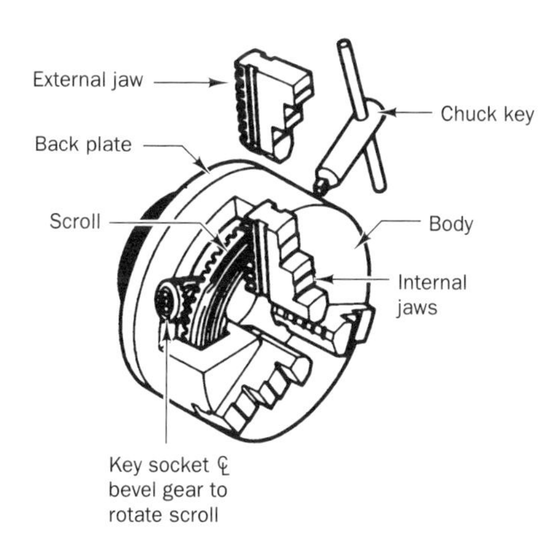

Figure 1.30 Three-jaw self-centring chuck

Three-jaw self-centring chucks can be used for turning external surfaces, boring and end facing. They should not be used with work that is not truly cylindrical such as hot-rolled black bar, or hexagonal, as this will put unequal load on the jaws and the workpiece may not be securely gripped.

Four-jaw independent chuck
With this type, each jaw moves independently on a screw thread and has its own chuck key socket. Four-jaw chucks are used for gripping square bar, hot-rolled black bar and irregularly shaped casting and forgings (Fig. 1.31).

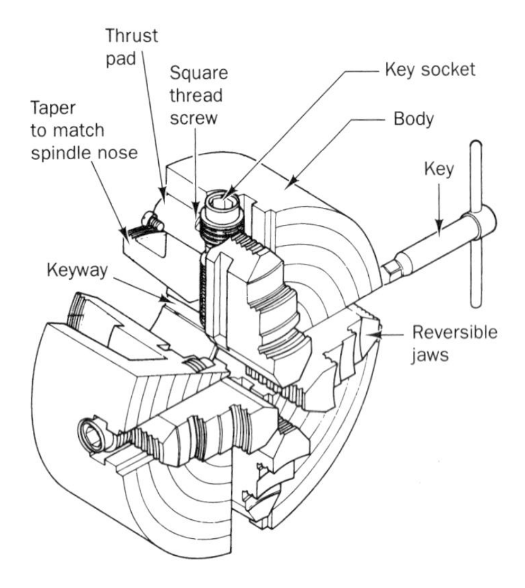

Figure 1.31 Four-jaw independent chuck

Because each jaw can be tightened separately, the independent four-jaw chuck has greater gripping power than the three-jaw self-centring type.

Face plate
This allows irregularly shaped castings or forgings, which may be too large for a four-jaw chuck, to be securely held (Fig. 1.32).

The face plate enables diameters and faces to be turned which are perpendicular or parallel to a pre-machined surface. The pre-machined surface locates against the face plate or on the angle plate. For through-boring operations, the workpiece should be separated from the face plate by parallel bars.

When the workpiece is irregularly shaped or eccentrically mounted, a considerable out of balance force may be present. This can cause vibration, poor surface finish and harm to the machine bearings. To prevent this, balance weights must be attached to the face plate by trial and error, until it can be set in any position without swinging round, due to gravity.

Catch plate carrier and centres
This is a very old and accurate method of turning long and cylindrical workpieces (Fig. 1.33).

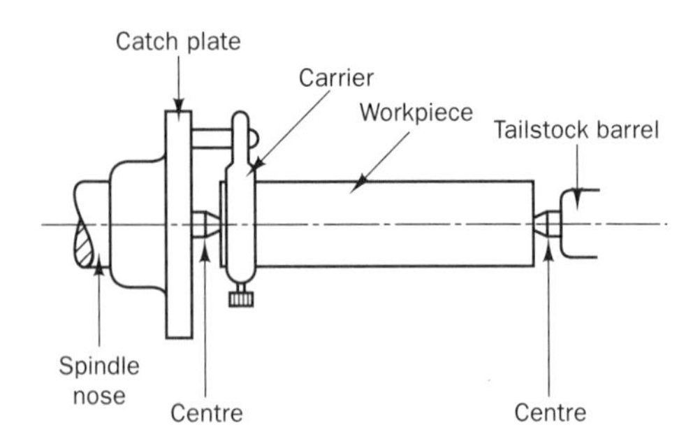

Figure 1.33 Workholding between centres

The workpiece is centre drilled at each end and has the carrier clamped to it. It is then held between centres and rotated by the catchplate pin which drives the carrier. An advantage of the method of workholding is that the work can be taken out, reversed and replaced in the machine without loss of concentricity. It can also be taken out for other machining operations. Disadvantages of the method are that no boring operations, and only limited end facing, can be performed.

Having secured the cutting tools and workpiece in a lathe, it is essential to follow the correct procedures for the following:

- Setting the spindle speed
- Feeding the cutting tool
- Supporting the free end of a workpiece
- Cooling the workpiece

Setting the spindle speed
As a general guide, the cutting speeds for turning different materials with high-speed steel (HSS) cutting tools are given in Table 1.9.

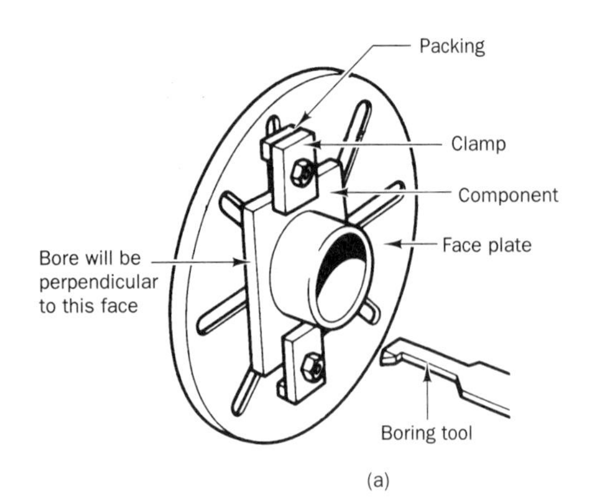

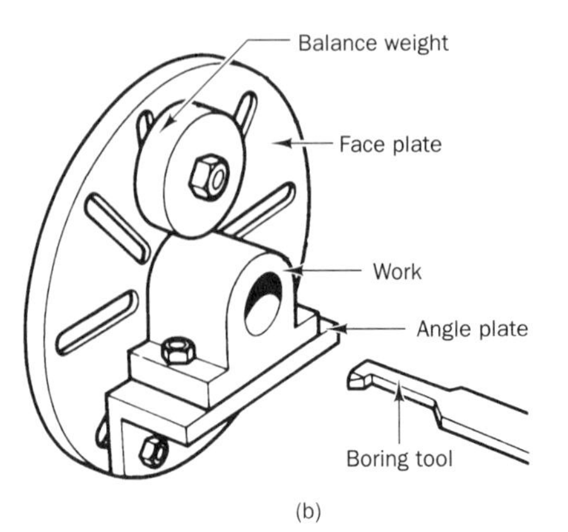

Figure 1.32 Use of face plate: (a) balanced work; (b) unbalanced work

Table 1.9 Cutting speeds for turning

Material being turned	Cutting speed (m/min)
Aluminium	70–100
Brass	70–100
Bronze (phosphor)	35–70
Cast iron (grey)	25–40
Copper	35–70
Steel (mild)	35–50
Steel (medium carbon)	30–35
Thermo-setting plastic	35–50

The spindle speed to which the machine must be set is measured in revolutions per minute, and depends on the diameter of the workpiece. The greater the diameter, the greater the surface speed of the workpiece and the lower will be the spindle speed needed to give the recommended cutting speed in metres per minute. The formula already given for drilling speed gives the rotational speed required for a given workpiece diameter.

Having calculated the required spindle speed, the nearest value from the range available is selected using the speed change levers on the lathe gearbox.

Feeding the cutting tool

To remove material as quickly as possible the depth of cut should be as large as the machine can handle. As a general rule, a deep cut coupled with a fine feed rate is the best combination for fast metal removal, long tool life and a good surface finish. This assumes, of course, that the tool is correctly ground and set in position and that the workpiece is securely held.

Most centre lathes have provision for automatic carriage feed. This is driven from the power traverse shaft or 'feedshaft'. The drive can also be directed to the cross-slide for automatic cross-traverse on many centre lathes. The lead screw, which runs along the side of the bed with the power traverse shaft, is also used to give automatic carriage feed, but it is only used for screw-cutting operations. The engagement levers for automatic feed are situated on the apron.

Supporting the free end of a workpiece

All except the shortest workpieces need to be supported at the tailstock end. They are supported on a centre which is held in the tailstock spindle but first they need to be centre drilled.

The centre drill is held in a Jacobs chuck in the tailstock and produces a hole in the workpiece which is part parallel and part tapered. The supporting centre may be a stationary or 'dead' centre, or a 'running' centre. If a stationary centre is to be used, the centre drilled hole should be greased to prevent overheating. The use of a 'running' tailstock centre, which rotates with the workpiece, will eliminate friction and overheating completely.

Cooling the workpiece

All except the smallest centre lathes are equipped with a coolant reservoir and delivery pump. This supplies coolant through an adjustable pipe and control tap to the workpiece. There are many types of coolant for different cutting processes and materials, but their functions are generally the same. They are intended to:

- Carry heat away from the workpiece and cutting tool.
- Lubricate the chip–tool interface to reduce tool wear.
- Prevent chip particles from becoming welded to the tool face, forming a built-up edge.
- Wash away chips (swarf).
- Improve the surface finish and prevent corrosion of the workpiece and machine.

Emulsified or soluble oils are the most common coolant for general machine shop use. They contain oil, detergent and disinfectant and, when mixed with water, they have a white milky appearance. The oil content gives the coolant its lubricating and corrosion protection qualities. The water content has the cooling effect and the detergent helps the water and oil to mix or 'emulsify'. The disinfectant kills bacteria which might breed in the coolant, causing the emulsion to break down and possibly spread infection.

The coolant should be directed onto the cutting area of the workpiece in a steady stream from which it filters back to the reservoir beneath the lathe bed. Care should be taken to ensure that the return filter does not become blocked with swarf.

Cast iron may be machined without a coolant. Its free graphite content gives it self-lubricating properties and less heat is generated compared to other metals. The graphite does, however, appear as black dust and lathes used regularly for turning cast iron should be fitted with dust extractors.

All lathes should be fitted with guards. A minimum requirement with centre lathes is that the guard should cover the rotating chuck or face plate. On more advanced automated types of lathe, the complete workpiece may be guarded. After fixing a workpiece in a chuck the chuck key should be removed and the guard placed in position before engaging the spindle clutch. The guard should not be removed until the clutch has been disengaged and the workpiece is stationary.

Milling

Milling machines use multi-toothed cutters for removing material. The devices used to hold the cutters are:

- Arbors – Arbors are used to hold the cutters on both horizontal and vertical milling machines. The arbor locates in a taper in the machine spindle. It is secured in position by a draw bolt and driven by two dogs on the spindle nose. The method of mounting the cutter on the arbor varies with the type of machine.

  The horizontal milling machine has a long arbor with a keyway running along it. The cutter is keyed to the arbor and positioned with spacing collars on each side. A nut on the end of the arbor is tightened to hold the spacing collars and cutter tightly together. The free end of the arbor is supported by the overarm steady.

  The cutter and the overarm steady should be positioned with as little overhang as possible on each side of the cutter and, in some instances, two steadies may be used. This reduces the likelihood of inaccuracies caused by arbour flexing and bending when cutting.

  On the vertical milling machine, face mills and shell end mills are mounted on a 'stub' arbour or held in position by a retaining screw. When fitting an arbor to a milling machine spindle, the taper and flange surfaces should be absolutely clean to ensure true running. Unlike the morse taper on drill shanks, the taper on milling machine arbors is for location only, and the dogs which engage in the slots on the arbor flange provide the drive. Care should be taken to see that the arbor is properly located before tightening the draw bolt.

- Collet chucks – The end mills used on the vertical mill have a parallel shank which is threaded at its end. This locates in an 'antilock' collet chuck which is so designed that the cutting forces tend to increase the grip of the collets on the cutter shank.

Milling cutters have a number of extremely sharp cutting edges and care is needed when handling them. Stout leather gloves give the best protection.

Milling machines have an intermittent cutting action and exert large cutting forces on the workpiece. This needs to be held very firmly and positively to prevent movement. As with drilling machines the most common methods are as follows:

- Clamping to the machine worktable – The methods used are similar to those described for drilling operations where large, or irregularly shaped components are clamped directly to the T-slots in the machine worktable or to an angle plate which is itself bolted down. At least two clamps should be used and be positioned so as to give maximum restraint to the workpiece. As with drilling operations, V-blocks can be used to support cylindrical workpieces.
- Clamping in a machine vice – The machine vices used on milling machines must be of a heavy duty type because of the high cutting forces involved. Heavy duty vices with a swivel base are often used which can be set at any angle in the plane of the work table.

  Components held in a machine vice for milling should be supported on parallel bars to ensure that the upper and lower machined surfaces of a component are parallel. While tightening up the vice, a blow to the workpiece from a hide hammer will help to keep it sitting tightly on the parallel bars. If the vice is badly worn, with excessive play in its slideways, there is a tendency for the moving jaw to lift as it is being tightened. This makes it very difficult to keep the workpiece in contact with the parallel bars and the vice should not be used for precision work.

Having secured the cutters and the workpiece in a milling machine, it is essential to follow the correct procedures for:

- Setting the spindle speed
- Feeding the worktable
- Cooling the cutter and workpiece

Setting the spindle speed

As a general guide, the cutting speeds for milling different materials with high-speed steel milling cutters are given in Table 1.10.

Table 1.10 Cutting speeds for milling

Material being milled	Cutting speed (m/min)
Aluminium	80–110
Brass	70–100
Bronze (phosphor)	35–70
Cast iron (grey)	25–40
Copper	40–70
Steel (mild)	35–50
Steel (medium carbon)	30–35
Thermo-setting plastic	35–50

The required spindle speed for a given diameter of milling cutter is calculated using the same formula as for turning and drilling.

Having calculated the required spindle speed, the nearest value from the range available is selected using the speed change levers on the milling machine gearbox.

Feeding the worktable

With horizontal milling machines it is important to feed the worktable and workpiece under the cutter in the correct way. There are two methods of horizontal milling. They are up-cut milling and down-cut milling (Fig. 1.34).

Up-cut, or conventional milling, should always be used unless the machine is specially designed or equipped for down-cut milling. Up-cut milling has the disadvantage that it tends to lift the workpiece from the worktable and that the cutting teeth tend to rub on the surface of the work before starting to bite. These are, however, outweighed by the fact that with all except new and specially equipped machines, there is backlash in the worktable feed mechanism.

If down-cut milling is attempted, there is a tendency for the workpiece to be dragged under the cutter and for the cutter to climb onto the workpiece. This can result in excessive vibration and possible damage to the arbor.

Down-cut milling should not be used unless the machine is specially designed for it or unless it is fitted with a backlash eliminator in the worktable drive mechanism.

Cooling the cutter and workpiece

As with turning and drilling, emulsified oil coolants are used for both horizontal and vertical milling operations. Care should be taken to ensure that the return drain from the worktable and the filter do not become blocked with swarf particles. Heavy duty milling machines, where the cutting action is severe, may use soluble oil in undiluted form.

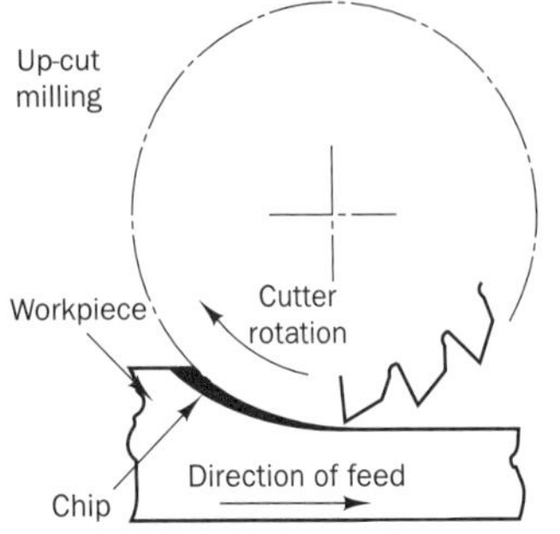

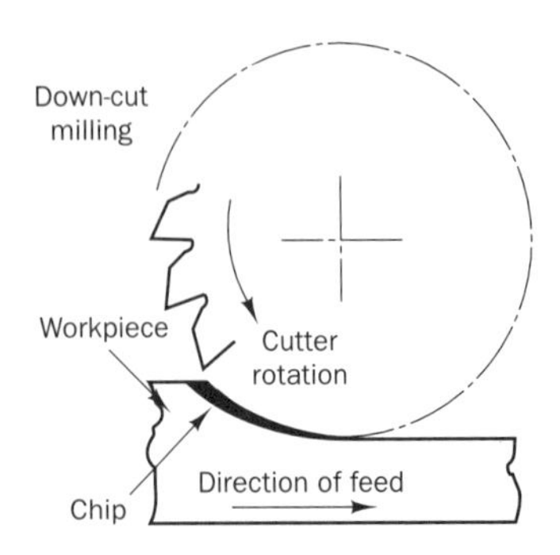

Figure 1.34 Up-cut and down-cut milling

The multi-toothed cutters used on milling machines, together with the swarf chips thrown out by the cutter, make milling a more dangerous machining process than either drilling or turning. It is essential that cutter guards are in position before engaging the spindle clutch and that the guard is not removed until the spindle is once again stationary.

Self-assessment tasks

1. What are the two basic methods of securing twist drills in a drilling machine?
2. How should cylindrical components be supported on a machine worktable?
3. What is the required spindle speed for a 10 mm diameter twist drill which is to have a cutting speed of 30 metres per minute?
4. What precautions should be taken when hand-feeding a drill?
5. How you would set a turning tool at the correct height?
6. How are drilling operations carried out in the centre lathe?
7. What kind of workpieces may be held in a three-jaw self-centring chuck?
8. Give one advantage and one disadvantage of turning between centres.
9. What is the purpose of the leadscrew which runs along the side of the bed of a centre lathe?
10. How is the free end of a cylindrical workpiece supported in the centre lathe?
11. What is the purpose of the overarm steady on a horizontal milling machine?
12. How are face mills and shell end mills mounted in the vertical milling machine?
13. Why is up-cut milling the recommended method of cutting for the majority of horizontal milling machines?
14. How should components be supported when held in a machine vice on a milling machine?
15. What is the reasons for using a coolant in machining processes?

Joining and assembly techniques

Product quality often depends on the care which has been taken when joining and assembling materials and components. Joining and assembly processes each have their own particular procedures and techniques. The most common processes are as follows:

- Screwed fastening
- Riveting
- Soft soldering
- Hard soldering
- Welding
- Adhesive bonding

Screwed fastening

Where metric and non-metric screwed fasteners are in use they should be kept separate and clearly labelled. Only those of the specified thread size and length should be used.

Good quality open-ended, ring and socket spanners should be used so as not to damage the faces of hexagonal nuts, bolts and screw heads (Fig. 1.35). Care should be taken not to over-tighten small-diameter fasteners, as they can quite easily be sheared across. When torque settings are specified a suitable torque wrench should be used.

Before joining components with screw fasteners, care should be taken to ensure that the joint faces are clean and that any burrs around machined faces and drilled holes are removed. Blind tapped holes should be inspected to ensure that no swarf remains in them from the tapping process.

The order in which screwed fasteners are tightened is sometimes important. When they lie on a pitch circle it is good practice to follow the sequence of tightening those which lie diametrically opposite. If the fasteners lie around a rectangle, it is good practice to start with those at the centre of the larger sides and work alternately outwards, tightening up those on the shorter sides last of all (Fig. 1.36).

Screwed fastenings are widely used to make electrical connections. The regulations which govern electrical installation work require connections and joints to be mechanically and electrically sound. It is recommended that joints in non-flexible cables should be made by soldering, brazing, welding or mechanical clamping. The joints should be appropriate to the size of the conductor.

Light duty terminations used in domestic installations, appliances and accessories make wide use of the screwed terminals shown in Fig. 1.37.

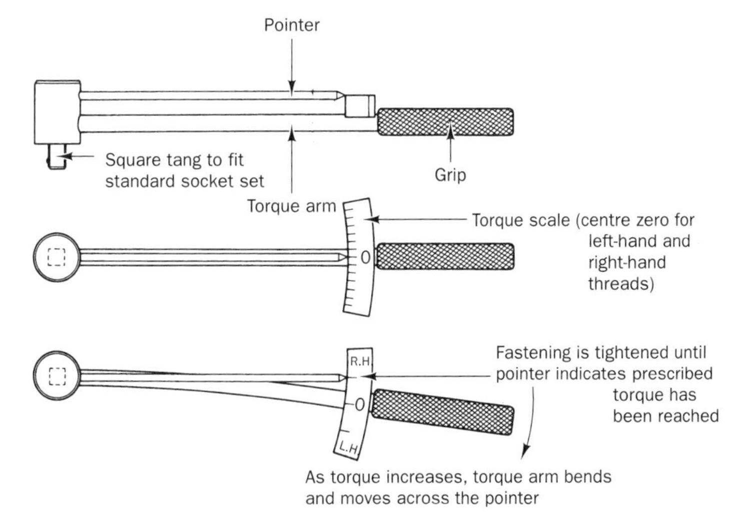

Figure 1.35 Typical torque spanner

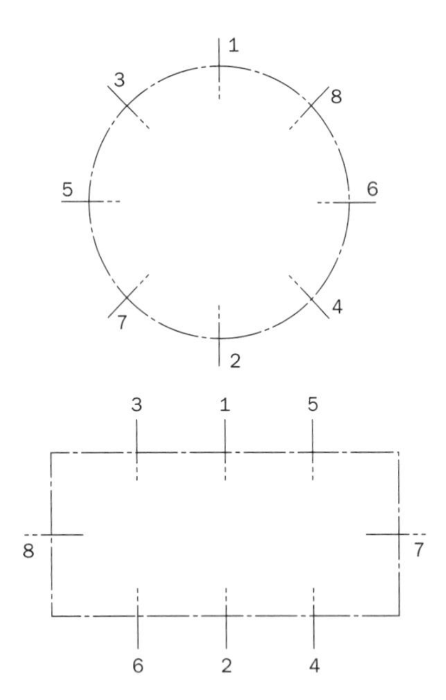

Figure 1.36 Tightening sequence

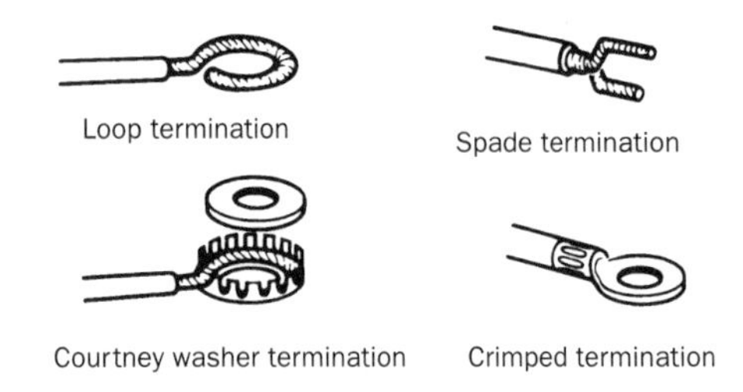

Figure 1.37 Light duty terminations

Crimped terminations are fitted on the end of a conductor after stripping off a suitable length of insulation. They are then compressed onto the conductor using special pliers. Loop, spade and other types of terminal are available for crimp fitting.

Heavy duty terminations (Fig. 1.38) are used in power distribution and supply applications. Both copper and aluminium are used for heavy duty cables. The soldered lug termination is more suited to copper which can easily be soldered using an appropriate flux and heat source. The crimped lug termination has found wide acceptance for use with both copper and aluminium cables. It is easy to fit using a special compression tool and there is no possibility of damage to the cable from heat or a corrosive flux.

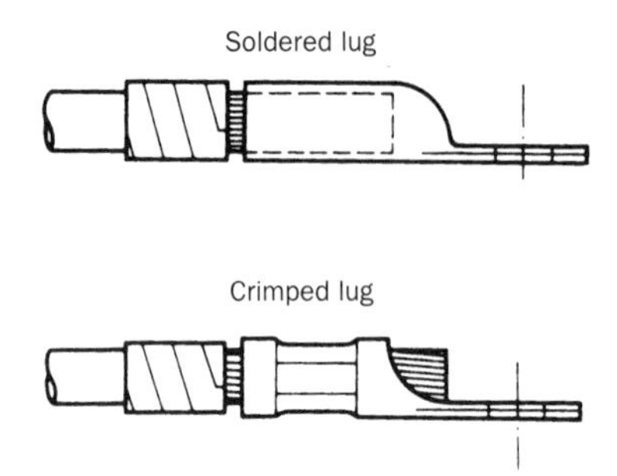

Figure 1.38 Heavy duty terminals

Riveting

Only rivets of the specified material, diameter, length and head shape should be used. Special care is required with aluminium rivets which may age harden if not used within a given time.

As with screwed or bolted joints, care should be taken to ensure that the joint faces are clean and that their edges and drilled holes are free from burrs.

When the rivets have been inserted in a joint, a drawing-up tool is used to ensure that they are seated correctly. A compressive force from a ball pein hammer or pneumatic riveting tool is applied to swell the shank and rough-form the head. A rivet snap of the correct shape is then used to finish-form the head.

In joints where pop-rivets are used, the same careful preparation of the joint faces is required. Pop-rivets of a sufficient length should be used, care being taken to use the sealed type if a liquid- or gas-tight joint is required.

Soft soldering

Before joining metal surfaces using soft solder, they must be carefully cleaned, first, by wiping off oil or grease and then by removing metal oxide using wire wool or emery cloth.

Flux should then be applied to the surfaces after which they must each be tinned or 'wetted' with the solder. This is done using a soldering iron whose copper bit has been heated in a gas flame, fluxed and loaded with molten solder. This is then applied to the joint surfaces until they are covered with an even film of solder as shown in Fig. 1.39.

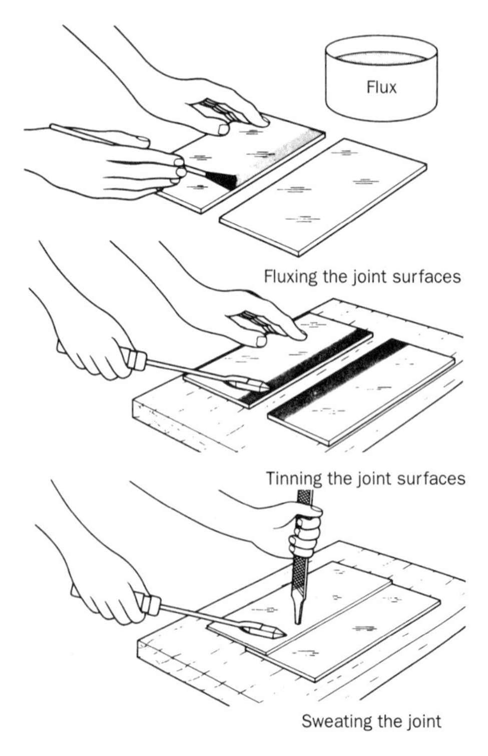

Figure 1.39 Soft soldering procedure

The tinned surfaces are then placed together and heated using the hot soldering iron while at the same time applying pressure. When molten solder from the tinned surfaces is seen to appear at the edges of the joint, the soldering iron is drawn along until the surfaces are completely fused together. This operation is known as 'sweating' the joint.

The pressure should not be removed until the solder has solidified, after which any active flux residue should be washed off and a rust inhibitor applied to the area surrounding the joint.

The lugs for soldering on electrical cables are generally ready loaded with solder. The cable insulation should be stripped back and a passive resin flux applied to the bare copper. This is inserted into the terminal lug. Heat is applied to sweat the joint until molten solder is seen emerging from the edge. Overheating should be avoided as the remaining insulation may become damaged.

Electronic components are usually supplied with their leads and terminals ready tinned. Tinned copper wire, or copper printed circuit board, is used for making the circuit. Having placed a component in position, heat is applied using an electric soldering iron whose bit is loaded with sufficient resin-cored solder to make the joint.

The bit should be removed as quickly as possible after making the joint to prevent damage to the component. Pliers can sometimes be used to grip the leads to sensitive components while they are being soldered in order to conduct excess heat away.

Care should always be taken when working with hot materials and corrosive liquids. Heat-resistant gloves should be worn for handling hot workpieces and eye protection should be worn when using active fluxes as spitting might occur when they are heated.

Hard soldering

As with soft soldering, the surfaces to be joined by hard soldering should be free of oil and grease and cleaned with wire wool, a wire brush or emery cloth to remove the surface oxide. They are then coated with a flux paste and assembled on a brazing hearth, surrounded by fire bricks, as shown in Fig. 1.40.

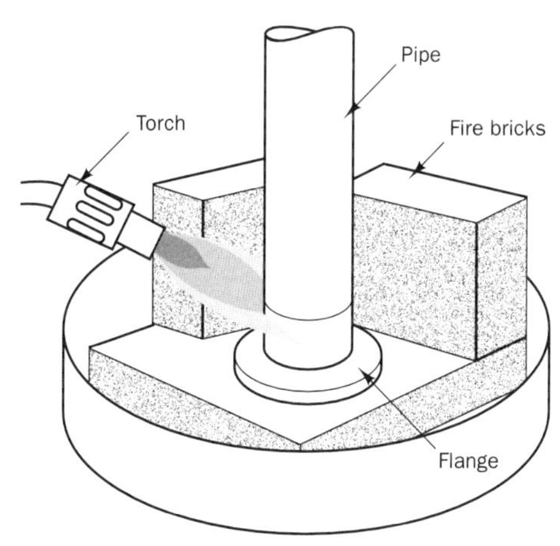

Fire bricks or other suitable insulating materials are packed around the component to be brazed. This helps to contain and reflect the heat supplied by the torch.

Figure 1.40 Hard soldering

Heat is then applied from a blow torch and, as the temperature of the work rises, the flux will be seen to be cleaning and protecting the joint surfaces. The hard solder, in the form of rod or wire, can then be touched on the edge of the joint. If the temperature is correct, it will melt and be drawn into the joint by capillary action. After the work has cooled down the flux residue, which sets as a glassy film, may be cleaned away with a wire brush or emery cloth.

The temperatures required for hard soldering are much higher than for soft soldering. Heat-resistant gloves should be worn during the process and when moving the hot workpiece. It is also advisable to wear eye protection against sparks.

Welding

Welding technology is a wide-ranging subject. It covers a variety of procedures and methods for welding different materials of various thicknesses and with different types of joint.

The edges to be joined should be free of any scale or rust. Depending on the thickness of the material, they may also need to be ground to an angle to assist weld penetration. The procedures and techniques which should then be followed depend on which of the two main types of welding are being used. These are as follows:

- Oxy-acetylene welding
- Metallic arc welding

Oxy-acetylene welding

Before using a new gas cylinder for oxy-acetylene welding, the valves should be opened momentarily to blow away any dirt or dust from the internal screw thread. The screw thread on the regulator and in the cylinder should then be wiped clean to ensure that no oil or grease is present before they are assembled. This is because oil and grease can suddenly burst into flames in the presence of high-pressure oxygen by a process known as spontaneous combustion.

The makers of oxy-acetylene equipment supply charts which recommend the nozzle size and line pressures which should be used for different material thicknesses. Having selected the correct nozzle, and set the line pressures, the gas flame can be lit. This is done by first turning on and lighting the acetylene and then slowly turning on the oxygen supply.

Two inner cones will be seen in the flame and the oxygen supply should be increased until the outer cone disappears to give a sharply defined inner cone. This is recommended flame setting for welding mild steel. It is known as a neutral flame in which the oxygen supply is just sufficient to burn all of the acetylene (Fig. 1.41).

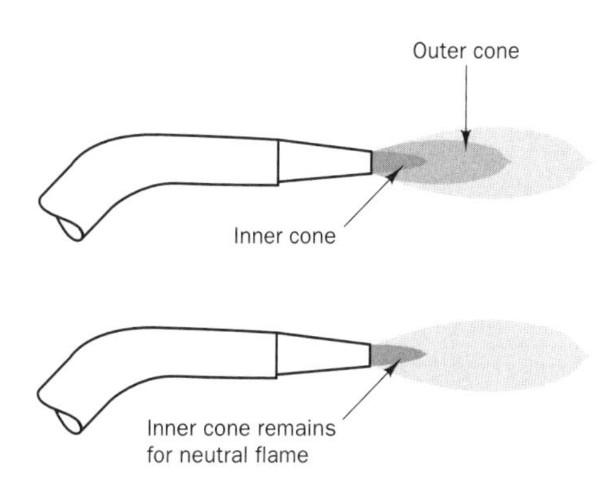

Figure 1.41 Obtaining neutral flame

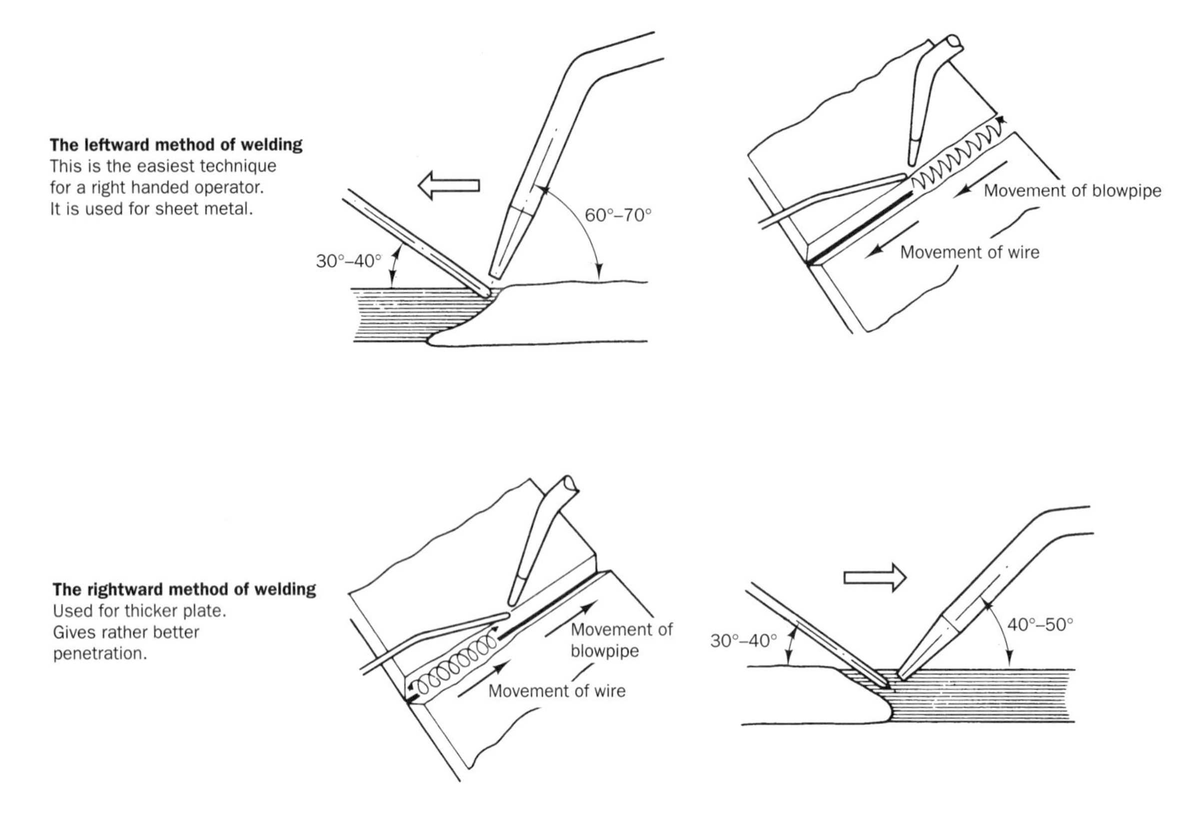

Figure 1.42 Oxy-acetylene welding techniques

With oxy-acetylene equipment the recommended techniques are shown in Fig. 1.42.

Leftward welding is recommended for plate thicknesses up to 6 mm and rightward welding for greater thicknesses.

Metal arc welding

The makers of arc-welding equipment supply data which gives the recommended current settings and sizes of filler rod for welding different plate thicknesses.

Metal arc welding is a highly skilled operation. The first technique to be mastered is striking the arc and this itself takes a great deal of practice. When the arc has been successfully struck, the recommended technique is shown in Fig. 1.43.

After completing the weld a coating of slag from the flux remains on the joint. This must be chipped off with a sharp-pointed hammer.

With both oxy-acetylene and metallic arc welding the speed of travel is critical. If the speed is too fast, the weld will not penetrate through the material to fill the joint. If the speed is too slow, the parent metals around the joint can be damaged. The correct speed for different plate thicknesses can only be achieved with practice.

Extremely high temperatures are created during welding processes with the added dangers of molten metal, sparks and very bright light. It is thus essential to wear the approved protective clothing and eye protection.

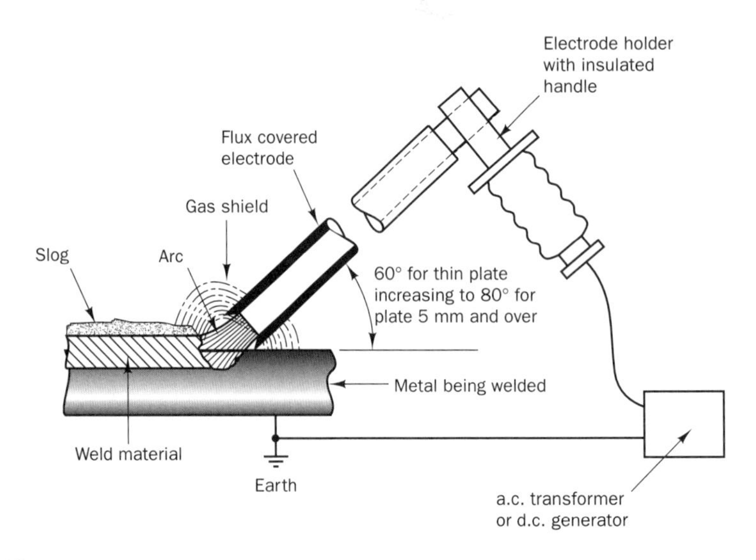

Figure 1.43 Electric arc welding

Adhesive bonding

Surfaces to be joined by adhesives must be absolutely free of dirt, oil or grease. They may need to be roughened to enable the adhesive to 'key' into the material.

Adhesives should be chosen which are appropriate for the materials being joined, their loading conditions and service environment. It is important to follow the adhesive manufacturer's recommendations and instruction when preparing the materials and applying the adhesive.

Thermoplastic adhesives should not be used where a joint is subject to heat as the adhesive may soften and fail. Solvent-based thermoplastics should not be used for joints with a large surface area unless the materials are porous, as the solvent must be allowed to evaporate. Thermo-setting adhesives give strong bonds but tend to be brittle and should not be used where vibration and flexing are present.

Where it is intended to replace mechanical fixings such as screws or rivets with an adhesive, the joint may need to be redesigned. Adhesives tend to be strong in tension and shear but weak when the surfaces are subjected to peeling and cleavage forces (Fig. 1.44).

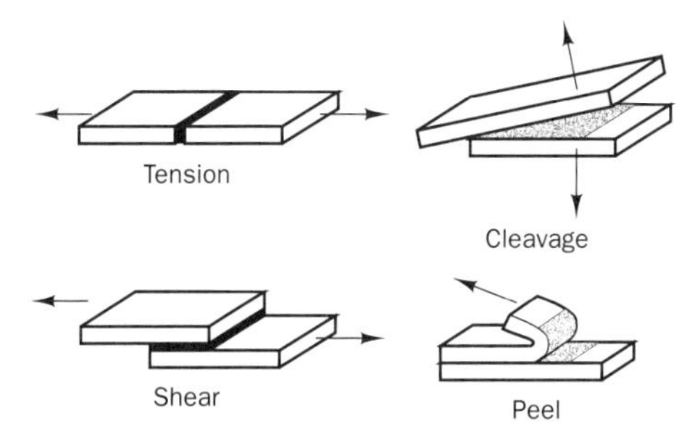

Figure 1.44 The stressing of bonded joints

Hand protection should be worn when using adhesives. Many are difficult to remove from the skin and can cause irritation. Contact with cyanoacrylate adhesives, or 'super-glues', should especially be avoided. They react chemically with the skin, bonding it to objects and bonding the fingers together. It is also important for the work area to be well ventilated especially where solvent-based adhesives are being used.

Self-assessment tasks

1. How can screwed fastenings be tightened to the correct torque setting?
2. How is a crimped termination fitted to the end of an electrical conductor?
3. What kind of tool is used to ensure that a rivet is seated properly before forming the head?
4. What is a rivet snap?
5. How are the surfaces of a soldered joint prepared before they are placed together?
6. How can sensitive electronic components be protected to prevent them from being damaged by the heat while being soldered in position?
7. Why are fire bricks packed around components which are to be joined by silver soldering or brazing?
8. What is meant by a neutral flame in oxy-acetylene welding?
9. Up to what plate thickness should the leftward method of oxy-acetylene welding be used?
10. What kind of service conditions are unsuitable for the use of thermoplastic adhesives?

Heat treatment techniques

Heat treatment processes are carried out to improve and enhance the properties of engineering materials. The most common heat treatment processes are:

- Annealing
- Normalising
- Quench hardening
- Tempering
- Case hardening

Annealing

Annealing is a process carried out on material which has been cold worked. Cold-working processes such as drawing, extrusion, rolling and pressing deform a material beyond its elastic limit and in so doing they deform its crystal or grain structure (Fig. 1.45).

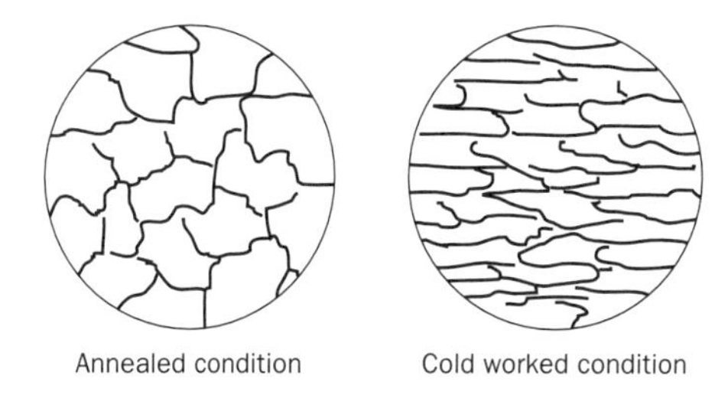

Figure 1.45 Effects of cold working on grain structure: (a) annealed condition: (b) cold-worked condition

As the grains become distorted, the material becomes harder. It is said to become 'work hardened', which makes it more difficult to deform. If further cold working is required, the material must be softened to restore its malleability and ductility. This is achieved by the annealing process in which the material is heated to its recrystallisation temperature. At this temperature new crystals or grains start to form at the points where the distorted grains are most stressed.

The material is held, or 'soaked' at the recrystallisation temperature for a period of time which depends on the amount of cold working and the use to which it will be put. If the soaking period is prolonged the new grains will feed off each other and grain growth will occur. This can make the material too soft for use. Plain carbon steel, copper, aluminium and cold-working brass can be softened by annealing.

With plain carbon steels, the material is heated to a cherry red colour which corresponds to the temperature band shown in Fig. 1.46.

The furnace is then turned off and the material is allowed to cool slowly in the 'dyeing furnace'. Other materials, such as aluminium, copper and cold-working brass, may be quenched when recrystallisation is complete. Their annealing temperatures are as follows:

- Pure aluminium 500–550 °C
- Pure copper 650–750 °C
- Cold-working brass 600–650 °C

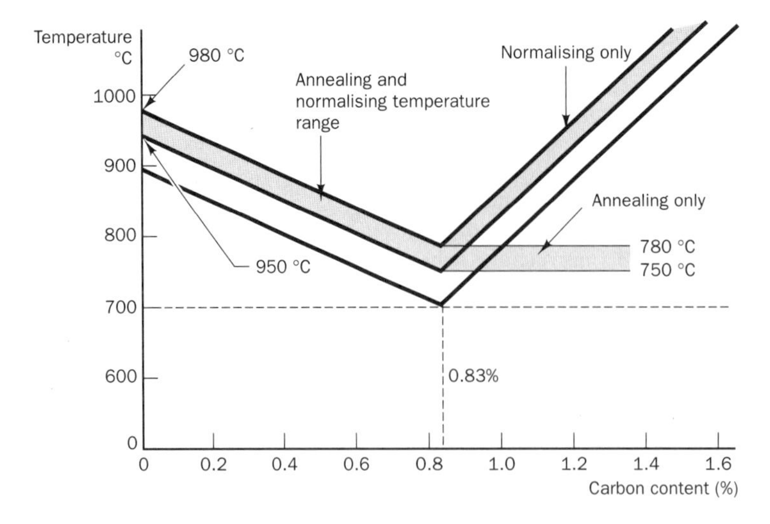

Figure 1.46 Heat treatment graph for plain carbon steels

Normalising

Hot-worked steel components such as those which have been hot formed to shape by forging or stamping often contain internal stresses which result from uneven cooling. This can cause distortion when the component is machined. The normalisation process is used to remove internal stress and refine the grain structure.

Plain carbon steels with up to 0.83% carbon content are heated to the same temperatures as for annealing, while those with above 0.83% carbon content are heated to within the range shown in Fig. 1.46 above. The components are then removed from the furnace and allowed to cool down in still air. Because the rate of cooling is quicker than for annealing, the grain structure is finer, leaving the material stronger and free from stress.

Quench hardening

This process is carried out on plain carbon steels of above 0.3% carbon content by heating them to within the same range of temperature as for annealing. They are then quenched in oil or water depending on the carbon content and the degree of hardness required.

Water quenching gives the maximum hardness. Oil quenching is less violent and leaves the steel a little less hard and brittle. Oil alone should be used for high carbon steels with above 0.9% carbon content as water quenching is too violent and causes cracking. Mild steel with a carbon content below 0.3% cannot be quenched hardened.

Tempering

Quench-hardened plain carbon steel is too hard and brittle for direct use. Tempering is a process which removes some of the hardness and increases the toughness of the steel. The hardened components are reheated to between 200 °C and 600 °C depending on their final use, as listed in Table 1.11. They are then quenched in oil or water.

Table 1.11 Tempering temperature

Component	Temper colour	Temperature (°C)
Edge tools	Pale straw	220
Turning tools	Medium straw	230
Twist drills	Dark straw	240
Taps	Brown	250
Press tools	Brownish-purple	260
Cold chisels	Purple	280
Springs	Blue	300
Toughening (crankshafts)		450–600

In industry, tempering furnaces are used which are set and maintained at the required temperature. Tempering can also be carried out on single components in the workshop by first polishing them and then heating them in a gas flame. The oxide colour films which spread over the polished surface give an indication of temperature and the point at which the components should be quenched.

Case hardening

Mild steel has too low a carbon content to be quench hardened. Case hardening is a process in which its surface hardness is increased while leaving the core in a soft and tough condition.

Components for case hardening are first carburised. This involves soaking them in a carbon-bearing material at the same temperature as that used for annealing. Over a period of time the carbon soaks into the steel to raise its carbon content near to the surface. The depth of penetration depends on the time of soaking.

Pack hardening is a traditional carburisation process where a solid carbon-bearing material is used. This is in the form of a powder which can be made from charcoalwood, bones and leather scrap with added chemicals to help the carbon to penetrate.

The components for carburising are packed with the powder in metal boxes and heated in a furnace until the depth of the high carbon case is sufficient. After removal, they may be given further heat treatment to refine the grain size of the core and to harden and temper the high-carbon case.

Chemical treatment techniques

Chemical treatment may be used to prepare materials and components for further processing, for material removal or for surface finishing. Some of the most common chemical processes are:

- Degreasing
- Pickling
- Etching
- Eletroplating
- Anodising

Degreasing

Materials and components sometimes need to be cleaned before use. Oil and grease can be removed by washing with paraffin or white spirits but these may not leave a completely clean and dry surface. Chemical solvents such as trichloroethylene are more effective but give off toxic fumes. They should only be used with specially designed degreasing equipment which is enclosed and situated in a well-ventilated area.

Pickling

Metals which have been hot formed by rolling, drawing or forging are often covered with an oxide film. This is particularly true of steel components which are covered with a flaky coat of black iron oxide, known as 'millscale'. This can be removed by pickling the material in sulphuric acid, phosphoric acid or a mixture of the two. Phosphoric acid has the advantage of leaving a layer of iron phosphate on the steel surface which protects it from rusting.

Etching

Chemical etching may be carried out to prepare a surface for receiving a finishing coat of paint or other protective material. Alternatively, it may be carried out to remove surplus material as occurs in chemical milling and in the production of printed circuit boards.

Various etching solutions are used depending on the material to be etched. In the case of printed circuit boards, where copper is the metal being removed, ferric chloride is a common etchant.

In both printed circuit board production and chemical milling, the areas which are to be left unetched are masked off with a material which is resistant to the etchant.

Protective clothing and eye protection should be worn when carrying out etching processes since the chemicals used may cause irritation to the eyes and skin. Adequate ventilation is also necessary as toxic fumes may be given off.

Electroplating

This is an electrochemical process in which a component is given a coating of another metal for protection or for decorative purposes. A direct current power supply is required together with a suitable chemical solution, known as an electrolyte.

The metal component to be plated is the cathode, which is connected to the negative terminal of the supply. A bar or slab of the plating metal is the anode, which is connected to the positive terminal of the supply. Both are immersed in the electrolyte as shown in Fig. 1.47.

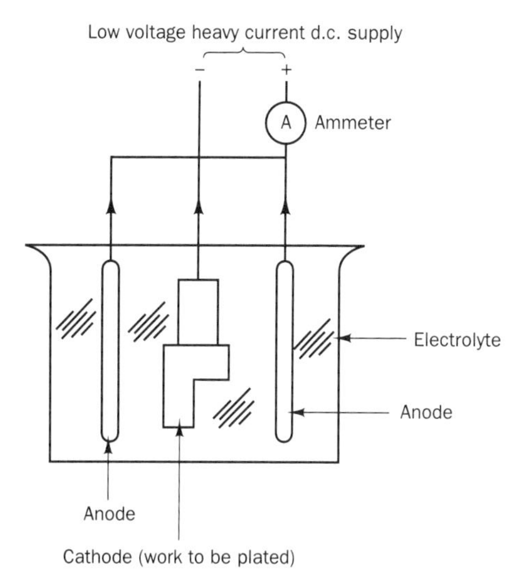

Figure 1.47 Electroplating

Over a period of time the anode dissolves into the electrolyte and is deposited on the cathode. The process can be used for plating with copper, nickel and zinc. With zinc it is known as electrolytic galvanising. An alternative is for the anode not to dissolve and for the electrolyte alone to supply the plating metal. Chromium plating is done in this way, but over a period of time the electrolyte becomes weaker and must be renewed.

Anodising

This is an electrochemical process carried out mainly on aluminium. Aluminium forms a thin oxide film on its surface which protects it against further attack. Anodising is a means of artificially thickening the film to give increased protection to components which will be exposed to the environment.

The aluminium component is placed in an electrolyte of dilute acid and is connected to the positive terminal of the direct current supply. The arrangement is the reverse of electroplating and, over a period of time a thick layer of aluminium oxide is built up on the surface of the component. The aluminium sections used for window frames are treated in this way.

Surface-finishing techniques

The type of surface finish given to an engineered product depends on its application and service conditions. The surface finish may be for protection, decoration or a combination of the two. Where a high degree of precision is required, specialist machining operations may be needed to produce a surface finish which is very smooth. Some common surface-finishing techniques are:

- Grinding
- Polishing
- Painting
- Plating

Grinding

Grinding is an abrasive material removal process. It is used on components which must be produced to a close dimensional tolerance and have a smooth surface finish.

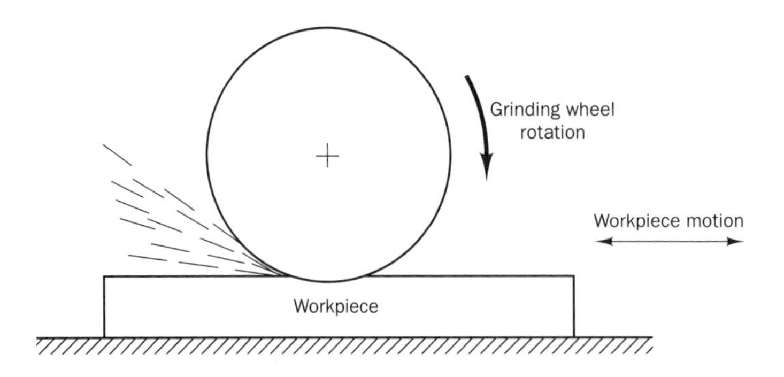

Figure 1.48 Surface grinding

Smooth flat surfaces are produced by surface grinding. A surface-grinding machine is rather like a horizontal milling machine but with a grinding wheel instead of the cutter (Fig. 1.48).

Smooth cylindrical surfaces are produced by cylindrical grinding. A cylindrical grinding machine is rather like a lathe but with a grinding wheel in place of the cutting tool (Fig. 1.49).

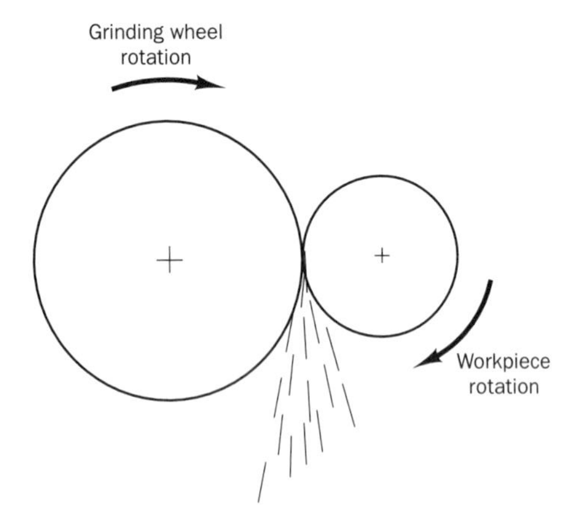

Figure 1.49 Cylindrical grinding

Polishing

Like grinding, polishing is an abrasive material removal process. The abrasive substance may be in the form of a paste or it may be suspended in a liquid. Surfaces which require polishing should already be in a smooth condition as very little material is removed during the process.

The abrasive substance is loaded onto a pad or 'buff' which rotates on the spindle of a polishing machine. Polishers may be of the hand-held portable type which is held against the workpiece, or the stationary machine type where the workpiece is held against the rotating buff.

For all kinds of grinding and polishing processes suitable protective clothing and eye protection should be worn.

Painting

Engineered products may be painted for protection, decoration or a combination of the two. Before paint is applied to a surface, the surface must be properly prepared. All traces of oil, dirt and grease must be removed, together with any rust or scale. Surface preparation processes include degreasing with a suitable solvent, pickling in acid and sand blasting.

Paint consists of fine solid particles, or pigments, suspended in a liquid. The pigment gives colour to the paint and the liquid dries or sets when it is applied to a surface. The main types of paint used on engineered products are:

- Simple oil paints and varnishes – These are air drying and mainly based on natural oils such a linseed oil. They may also contain resins, bitumen and asphalt which give strength and added protection. They find general use indoors and outdoors, as a decoration and protection for wood and metalwork.
- Stoving paints and enamels – These are resin based and dry quickly with the application of heat from an infrared source. The paint film is hard wearing and adheres well to metal. It is used for cookers, refrigerators, motor vehicle bodies and many items of laboratory and industrial equipment.
- Catalytically drying paints – These are resin based with an added hardener and solvent to keep the resin liquid form before use. On application the solvent evaporates and the resin quickly hardens as a chemical reaction takes place with the hardener. The process is similar to that which occurs in the production of thermo-setting plastics. The final paint film is hard and tough and gives better protection to wood and metal than oil-based paints.
- Lacquers – These include cellulose paints which dry entirely by the evaporation of a solvent. They give a thin hard-wearing film with a very smooth surface finish. This can be further enhanced by polishing with a fine abrasive. Shellac varnish is another lacquer which dries purely by evaporation of a solvent. It is used by the electrical industry as an insulation for the coils and laminations in motors and transformers. They are mainly applied by a spray gun or from an aerosol can and are widely used on furniture and in motor vehicle body finishing and repair.

The application of paints by brush or spray requires certain safety precautions. The work area should be well ventilated and suitable protective clothing and fire-fighting equipment should be provided.

Plating

The plating of engineered products with metals such as zinc, tin and cadmium protects them against corrosion. Plating with a hard metal, such as chromium, also gives a wear-resistant surface and may enhance the appearance of a product. In addition to electroplating which has been described above, plating can be applied by the following methods:

- Hot dipping – Steel is plated with zinc and tin by this process. With zinc, it is called 'hot-dip galvanising'. The steel component is first cleaned and then coated in a flux. It is then dipped in a bath of molten zinc and quenched in water.

In service the zinc is said to give 'sacrificial' protection to the steel. This means that even if the zinc coating is damaged and the steel is exposed, it will not corrode. If moisture is present, electrolytic corrosion will occur, but it is the zinc which corrodes away at a very slow rate, and not the steel.

Sheet steel is plated with tin to produce 'tinplate'. This is widely used for food canning on account of the tin being resistant to any chemical reaction with the food product.

The tin is fed in a continuous strip through a flux and then through a bath of molten tin. It then passes through rollers which give the tin coating a uniform thickness.

Unlike zinc, tin does not give sacrificial protection to steel. If it is damaged, and moisture is present, electrolytic corrosion occurs and it is the steel rather than the tin which corrodes.

- Hot spraying – This is also called 'flame spraying'. The coating metal is fed into a blowpipe where it is melted and emerges as a jet of molten globules. Zinc, aluminium and cadmium can be sprayed in this way. The advantage is that it can be applied to large irregularly shaped objects in the same way as paint spraying.
- Powder bonding – Zinc may be plated on steel components by a process known as Sherardising. The components are placed in a revolving cylinder filled with zinc dust and zinc oxide. At temperatures between 250 °C and 450 °C the dust particles adhere to the metal which becomes coated with a very thin film of zinc alloy. Aluminium may be plated by a similar process known as Calorising.

Self-assessment task

1. How should steel components be cooled down in the annealing process?
2. Why do some steel components require normalising before they are machined?
3. What is the percentage of carbon which steel needs to have so that it can be quench hardened?
4. Why should high-carbon steels be quenched in oil rather than water?
5. What effect does tempering have on components which have been quench hardened?
6. What is case hardening?
7. Which chemical treatment process is used in the production of printed circuit boards?
8. To which of the d.c. supply terminals is a component connected when being electro-plated?
9. How is the abrasive material applied to a component during the polishing process?
10. What is hot-dip galvanising?

Selection of processes and specific techniques for making electromechanical products

Having read the earlier sections of this unit, the student should be able to identify the processes used in making electromechanical products and give examples of products where the ranged processes and techniques have been used.

The student should also be able to identify the most appropriate processes and specific techniques required to produce a given electromechanical product together with the associated safety procedures and equipment.

Portfolio suggestions for given electromechanical products are:

- Adjustable desk lamp
- Battery charger
- Torque wrench
- Lawn sprinkler
- Alarm system

The identification, selection and reporting activities will need to be planned as a series of tasks. A suggested plan is:

- Task 1 – Write short notes on each of the ranged processes giving two examples of products where each process is used. The ranged processes are:
  - material removal
  - joining and assembly
  - heat treatment
  - chemical treatment
  - surface finishing

  The products chosen as examples should involve different specific techniques.
- Task 2 – For a given electromechanical product, write notes which outline the selection of the most suitable processes for making the product, considering all of the selection criteria in the range. The selection criteria are
  - material properties
  - quality
  - tolerances
  - quantity
  - cost

  If alternative processes could be used, explain why they were not selected.
- Task 3 – For each of the selected processes, write notes which identify the specific techniques which are most suitable, considering all of the above selection criteria. If alternative techniques could be used, explain why they were not selected.
- Task 4 – For each of the selected processes and techniques, write notes which identify any associated safety procedures and equipment.

Identification of relevant safety procedures and equipment for the processes selected

Safety procedures which are specific to material removal, assembly and finishing processes have already been described. These include the correct methods of work and tool holding, and the approved working practices and techniques.

General procedures

The following procedures apply equally to all areas of processing:

- Personal hygiene
- Personal conduct
- Reporting accidents
- Evacuation of the workplace

Personal hygiene

It is advisable to rub a barrier cream into the hands and lower arms before starting work. This prevents dirt from entering the pores of the skin and becoming engrained. Barrier creams have antiseptic properties which protect the skin from germs. In addition, they are water-soluble so that they are easily washed off, together with the dirt, at the end of a work period.

The hands should always be washed before and after using the toilet and before meal breaks. Barrier cream should then be re-applied before commencing work. It is not advisable to use solvents to clean the hands as these are often toxic and may cause skin irritation.

Personal conduct

Employees should conduct themselves in a responsible manner. Horseplay, practical jokes which involve tools and equipment, shouting and throwing things have no place in an industrial environment. Such actions can lead to damaged equipment, wasted materials and serious accidents.

Pressure hoses, compressed air lines, electrical equipment and dangerous chemicals are potentially lethal and on no account should they be used for other than their intended purpose.

'No smoking' instructions should be obeyed and on no account should employees be present in the workplace while under the influence of alcohol or illicit drugs.

Reporting accidents

All accidents, however minor, should be reported to the work area supervisor. First aid treatment should be given for personal injuries. The nature of the accident and the treatment given should then be recorded.

Evacuation of the workplace

In the event of fire, escape of toxic substances or any other kind of emergency situation, it may be necessary to evacuate the workplace. It is the duty of an employer to instruct employees on how to leave the workplace quickly and safely.

It is the duty of the employee to become familiar with these instructions so that a quick and orderly evacuation can take place. Practice evacuation drills should take place periodically in association with the emergency services to test the procedures and, where possible, to improve them.

The instructions should include the procedures to be followed for the following:

- Raising the alarm – This should be done immediately a fire or other emergency situation is discovered. Employees should be aware of the location of alarm points in the work area and how to operate them. This often involves breaking a glass cover which releases a spring-loaded button to trigger the alarm.
- Responding to the alarm – On hearing the alarm, and before leaving the workplace, process equipment should be turned off and isolated from the power supply.
- Leaving the building – The evacuation route should be marked by green arrows in corridors, passageways and stairs. Employees should follow this route quickly but without running and without panic.
- Assembling outside the building – After leaving the building, employees should meet at the specified assembly point. A role call should then be taken by a supervisor, or other appointed officer, to ensure that everyone is present.

Safety equipment

The safety equipment which an employer must provide depends on the type of work which is taking place and the likely hazards which may occur. It includes:

- Protective equipment
- Emergency equipment
- First-aid equipment
- Safety equipment

Protective equipment

This includes the barriers and guards which an employer is legally required to provide to protect employees from hot surfaces, hazardous substances and the moving parts of machinery.

Guard rails, coloured with diagonal black and yellow stripes, are often positioned permanently around furnaces, chemical processes and machines. The colour indicates that it is hazardous to step beyond the fence. Portable barriers may also be used to prevent employees from entering temporarily hazardous areas such as where maintenance work is in progress.

The belt drives, chain drives, gear trains and power transmission shafts of machines must be guarded so that they cannot be touched while in motion. It is an offence to have or sell machinery which is not adequately guarded or for an employer to put it in use.

If the guards need to be removed for maintenance work, the machine should be stopped and isolated from its power supply. If it is electrically driven, the key should be removed from the isolating switch and, if one is not provided, the fuses should be removed.

Machines may be fitted with different types of cutter guard, or workholding guard, depending on their use. It is the duty of technicians and operators to make sure that the guards are undamaged and working correctly and to use them in the proper manner. It is an offence to remove or tamper with the guards. If they are found to be faulty, the matter should be reported and rectified before using the machine.

Emergency equipment

An emergency can arise due to the outbreak of fire, explosions, escape of toxic fumes and harmful high-energy radiation. Following evacuation of the workplace, specialist equipment in the form of respirators, protective clothing and monitoring equipment is required for the emergency teams who must restore the situation.

In general engineering workshops and offices the most likely emergency situation is the outbreak of fire and employers are required by law to provide appropriate fire-fighting equipment.

In the event of fire, evacuation is the best course of action, leaving the emergency services or the employer's trained emergency team to fight the fire. If it is judged that there is no danger of a fire spreading and that it can be quickly put out, it is essential to choose the correct type of fire extinguisher. Employees should be aware of where these are positioned, the types of fire for which they should be used and the way to operate them.

A fire needs three things in order to continue burning: it needs a supply of fuel or flammable material; it needs oxygen in the form of an air supply; and it needs a heat source or high temperature to ignite the fuel and continue the combustion process. If the fuel is removed, or the air supply cut off, or the temperature lowered sufficiently, the fire will be contained and extinguished. The following types of extinguisher are those usually found in the workplace:

- Hose reel and pressurised water extinguishers – These are coloured red. They are suitable for combustible materials such as wood, paper and textiles. They should not be used on burning liquids as this tends to spread the fire.
- Carbon dioxide extinguishers – These are coloured black. They are suitable for fighting fires in electrical appliances or equipment. Here it is important not to use water where there may be live conductors and to isolate the equipment as quickly as possible.
- Foam extinguishers – These are coloured cream. They are suitable for fires which involve flammable liquids such as oils, petrol, paint and solvents. The foam has the effect of cutting off the oxygen supply to the fire.
- Vaporising liquid extinguishers – These are coloured green. They provide a vapour which also cuts off the oxygen supply to a fire. The vapour is however toxic, and this type of extinguisher should be not used in confined spaces. They are useful for containing small fires outdoors in motor vehicles or electrical equipment.
- Dry powder extinguishers – These are coloured blue. With this type of extinguisher, the powder not only helps to smother the fire, it produces carbon dioxide gas when heated, which further cuts off the oxygen supply. They are useful for all flammable liquid fires and particularly those in kitchens and food stores. The powder is non-toxic and easy to remove with a suction hose or vacuum cleaner.
- Fire blankets – These are made of fire-resistant synthetic fibres and are placed over a fire to cut off the air supply. They are also useful for wrapping round a person whose clothes are on fire.

First-aid equipment

The amount of first-aid equipment and the number of employees trained to provide first aid varies with the size of an industrial establishment. As an absolute minimum a small engineering firm should have a first-aid station equipped with the solutions and dressings needed to treat minor cuts, bruises and burns.

Bandages, slings and splints should also be provided to immobilise suspected fractured limbs and a stretcher should be available to remove an accident victim to a safe area. In areas where toxic or caustic chemicals are in use, the first-aid station should be equipped with the solutions necessary to neutralise them and irrigate the affected parts of the body, particularly the eyes.

Safety wear

Large engineering firms often supply all items of safety wear free to employees. Smaller firms may require employees to provide certain items of clothing or footwear themselves, or supply them at a small charge. Items of safety wear include the following:

- Protective clothing – A one-piece boiler suit is the most appropriate item of protective clothing for general engineering work. The boiler suit should be close fitting so as not to become entangled in machinery and it should be changed and washed regularly.

  Work with corrosive liquids, molten metal and heavy duty welding equipment often requires additional body and leg protection against splashes and sparks. A variety of corrosion and heat-resistant jackets, aprons and gaiters are available for use with these processes.
- Head protection – Long hair is a serious hazard as it can easily become entangled in a machine. It should be contained by a close-fitting cap.

  When working in an area such as a construction site or where overhead cranes are in continual use, a safety helmet should be worn. These are light in weight but have a high resistance to impact. They are also non-flammable and should the head come into contact with a live electrical conductor, they give protection against electrical shock.
- Eye protection – Eye protection is essential for material removal process such as turning, milling, drilling, grinding and polishing. It is also essential when chemicals such as soldering fluxes are being used to protect against splashing and spitting. A variety of safety spectacles, goggles and visors are available for use with these processes.

  Welding processes require more specialist eye and face protection, as sparks, heat, bright light and ultraviolet radiation, may be given off. Special goggles with heavily tinted lenses are used for oxy-acetylene welding but on no account should these be used for arc welding where the light given off is much brighter.

  Hand held shields or visors which cover the full face, and filter out the harmful rays, should be used for arc-welding.
- Face masks and respirators – A face mask or respirator should be worn when working in the presence of dust or fumes such as when using grinding or paint-spraying equipment. There are many patterns but most of them have filter elements which should be replaced regularly.
- Hand and arm protection – A variety of gloves and gauntlets are available for protecting the hands and arms against heat, sharp material and corrosive liquids.
- Foot protection – Lightweight fashion and sports shoes should not be worn in the workplace. They offer no protection against falling objects or penetration by sharp objects. Safety shoes or boots should be worn which have steel toecaps and reinforced non-slip soles.

More detailed information on safety equipment, procedures and systems can be obtained from the British Standard Specifications BS 3456, BS 4163, BS 5304, BS 5378 and PD 7304.

Self-assessment tasks

1. Why is it advisable to apply a barrier cream to the hands before starting work?
2. What is the correct procedure when an operator suffers a minor cut or graze?
3. What action should employees take on hearing the emergency alarm?
4. What colours are the guard rails painted which are positioned around hazardous equipment?
5. What procedure should be followed if the guards have to be removed from a machine to carry out maintenance operations?
6. What is contained in a fire extinguisher which is coloured cream and what kind of fires should it be used with?
7. Which type of fire extinguisher is most suitable for fighting fires in electrical equipment and what is its colour?
8. What particular items of first-aid equipment should be provided in areas where toxic or caustic chemicals are in use?
9. What item of safety wear should be worn when lifting or moving heavy objects?
10. What items of safety wear should be provided for oxy-acetylene welding operations?

1.3 Production of an engineered product to specification

Having decided which production processes are the most appropriate for a product, the next step is to draw up detailed production plans. These contain the sequence in which the processes are to be carried out and the specific techniques which will be used to complete each process. The production plans should also contain the required quality specifications and details of the tools and equipment which are required.

Safety is a prime consideration in any production operation. Where necessary, the production plans should contain details of the safety equipment which is to be used and the safe working procedures which are to be followed. It is the responsibility of the employer to provide a safe working environment and adequate training. It is the responsibility of the employee to use safety equipment and follow safe working procedures.

Having finalised the production plans and provided the materials, the tools, the equipment and the training, production can proceed. The product can then be produced as safely and as economically as is possible.

Topics covered in this section are:

- Identification of the sequence of processes necessary to produce an electromechanical engineered product to specification.
- Selection of the specific techniques needed to perform each of the processes identified in the sequence.
- Selection of materials and components to produce the engineered product to specification.
- Carrying out processes to produce the product to specification.
- Correct use of relevant safety procedures and equipment.
- Maintaining tools, equipment and the working area in good order during and after processing.

Identification of the sequence of processes necessary to produce an electromechanical product to specification

In planning the sequence of processes required to make a product it is often helpful to use a block diagram, or flow chart (Fig. 1.50). This can show the sequence of processes required to make the component parts of a product and how the component parts are brought together for assembly and finishing.

The flow chart is sometimes also called a production process layout, and may contain the estimated time required for each stage. Further information on flow charts and block diagram formats can be obtained from the British Standards Specification BS 4058.

Selection of the specific techniques, materials, components, tools and equipment to produce the engineered product to specification

Having decided on the best sequence of processes, the specific process techniques must be selected which will be used to meet the product specifications. The product specifications are to be found in the engineering drawings and memoranda which have been received from the design engineers. The specifications may include the following.

- Material and material properties
- Product dimensions and tolerances
- Surface finish or texture
- Test readings and results
- Production quantities

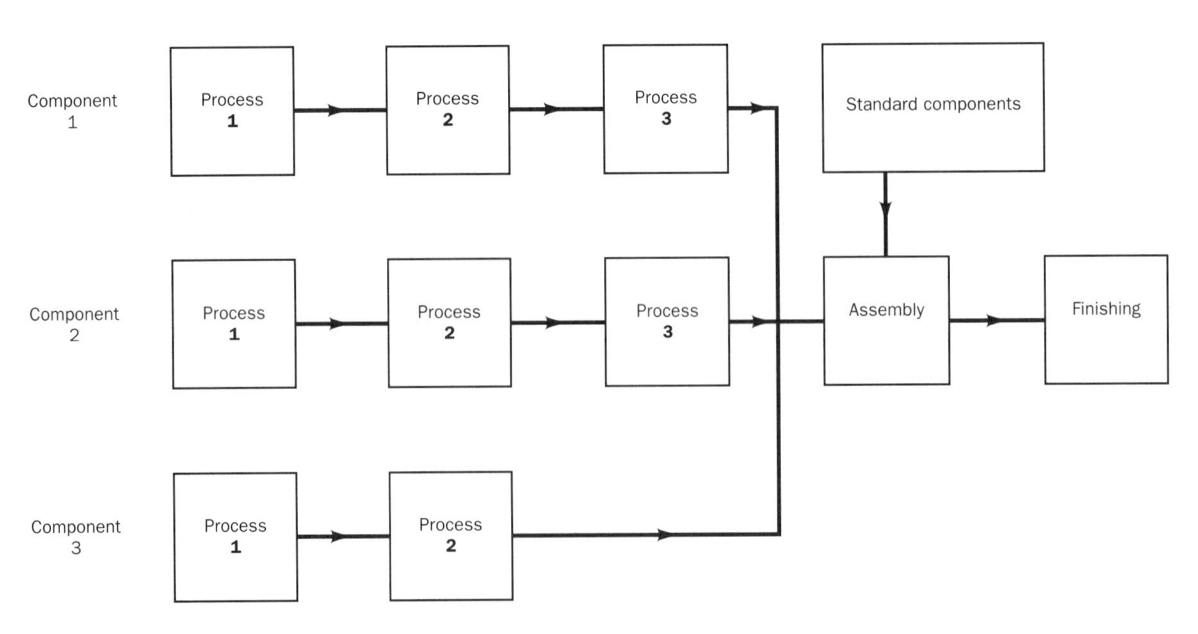

Figure 1.50 Flow chart

Material and material properties

In addition to the description of the material, the specification may give a British Standards Specification number. A typical example given on a drawing might be:

Medium carbon steel BS 970:080M40

The drawing or memoranda from the designers might also contain heat treatment and surface hardness specifications. A typical example might be:

Harden by oil quenching from 850 °C and temper at 250 °C
Surface hardness VPN 800

The term VPN is the Vickers pyramid hardness number.

Product dimensions and tolerances

The dimensions which are needed to make a product are given on its engineering drawings. The dimensions are usually given from one of its surfaces which is known as a datum face (Fig. 1.51), or from a datum line or point. This is to avoid multiple, or 'cumulative' errors occurring during production.

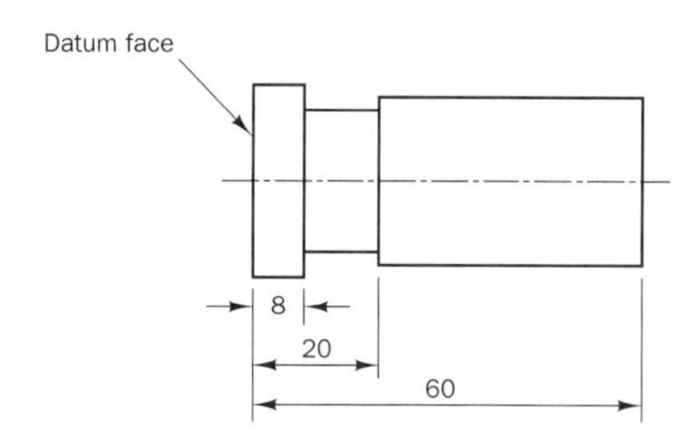

Figure 1.51 Dimensioning from a datum face

The overall dimensions may not be given, but in such cases they can be obtained by adding together the intermediate dimensions. Diameters are always prefixed by the symbol ϕ.

A toleranced dimension defines the allowable limits of the size of a feature on a product, i.e. the dimensional accuracy to which it must be produced. Some typical toleranced dimensions are given in Fig. 1.52.

The tolerances often dictate the choice of the production process and process technique which must be used. Depending on the size of a workpiece, machining limits of ±0.03 mm should be within the capability of well-maintained

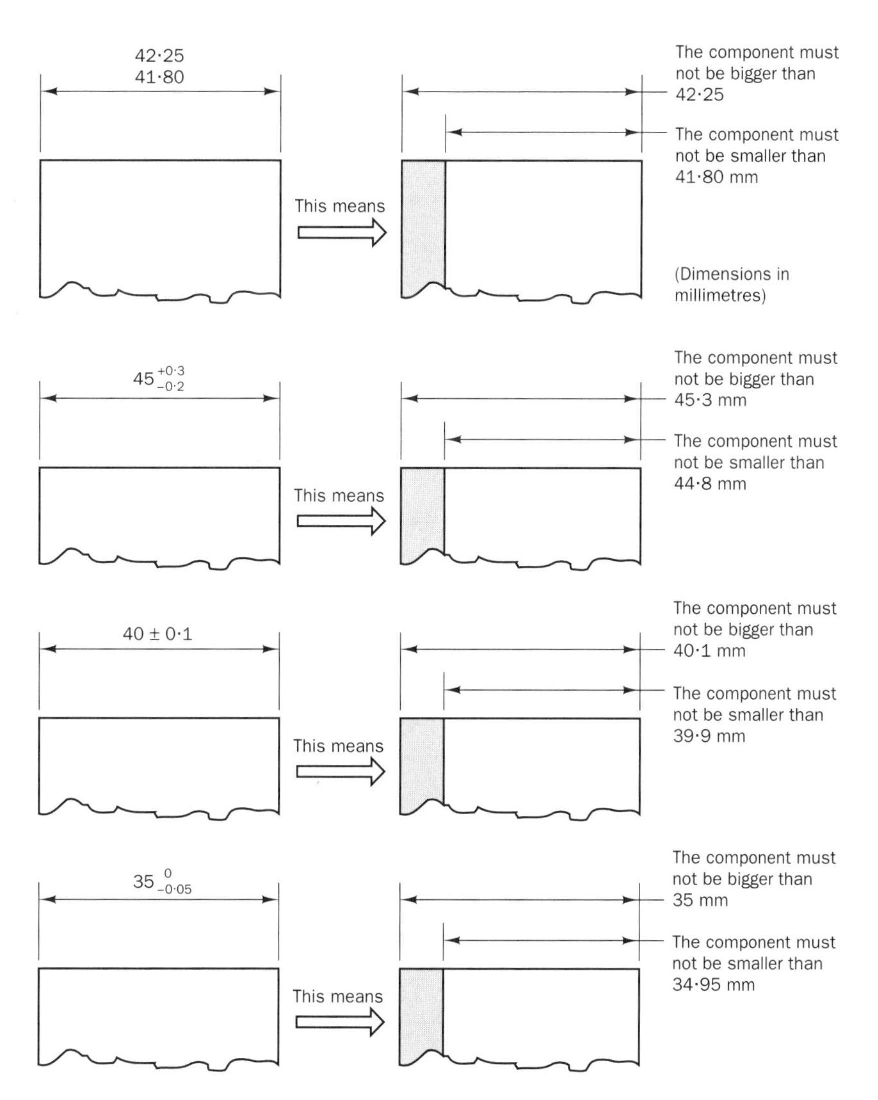

Figure 1.52 Toleranced dimensions

lathes and ±0.1 mm within the capability of milling machines. Limits of ±0.008 mm, however, might indicate that grinding is the most appropriate material removal process.

The dimensions for which no tolerance is given are sometimes called open dimensions. Guidance as to the degree of accuracy required with these may be given on the title block of the engineering drawing, for instance:

All dimensions to be ±0.25 mm unless otherwise stated

Further information on engineering drawings, constructional drawings, circuit and other diagrams, units, symbols dimensions and tolerances can be obtained from British Standards Specifications BS 5070, BS 1192, BS 308, BS 5775 and BS 4500.

Surface finish and texture

The engineering drawings for a product specify the required surface roughness or texture of the surfaces which are to be machined, chemically treated, plated, etc. They are indicated by a symbol such as that shown in Fig. 1.53(a).

The number specifies the required degree of surface roughness measured in micrometres (μm), where 1 μm is a thousandth of a millimetre. This may be accompanied by additional information which specifies the finishing process to be used (Fig. 1.53(b)).

Here the limits of surface roughness are also specified, i.e. between 0.2 μm and 0.8 μm. The symbol need not be shown on the surfaces of components which are to be processed all over. Here it is sufficient to enter the specification in the title block of the drawing as shown in Fig. 1.53(c).

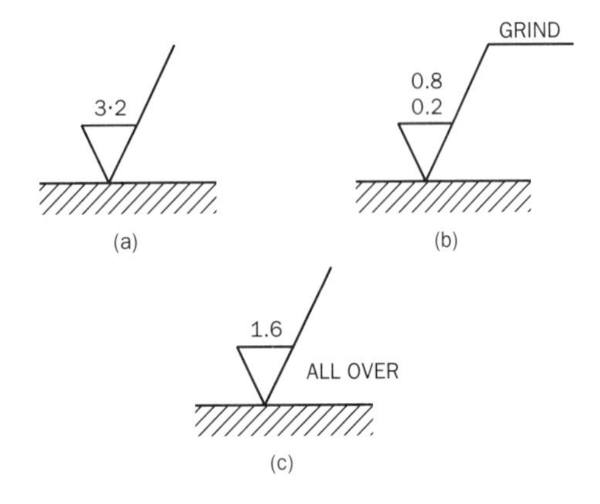

Figure 1.53 Surface finish and texture

Test readings and results

The specifications for an engineered product may require that quality control checks are carried out at key stages in its production using specialist test equipment. The required test readings and results may be given on the engineering drawings or circuit diagrams or listed in memoranda from the design or quality control engineers.

Production quantities

The quantity and rate at which a product is required often governs the choice of production processes and process techniques. The scale of production can range from single item production, or 'jobbing', through batch production to continuous production.

Small-scale production generally requires general purpose or flexible production systems which can be readily adapted to process different products. Large-scale repetitive batch and continuous production is often carried out on specialist automated production systems which are dedicated to one product.

Having taken into account the complete specification for a product, a decision can be made as to the best sequence of production processes and process techniques. Their details must then be documented and this is often done on a process planning sheet. In some firms this is also called a process specification sheet or production operations sheet. They are generally held on a database from which hard copies can be produced as required. A typical format is shown in Fig. 1.54.

With many products, a number of components are processed separately before coming together for assembly and finishing. In such cases it is usual to draw up separate process planning sheets for each component and also one for the assembly and finishing operations. The processes are usually numbered 10, 20, 30, etc. This enables any extra processes, which may be needed, to be slotted in as 15, 25, etc.

- Materials/components – This column should specify the grade of material required and its form of supply, e.g. casting, forging, bar, sheet, granules, etc. For assembly operations the component parts should be listed with a part number or a standard value, e.g. M8 × 1.25 setscrew or 220k 1/4 watt resistor.
- Machine/equipment – This column should give a brief description of the machine or equipment type, e.g. pillar drill, vertical mill, tempering furnace, etc. In some cases it may also be appropriate to enter the location and inventory number of the machine or equipment.
- Tools/gauges – This column should list any jigs, fixtures, hand tools, gauges or items of test equipment required for the process.
- Quality indicator – This column should contain critical dimensions, hardness values, surface finish details or the readings which should be indicated on test equipment to show that the process has been successfully completed.
- Estimated time – This column should give the processing time required per piece or, in the case of processes such as heat treatment, the processing time per batch.

Process number	Process description	Materials/components	Machine/equipment	Tools/gauges	Quality indicator	Est. time
10						
20						
30						
40						

Figure 1.54 Process planning sheet

The product flow chart and the associated process planning sheets form part of the production process plan for a product. This contains all of the information needed to make the product and should not be confused with the production schedule.

A production schedule or production programme is issued to cover a period of time and states when, and in what quantities, the product is to be made. It contains the start and finish dates of the production processes for the product or for a range of products which follow a similar processing pathway.

Carrying out processes to produce the product to specification

Having read the previous sections of this unit, the student should be able to plan and carry out the production of an electromechanical product. Some portfolio suggestions were made for such products:

- Adjustable desk lamp
- Battery charger
- Torque wrench
- Lawn sprinkler
- Alarm system

The sequencing of the selected processes and the selection of suitable materials, components, tools and equipment will need to be planned as a series of tasks. These, together with the actual production activities, should be recorded in the student's activity log. A suggested plan is:

- Task 1 – Draw up a flow chart showing the sequence of processes required to produce the product. Log the activity
- Task 2 – Draw up a process plan which contains details of the selected techniques, materials, components, tools and equipment. Log the activity.
- Task 3 – Carry out the sequence of processes making correct use of relevant safety procedures and equipment. Ensure that each process is completed to specification particularly as regards dimensional tolerances and surface finish. Log each production activity.
- Task 4 – Maintain tools, equipment and the work area in good order during and after processing. Log each maintenance activity.
- Task 5 – If the tools and equipment used do not cover all the categories in the range – i.e. hand tools, power tools, machine tools – make general notes on any of the remaining categories.

Correct use of relevant safety procedures and equipment

Details of the safety procedures and equipment associated with engineering processes are given earlier in this unit. Employers are legally required to provide safety equipment and training in safe methods of working. Employees also have legal responsibilities. They are required to use the safety equipment correctly, follow safety procedures and generally conduct themselves in a safe and proper manner.

The law dealing with the health, safety and welfare of people at work is contained in The Health and Safety at Work etc. Act of 1974.

In addition to this there are many other Acts of Parliament and subsidiary regulations which are concerned with safety in the workplace. These include the Offices, Shops and Railway Premises Act and the regulations covering the use of abrasive wheels, the use of milling machines and the protection of eyes.

The Health and Safety at Work, etc., Act provided for creation of the Health and Safety Commission. This is made up of people from industrial management, the trade unions, local authorities and one independent member to represent the general public. Its duties are to oversee and control matters concerning occupational health and safety including the organisation and functioning of the Health and Safety Executive. This is a body which carries out workplace inspections to ensure that all laws and regulations are being complied with.

The inspectors have the power to serve improvement notices. These require that faults are remedied within a given time but allow the work to continue. When the fault is causing a serious risk of personal injury, the inspectors may issue a prohibition notice which closes the workplace down until the fault has been remedied. In addition to serving improvement and prohibition notices, the inspector may prosecute employers or employees who contravene the law and safety regulations.

The Health and Safety at Work, etc., Act also lays down the

- Employer's responsibilities
- Employee's responsibilities

Employer's responsibilities

Under the Act it is the duty of the employer to make the following provisions:

1. The employer must provide a safe place of work and a safe working environment. There must be adequate heating, lighting and ventilation. Stairways and floors must be made safe with easy access and exit and all necessary precautions should be taken against the risk of fire.
2. The employer must ensure that plant and equipment are properly maintained and fitted with the appropriate guards. Pressure vessels, lifting gear, electrical equipment and installations, etc., must be regularly inspected and all relevant process regulations must be followed.
3. The employer is responsible for using safe systems of working and safe methods of handling, storing and transporting goods. Appropriate safety equipment and clothing must be provided and lifting gear should have its safe working load clearly marked. Gas cylinders and flammable substances should be stored in well-ventilated compounds, away from the working area and shielded from direct sunlight and frost.
4. The employer must keep a register in the workplace of all accidents and dangerous incidents. Certain types of accident and incident must be reported to the inspectorate. Inspectors have the right to see the register on demand and investigate the cause of anything which has occurred.
5. The employer must provide information, instruction,

training and supervision. Employees must be trained to follow safe systems of working when handling materials or operating machinery. Young and inexperienced workers must be properly supervised. Employees must be made fully aware of any hazardous substances which are being used and be properly trained to handle them. They must also be properly trained in the use of emergency equipment and procedures.

Employee's responsibilities

Like the employer, the employees can be prosecuted for breaking safety laws and regulations. Employees have the following responsibilities:

1. Employees are legally bound under the Health and Safety at Work, etc., Act to cooperate with the employer to fulfil the requirements of the Act.
2. Employees have a legal responsibility to take reasonable care of their own health and safety. They must follow safe working procedures and make full use of the safety equipment and clothing provided.
3. Employees must act in a responsible manner which does not endanger other workers or members of the public. Equipment provided for health and safety must not be misused or interfered with. This is an offence which may lead to prosecution.

Maintaining tools, equipment and the working area in good order during and after processing

Maintenance activities should be carried out regularly to keep the work area, tools, equipment and process plant in safe working order. Poor maintenance can cause accidents and unscheduled production stoppages. It can also be a cause of poor product quality. Maintenance activities can be divided into two groups: routine workplace maintenance and scheduled maintenance.

Routine workplace maintenance

Before starting work, the workplace, which may be a workbench or machine, should be clear of waste material. The floor area should be clear of any obstruction or spillage and there should be no tools or equipment left from previous activities.

During the working period, the workplace should be maintained in a tidy condition. Waste material should not be allowed to accumulate, spillages should be cleaned away immediately and tools and equipment should be returned to the appropriate place when not in use.

Guidelines for the maintenance of commonly used hand, power and machine tools are as follows:

- Files – File handles should be inspected before use and replaced if they are damaged or badly fitting. File teeth should be cleaned with a wire brush or special file card if they become clogged or 'pinned' with waste material. File teeth are hard and brittle and easily damaged. After use files should not be stored separately on racks or boards as they are said to 'make bad neighbours' for other tools.
- Hacksaws – Hacksaw blades should be inspected before and after use and replaced if there are any cracks or broken teeth. When in use, the blade tension should be checked periodically and tightened if necessary.
- Punches and chisels – Punches and chisels should be inspected before and after use for sharpness and for mushrooming of the head. Where necessary they should be reground.
- Scrapers – Scrapers need to have a very sharp and smooth cutting edge. This should be kept in good condition by sharpening on an oilstone during and after use. As with files, loose or damaged scraper handles should be replaced immediately.
- Measuring and marking out tools – Measuring tools such as micrometers, vernier calipers and dial test indicators should be maintained in clean condition, and when not in use they should be kept in protective cases. They are precision instruments and should not be left around the workplace to come into contact with other tools and materials.

  Engineer's rules, squares, scribers, calipers and dividers should also be kept in a clean condition and be lightly oiled. They should never be used for purposes other than measuring and marking out. Periodically, the points of scribers and dividers should be ground up to maintain their sharpness.
- Power tools – Hand-held power tools such as angle grinders, die grinders, circular saws and power drills should be inspected before use for deterioration or damage to their electrical leads, outer casing and safety guards, where fitted.

  Dust, dirt and waste material should not be allowed to accumulate on them during use. Afterwards they should be stored in a dry place together with any accessories such as special attachments, spanners and chuck keys.
- Machine tools – Before using machine tools such as lathes, pillar drills and milling machines a check should be made on the condition of the guards, electrical connections and auxiliary systems such as the coolant supply and lighting.

  The levels in the lubricant and coolant reservoirs should be checked and replenished if necessary. Lubricating oil should also be applied to leadscrews, slideways and bearings as required.

  Swarf should not be allowed to accumulate during use and afterwards the slideways, worktables and coolant trays should be cleared of any remaining swarf or coolant.

Faulty tools and equipment should be reported immediately, especially if the fault is electrical or concerned with guards and toolholding or workholding devices. After finishing work, the workplace should be cleaned and left in a tidy condition ready for future activity.

Scheduled maintenance

Scheduled maintenance can prolong the working life of machines and equipment, prevent unscheduled stoppages, contribute to product quality and help to prevent accidents. Planned maintenance activities may be carried out by production workers themselves at specified times when production is halted. Alternatively, they may be the duty of a specialised maintenance team who operate outside production periods such as at night or at weekends.

The maintenance plan for a particular item of machinery or area of plant can be presented on a maintenance planning sheet (Fig. 1.55). This lists all the maintenance activities and states how often they should be carried out.

Maint. opn	Operation description	Materials/components	Equipment	Tools/gauges	Est. time	Frequency
10						
20						
30						
40						

Figure 1.55 Maintenance planning sheet

The maintenance operations are numbered 10, 20, 30 in the same way as the production processes on the production planning sheet. The operations may include cleaning, adjusting, replacing components, replenishing materials such as lubricants and testing to ensure that the machinery or plant is functioning safely in readiness for use.

The frequency column states how often a maintenance operation must be performed with the most frequent being operation 10 and the last one being performed the least often. A separate record or schedule should be kept which shows the maintenance operations completed to date and those which are due for attention.

Self-assessment tasks

1. What do terms such as BHN 250, or VPN 800, indicate on the engineering drawing of a component?
2. What does a toleranced dimension define?
3. Draw the symbol which would appear on an engineering drawing if the surface of a component is to be ground to a surface roughness of 3.2 microns.
4. Why are the dimensions of a component often given from a datum surface?
5. What is the difference between the production process plan and the production schedule for a high volume production component?
6. Which Act of Parliament contains the law relating to the safety and welfare of people at work?
7. What kinds of training is an employer legally required to provide for the workforce?
8. What are the general responsibilities of an employee as defined by the Health and Safety at Work, etc., Act?
9. What safety checks should an operator make before starting work on a machine?
10. What are the benefits which can come from operating a programme of scheduled maintenance?

1.4 Unit test

Test yourself on this unit with these sample multiple-choice questions.

1. A material that is both light in weight and a good conductor of electricity is

(a) resin-bonded fibre glass
(b) cast iron
(c) aluminium
(d) polyvinyl chloride (PVC)

2. High voltage electrical insulators are made from

(a) cast iron
(b) ceramics
(c) aluminium
(d) resin-bonded fibre glass

3. Steel is used for lifting chains because of its

(a) solvent resistance
(b) tensile strength
(c) electrical conductivity
(d) malleability

4. Soldering iron bits are made from copper because of its high

(a) electrical conductivity
(b) corrosion resistance
(c) ductility
(d) thermal conductivity

5. Standard values of electrical components are used wherever possible because they are

(a) hard wearing
(b) readily available
(c) heat resistant
(d) easy to handle

6. The type of material used to make grinding wheels is

(a) ceramic
(b) metal
(c) polymer
(d) semi-conductor

7. The component used to change alternating current to direct current in electric circuits is a

(a) resistor
(b) inductor
(c) diode
(d) capacitor

8. The machining process most often used to reduce the diameter of a round bar is

(a) drilling
(b) crimping
(c) milling
(d) turning

9. The process used to increase the hardness of steel components is

(a) heat treatment
(b) chemical treatment
(c) material removal
(d) welding

10. The process in which the technique of crimping is used is

(a) chemical treatment
(b) material removal
(c) joining and assembly
(d) heat treatment

11. Etching is a technique used in the process of

(a) joining and assembly
(b) material removal
(c) heat treatment
(d) chemical treatment

12. A technique used in material removal is

(a) tempering
(b) milling
(c) plating
(d) wiring

13. The heat treatment technique used to reduce the hardness and increase the toughness of a cold chisel is

(a) etching
(b) tempering
(c) soldering
(d) plating

14. The material removal technique used to produce the groove in the V-block shown in Fig. 1.T1 is

(a) turning
(b) drilling
(c) grinding
(d) milling

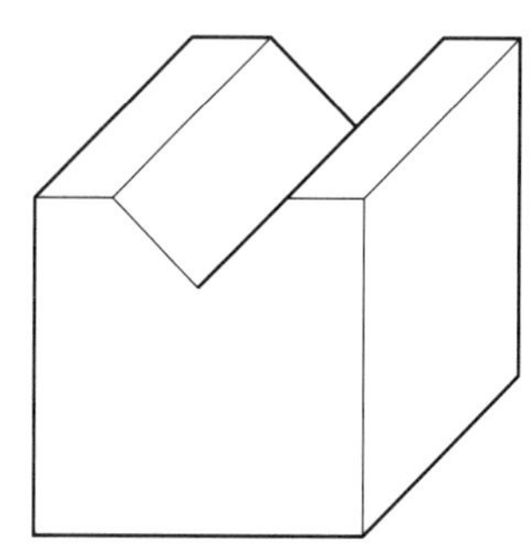

Figure 1.T1

15. Aluminium window frames are anodised to give them increased corrosion resistance. This is a technique of

(a) heat treatment
(b) material removal
(c) joining and assembly
(d) chemical treatment

16. When working on a grinding machine it is essential to wear

(a) a safety helmet
(b) safety boots
(c) rubber gloves
(d) safety goggles

17. A leather apron, heat-resistant gloves and a heavily tinted safety visor are the recommended items of safety equipment for

(a) turning
(b) arc welding
(c) milling
(d) soft soldering

18. The technique that produces components with a very smooth surface finish and close dimensional tolerances is

(a) grinding
(b) tempering
(c) drilling
(d) crimping

19. The method of joining steel plates to give the strongest joint is

(a) soldering
(b) riveting
(c) adhesion
(d) welding

20. The material removal technique that would be the best for producing the component in Fig. 1.T2 complete to specification is

(a) milling
(b) grinding
(c) turning
(d) drilling

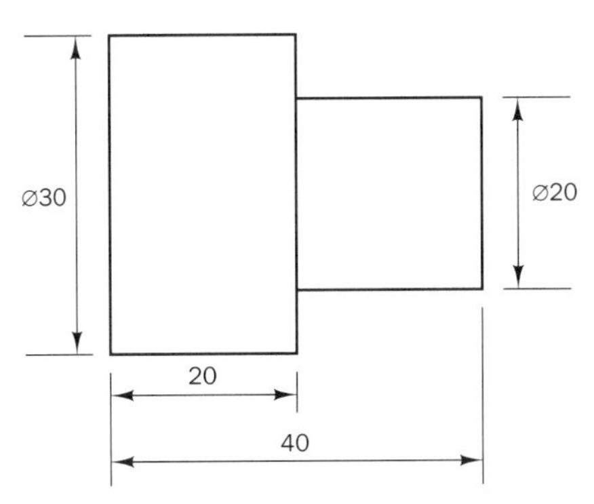

Figure 1.T2

21. The workplace and tools should be kept in good order to avoid

(a) material shortages
(b) accidents
(c) customer complaints
(d) component failure

22. When assembling heavy components it is essential to wear

(a) safety footwear
(b) ear protectors
(c) safety goggles
(d) strong leather gloves

23. A fire extinguisher suitable for fighting fires in electrical equipment is coloured black and contains

(a) carbon dioxide
(b) water
(c) powder
(d) foam

24. On no account should machine operators remove

(a) tools and cutters
(b) waste material
(c) machine guards
(d) finished components

25. To prevent skin infection and irritation from oils and greases it is the recommended practice to

(a) wear a leather apron
(b) use a barrier cream
(c) wear a face mask
(d) have good ventilation

Graphical Communication in Engineering

Colin Chapman

This unit introduces graphical communication within engineering. The chapter covers a variety of communication techniques appropriate to a wide range of applications. It is important to be familiar with many graphical methods in order to develop the ability to select and use a method of communication that is appropriate to its application. For example, sketches that are appropriate for the communication of ideas may be inappropriate for workshop drawings, and formal drawings will be inappropriate for presentation and marketing applications.

Graphical communication involves working with a wide range of media including those available through information technology. Use of IT software is particularly appropriate for charts, graphs and the presentation of information taken from spreadsheets and databases. Many software packages are available for both modelling and drawing using BS and ISO symbols and conventions. It is important to have access to appropriate hardware and software, suitable for the user and the task to be carried out. Some of the best and most highly regarded computer-aided design and draughting (CADD) packages offer too much sophistication for many students and simplified and more appropriate versions are often available.

The areas covered in this chapter are:

- Sketching, drawing and CAD for conceptualising and communicating ideas.
- Diagrams and charts for communicating information.
- Schematic and circuit diagrams.
- Engineering drawing.

After studying this unit you should be able to:

- Sketch freehand two- and three-dimensional geometric and engineering shapes and forms.
- Develop ideas in sketch form to satisfy product and/or system requirements.
- Prepare sketches and drawings using a range of techniques and media appropriate to a variety of applications and audiences.
- Produce spreadsheets, flow diagrams and bar charts in order to communicate information.
- Produce schematic diagrams for mechanical, pneumatic, hydraulic, electrical and electronic systems.
- Produce circuit diagrams for pneumatic, hydraulic, electrical and electronic circuits.
- Understand the application of both pixel-based and vector-based drawing software.
- Understand the application of standards and conventions within engineering drawing.
- Understand orthographic projection as applied to engineering drawing.
- Produce dimensioned and scale drawings of engineering details using both drawing instruments and CAD software.
- Understand the application of standard representations of common engineering features within engineering drawings.
- Read and interpret multi-component assembly drawings.

2.1 Sketching, drawing and CAD for conceptualising and communicating ideas

Sketches can be used to develop and clarify your own ideas and as a means of communicating those ideas to others. Either way it is important to be able to sketch clearly and quickly. Freehand sketches should enable you to use a piece of paper or a note pad as an extension of your brain and your thinking processes. It does not matter if things get refined later or even rejected completely, it is more important to get the ideas down on paper than it is to spend time making them look like works of art.

Some of Alec Issigonis' now famous 1950s sketches of the Morris Mini Minor were said to have been made on a restaurant table cloth (Fig. 2.1). These ideas are regarded as a milestone in small car design. Issigonis' designs were so successful that they continue to influence car design today.

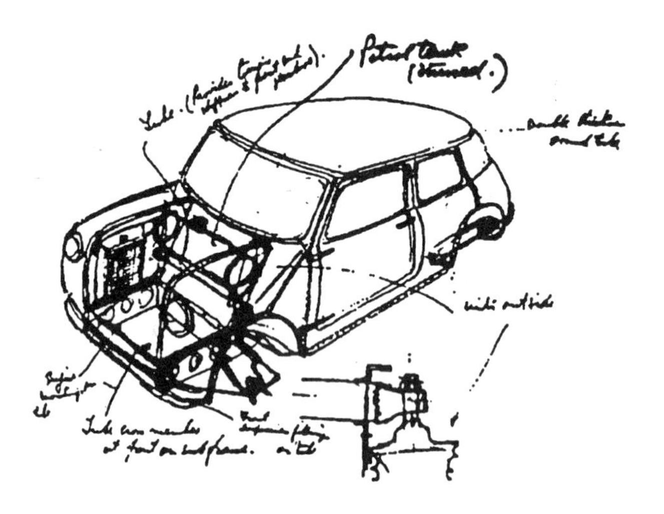

Figure 2.1 Original sketch of the Mini by Alec Issigonis

Freehand sketching

The ability to sketch can be greatly improved by practice and by learning a few techniques, but first let us consider the equipment to use. Professional graphic designers use a range of pens and pencils for making drawings, but for preliminary sketches, within the context of engineering design, softish pencils such as 'B' or '2B' are good and many designers use an ordinary ball-point pen. Few use felt or fibre-tip pens unless they wish to colour or render their drawings at a later stage. It is important to use the drawing instrument that flows easily and enables you personally to work easily.

To learn how to sketch fluently start by drawing simple, flat geometric shapes. Simple shapes are built up from either straight or curved lines. The key to drawing straight lines is to look where you are going and not at the point of the pencil or pen that you are using. Try drawing long straight lines vertically and horizontally to join dots at either end of an A4 sheet of paper, and look at the dot not at the pencil. Follow this by drawing right angles and work up to drawing squares and rectangles (Fig. 2.2).

Curves and circles can be constructed within squares based around the fact that a circle will touch a square in four places at a mid point along each side. Try drawing circles in squares and ellipses inside rectangles. Use a flexible wrist action to draw curves (Fig. 2.3). Don't push the pencil but turn the paper round so that it flows easily, and try to avoid drawing circles with points. Bring it all together by sketching some familiar three-dimensional objects (Fig. 2.4). Always begin with the overall outside shape and ensure that these proportions are correct, then identify the simple shapes that make up the complex one and work on it a little at a time.

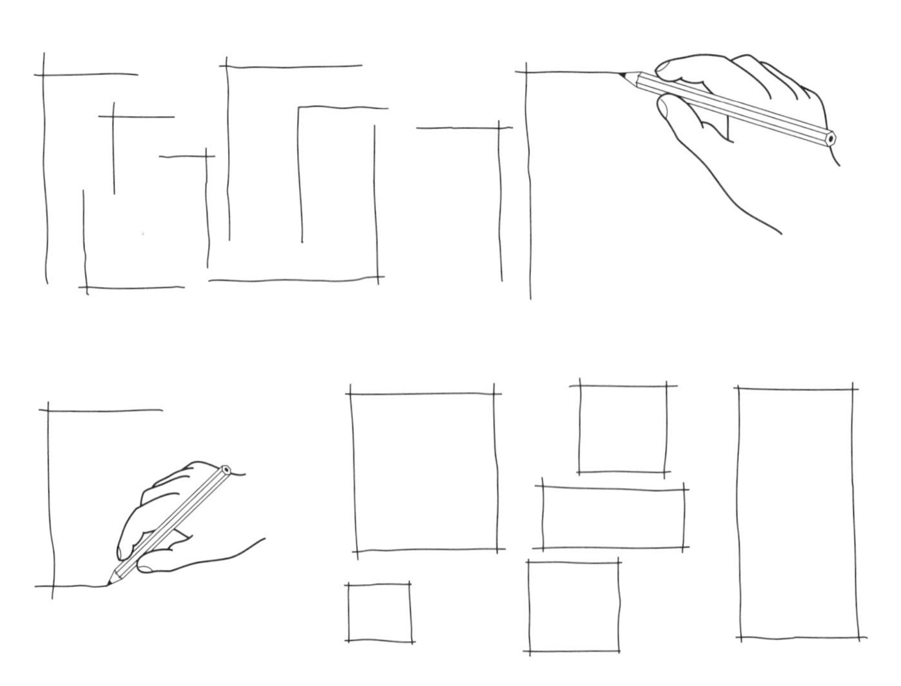

Figure 2.2 Sketches of right angles, squares and rectangles

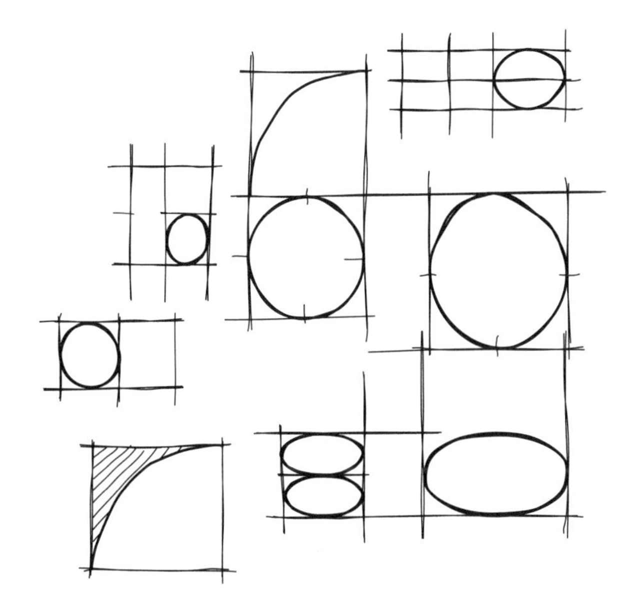

Figure 2.3 Sketches of curves and circles

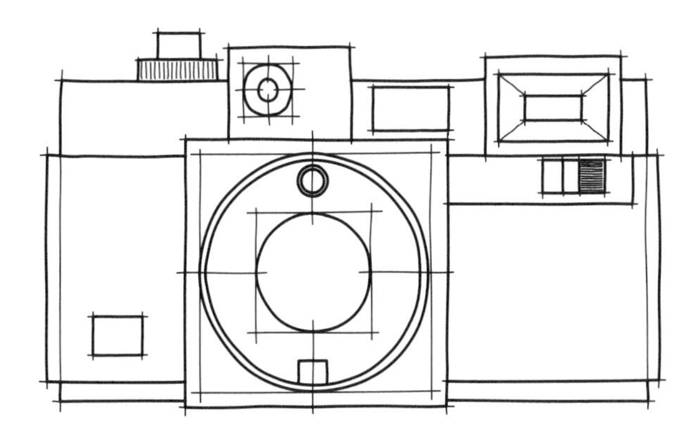

Figure 2.4 Sketch of a camera

Three-dimensional, or pictorial, sketches use a technique called 'crating'. This involves drawing an overall rectangular form, (a 'crate') that would just contain the object that you intend to draw, rather like a close-fitting fish tank. You can then draw the object inside the crate; this technique helps to keep the proportions correct and the sides parallel.

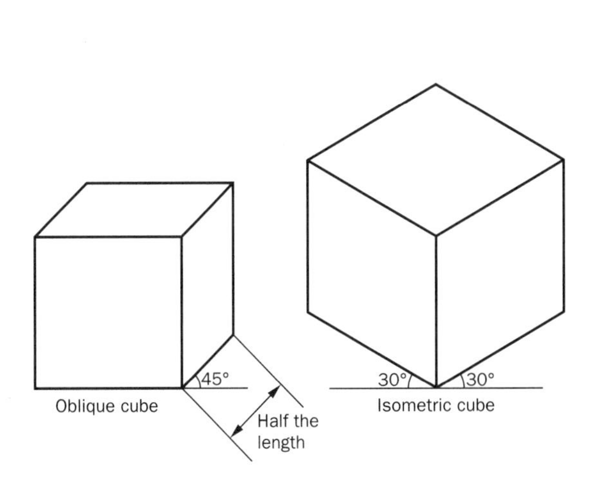

Figure 2.5 Oblique and isometric sketches

Pictorial drawings are most commonly 'oblique' or 'isometric'. Perspective drawing is another common form of pictorial drawing but it is unlikely to play a large part within the context of engineering. Oblique drawings have a 'correct' front face with right angle corners and side edges that slope back at 45°. In isometric drawing, which is the more popular of the two, the drawing is tipped forwards and the front and side edges slope at 30°. For oblique drawings the depth of the drawing has to be shortened in order for the object to look right (Fig. 2.5).

Circles and curves are dealt with as before but this time an outer square is drawn first (oblique or isometric). An oblique or isometric circle will touch the sides of the 'square' halfway along their lengths.

Developing and communicating ideas

Freehand sketching is the best method of exploring and communicating ideas. It is far easier to work on design detail in sketch form on paper than it is to work things out in your head or try to describe them. Initially you might generate and record ideas just for yourself but as they develop there will be a need to communicate them to others.

Use your initial sketches for reflection and development. Make yourself develop alternatives and variations. Design is about looking, reflecting and considering. It is also about working with others and bouncing ideas around – and this means communicating. Few designers operate on a 'eureka' basis. Issigonis didn't sketch on that table cloth for his benefit alone, he needed to communicate and share his ideas with colleagues.

Figure 2.6 shows some sketches which look at possible ways of locating the hinge pin of a cover. Clearly there needs to be more consideration given to materials and manufacturing processes before further design decisions can be taken. As designs develop they have to be informed by the possible manufacturing stages, but initially it is important that all possibilities are considered.

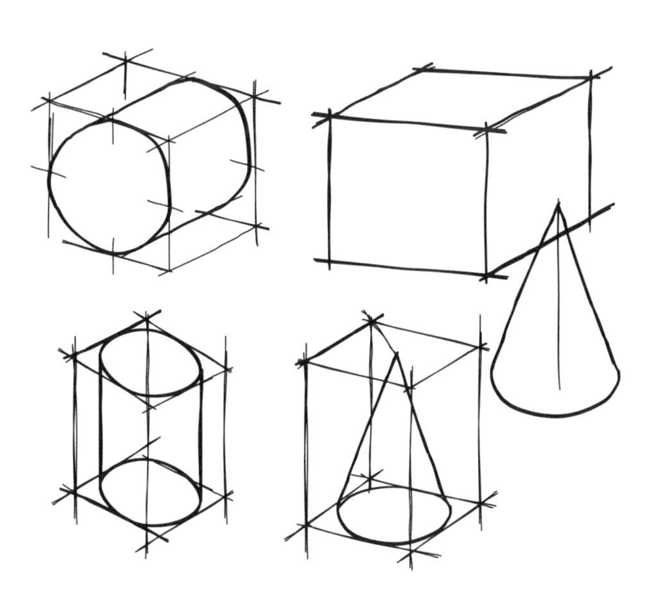

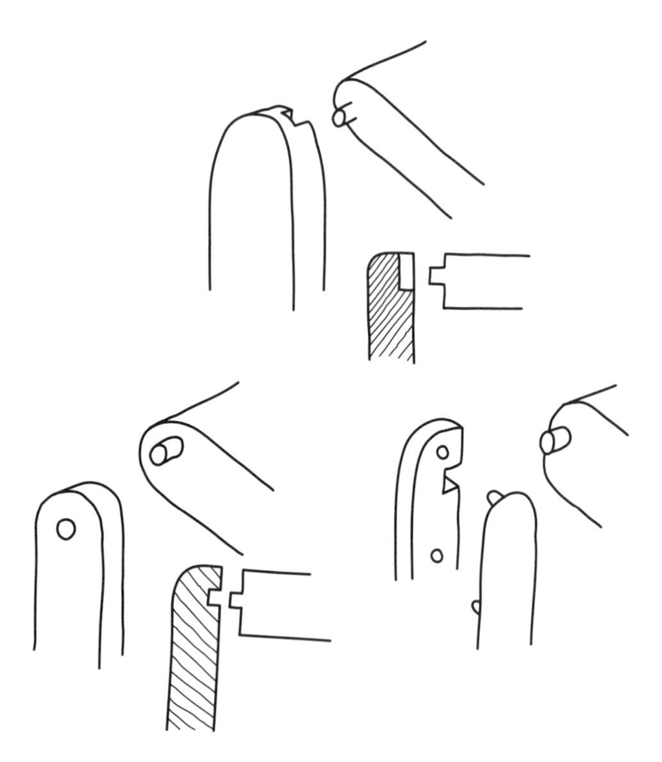

Figure 2.6 Sketches of hinge arrangement

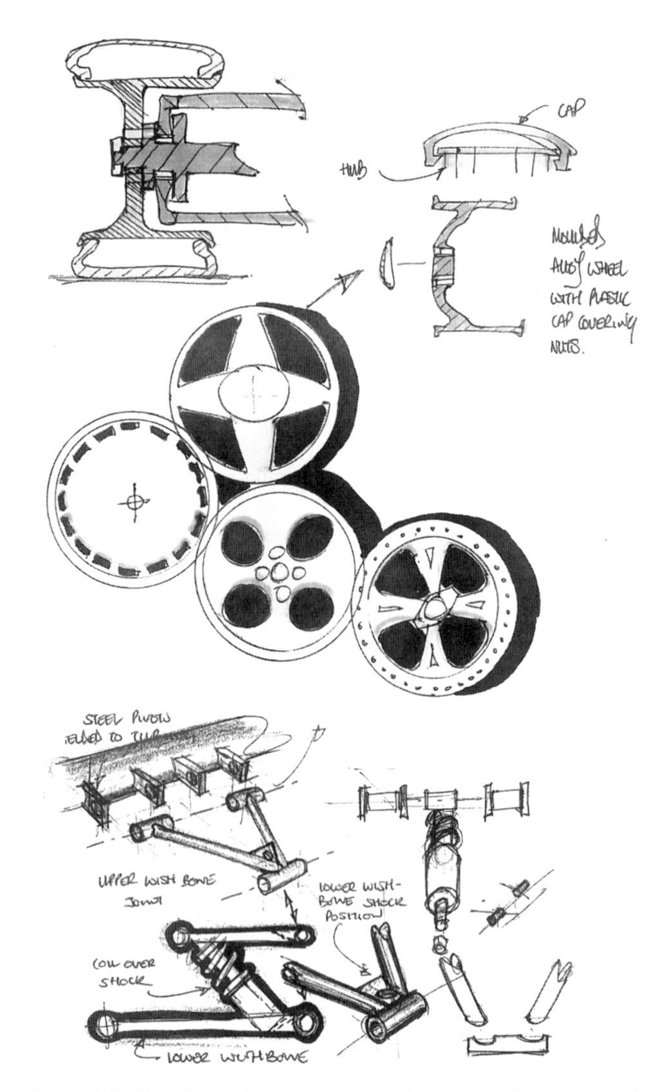

Figure 2.7 Sketches of a concept car (*reproduced courtesy of* Tom Morgan (De Aston School, Market Rasen))

Figure 2.7 shows examples of a designer's work exploring potential variations for sports car wheel hubs and also looking at the possible construction of a suspension unit. You can see that materials and processes are receiving consideration as well as form and appearance. Here two aspects of the same product have different priorities. While wheels are clearly a functional part of a car, their appearance also makes a statement about the type of car: sports car, family car, city car, etc. The suspension units are primarily functional and here the form follows the function.

Presentation drawings

All engineered products are becoming more complex and expensive, whether kitchen appliance, machine tool or military aircraft. Increasingly products are discussed with clients and even sold through the medium of a prototype model or presentation drawing, long before the actual product exists. Marketing departments need to create a demand before production is underway so that sales will peak faster; visual images are powerful inducements for clients and potential customers.

Presentation drawings can take many forms and have the advantage over the finished product that they can be used to highlight and even exaggerate particular selling features. For example the sleekness and suggestion of speed of a car, or the way that a new rail depot or bridge will fit within the local environment.

Figure 2.8 shows a sequence of illustrations of the stages in making a presentation drawing of a blow-moulded bottle for hand cream using marker pens.

1. The outline has been drawn on top of a feint pencil line using a fine fibre-tipped pen. A light coloured marker pen is used to add the first areas of colour to the cap and the body of the bottle. The front of the body is flat so this is left unmarked.
2. With a strip of tape placed down one side, the front can now be streaked across with quick movements. Streaking suggests a flat surface.
3. Further work with the same markers deepens the colours of the bottle and the cap on the left-hand side. This suggests depth of form and a right-hand light source.
4. Finally some black lines are added, also on the left, and white crayon is used to highlight the top edge of the bottle and separate it from the top. The finishing touch is a suggestion of shadow which is used to lift the product from its background.

Another presentation technique is airbrush rendering. This can be very time consuming as the majority of the work is involved in creating templates and masks. It is essential that the work is planned in the finest detail and that it is well practised (Fig. 2.9).

Computer-aided design (CAD)

Computer software can be used to aid the designer at many stages of the process of design and during the eventual transfer to manufacturing and process control. Design software is by its

Figure 2.8 Marker rendering

Figure 2.9 Airbrush rendering (*reproduced courtesy of* Dave Eastbury)

nature more sophisticated than computer draughting packages that may only support engineering drawing. The process of designing is a process of rearranging, visualising and making decisions and it should not be confused with inventing. If a design exists then it can be reworked and developed or imported and included. If it exists on a computer then the process can be done faster and more economically.

Consider the followings ways that CAD can assist the design engineer:

- Enables modification and reworking of existing designs.
- Provides access to libraries of standard features and components.
- By plotting the paths of moving features for clearance and collision detection.
- Aids design decision making by supporting the ability to visualise completed components and products.
- Enables engineering analysis of components subject to stresses and fatigue.
- Allows information to be shared electronically via communication networks such as e-mail, video conferencing and the Internet.

Access to libraries of standard components helps to speed up the process (Figs 2.10 and 2.11). The library file can also be used to show available stock and preferred components.

Parametric design software enables the design engineer to plot the path of moving parts to see how they could potentially interfere with other aspects of the design. Software of this nature is also used to design manufacturing cells to avoid clashes between robot operations.

The interpretation of computer generated images into three-dimensional forms enables the designer to better visualise and communicate the outcome. Figure 2.12 shows wire frame and solid model interpretations of a cast component. While the solid model is the easier to visualise, because it is closest to reality, it is expensive in terms of hardware requirements and processing time, making it slow to manipulate on the screen. Designers often stop at the wire frame stage. Wire frame modelling as a 'language' for communication is very popular and has the advantage of being able to see features behind and on the rear surface of components. By using colour to identify different components and surfaces wire frame assemblies become easy for the trained designer to interpret.

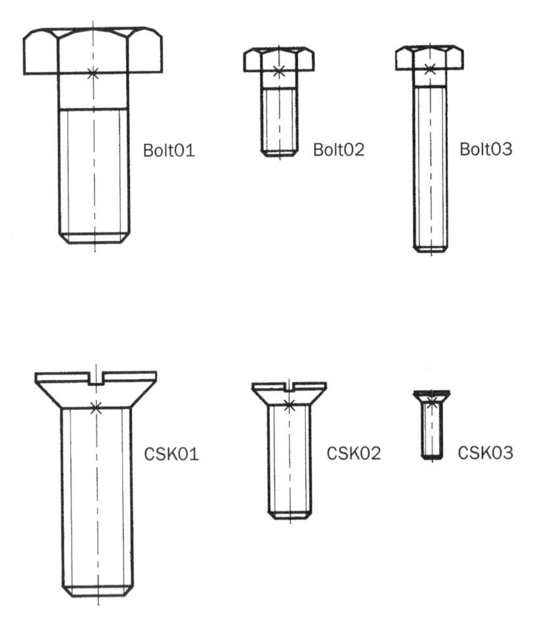

Figure 2.10 Extract from a library of standard bolts and screws

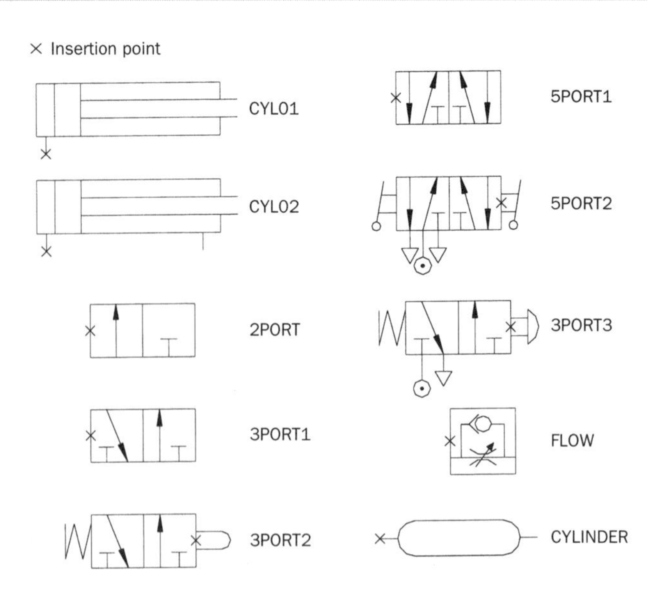

Figure 2.11 Extract from a library of standard pnuematic circuit symbols

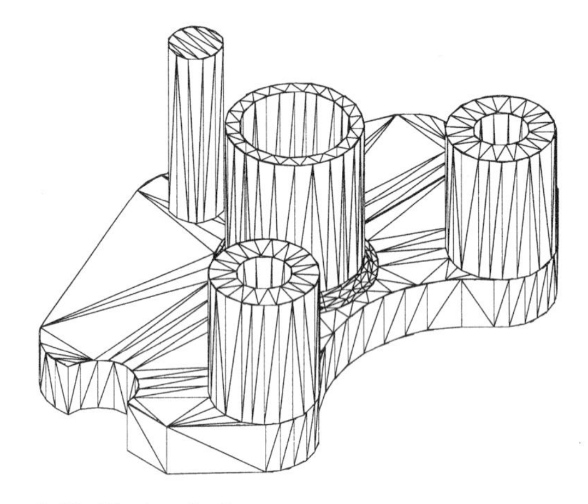

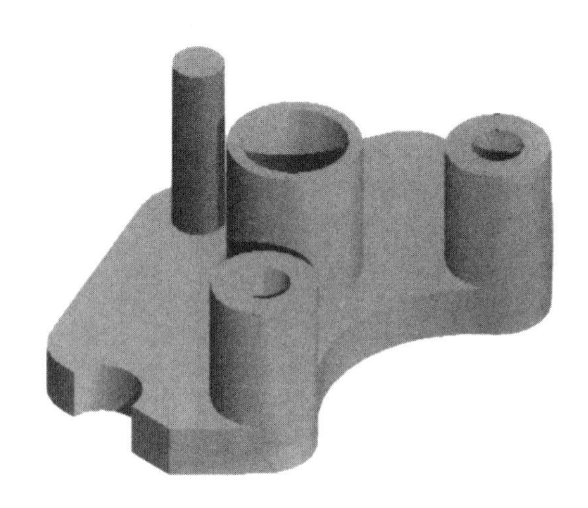

Figure 2.12 3D visualisations

Self-assessment tasks

1. Using the freehand sketching techniques described at the beginning of this section make some sketches of familiar rectangular items such as televisions, stereo systems, computer systems, etc. Look carefully at the proportions and sides that are parallel.
2. Take the best of your sketches from Task 1 and turn it into a three-dimensional oblique drawing. Do not use any drawing instruments.
3. Using isometric techniques, sketch a compact camera.
4. Using sketches and notes carry out a study of a simple mechanical device such as a stapler, can opener, pencil sharpener or hole punch.
5. For what purpose might a designer produce a presentation drawing?
6. Using any technique that you feel is appropriate, make a presentation drawing of a product that you are working on as part of your course.
7. Make a list of the advantages and disadvantages of using computer systems as an aid to designing and communicating design ideas.
8. What are the advantages of using computer generated wire frame modelling over other graphic and prototyping techniques?

2.2 Diagrams and charts for communicating information

Manufacturing industries, as they become international and develop international markets, become increasingly complex and have an ever greater need to communicate information. Non-verbal communication takes place within organisations, between organisations, and between organisations and their customers and suppliers irrespective of national boundaries or language. The mode of communication adopted will depend upon two key considerations.

- **Who** is it for? – 'The target audience'
- **What** does it need to say? – 'The message'

By addressing these questions it is possible to determine the most appropriate mode of communication: it could be a written report and in another case a picture. Quite often the most appropriate form lies between the two extremes and involves both graphic images and text.

Data charts

Manufacturing and planning data charts are used to display information that frequently needs to be referred to. Such charts are presented within handbooks or are produced with a wipe clean surface to be displayed on walls near to the appropriate process. Cutting speeds for machine tools, drilling sizes for screw threads and temperature settings for heat treatment are typical of the types of information that may be presented in this manner. The chart shown here (Fig. 2.13) shows the speed settings for drilling operations in columns (vertically) according to the cutting speed of the material, and rows (horizontally) according to the diameter of the drill.

Metric series	CUTTING SPEEDS Approximate							*Metric series*
ft/min	30	40	50	60	70	80	90	100
m/min	9	12	15	18	21	24	27	30
diam/mm	*Revolutions per minute*							
·5	5817	7756	9695	11634	13573	15512	17451	19390
1·0	2909	3878	4847	5817	6786	7756	8725	9695
1·5	1942	2589	3237	3884	4532	5179	5826	6474
2·0	1456	1942	2427	2912	3397	3883	4369	4854
3·0	970	1294	1617	1940	2264	2587	2911	234
4·0	728	970	1213	1455	1698	1940	2183	425
5·0	582	777	970	1164	1359	1553	1747	941
6·0	485	647	808	970	1132	1294	1455	617
7·0	416	555	693	832	970	1109	1248	386
8·0	364	485	606	728	849	970	1091	213
9·0	324	431	539	647	755	962	970	078
10·0	291	388	485	582	679	776	873	970
11·0	265	353	441	529	617	706	794	882
12·0	243	324	404	485	566	647	728	808
13·0	234	299	373	448	522	597	672	746
14·0	208	277	346	416	485	554	623	693
15·0	194	259	323	388	453	517	582	647
16·0	182	243	303	364	424	485	546	606
17·0	171	228	285	342	399	456	513	571
18·0	162	216	269	323	377	431	485	539
19·0	153	204	255	306	357	408	459	511
20·0	146	194	242	291	340	388	436	485
21·0	139	185	231	277	323	370	416	462
22·0	133	177	220	265	309	353	397	441
23·0	127	169	211	253	295	337	380	422
24·0	121	162	202	242	283	323	364	404
25·0	117	155	194	233	272	310	349	388

Figure 2.13 Typical data sheet

Graphs

Graphs are one of the most common graphical means of displaying information (Fig. 2.14). In its simplest form a graph will just show the relationship between two factors such as time and distance. Conventionally, when time is displayed upon a graph, it occupies the horizontal or '*x*' axis. The vertical axis is called the '*y*' axis. For example, vehicle A travelling at a constant speed appears as a straight line on a distance/time graph whereas vehicle B which is accelerating or changing its speed will generate a graph that is curved.

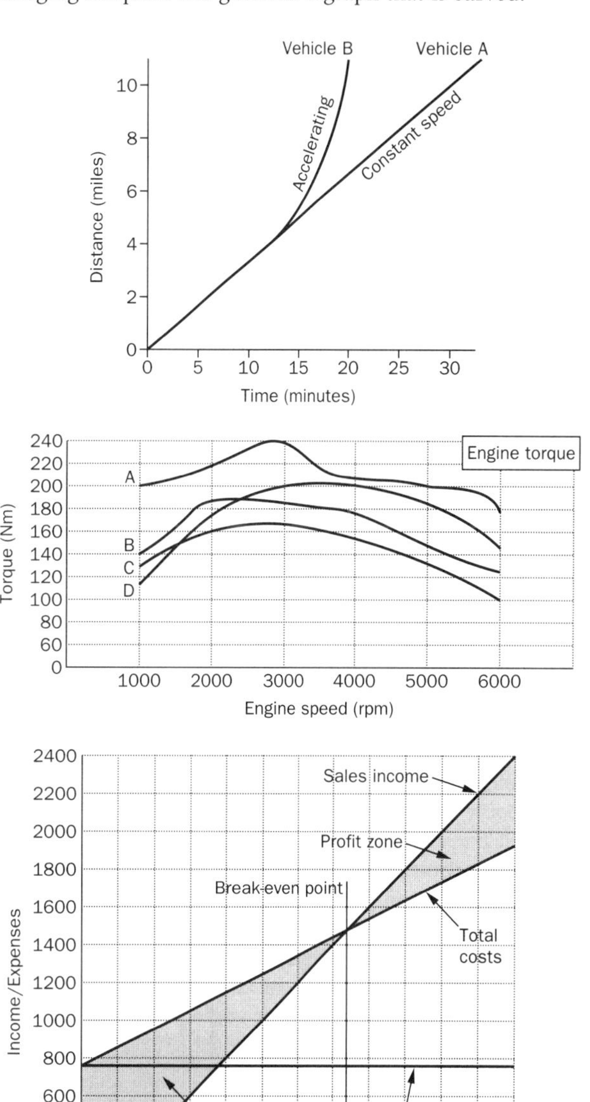

Figure 2.14 Examples of graphs

The other two graphs shown in Fig. 2.14 have different information to communicate. The graph showing engine performance has the torque versus engine speed of four engines (A, B, C, and D) plotted on it. The purpose of the graph is to compare the torque capability of the four engines.

The break-even analysis graph is used to determine one essential piece of information, the point where loss changes into profit: a most crucial point for any business.

Graphs are used extensively within newspapers and the media to show trends rather than for extracting detail. The upwards slope of a graph indicates increase whereas a downwards slope shows decline. They are not intended to be 'quantitative', this means that you can't read off the graph any numerical values as you could in the three graphs shown in Fig. 2.14.

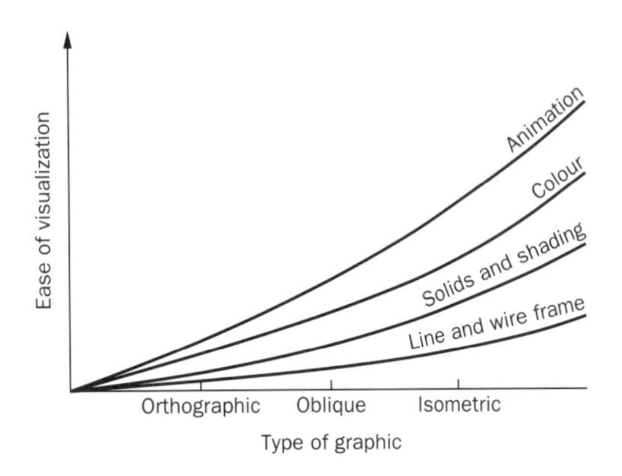

Figure 2.15 Types of graphical communication and their relative ease of visualisation

The graph shown in Fig. 2.15 has no quantifiable information on the axes but it does enable some important comparisons to be made that are particularly relevant to this unit.

Bar charts

Bar charts are used to display information in a more accessible form than a graph. In its simplest form a bar chart is more properly known as a 'histogram' which is really a graph that is drawn in steps rather than as a curve. The histogram shown in Fig. 2.16 is the sort that appears in holiday brochures showing the temperature of a resort at different times of the year. Again time, in this example in months, is shown conventionally along the *x*-axis.

Bar charts can be displayed in three-dimensional form and either vertically or horizontally. The bar chart in Fig. 2.16 shows how the UK workforce is divided up. The 'message' being conveyed is that manufacturing industry has to earn enough to cover the costs of other people within the economy particularly those in service industries.

Pictograms are another form of bar chart, using images to represent the message of the chart. In Fig. 2.17 the shape of a person is used to represent people employed within various aspects of manufacturing.

In Fig. 2.18 all of the 'bars' are the same size, representing 100 per cent of taxation, and the chart is used to compare how the taxation in a number of countries is made up from a range of taxation sources. You can see that as a percentage of total taxation income tax in the UK is similar to that in Japan, Italy and Germany, and less than that of Canada and the United States. The chart does not, however, say how much taxation is collected or how much income tax is paid in those countries as a percentage of income. It is important to remember that the message being portrayed is the chosen message of the originator of the information and not necessarily the whole story.

Pie charts

Pie charts are limited in that they can only show how things are divided up. Often when the aim is to draw attention to a particular aspect, that slice of the pie is drawn separately. The UK taxation example in Fig. 2.18 can be represented as a pie chart, as shown in Fig. 2.19; the 30 per cent that is income tax has been highlighted.

Computer-generated graphs and charts

Spreadsheets and databases usually have the facility to produce graphs from the data stored within them. Figure 2.20

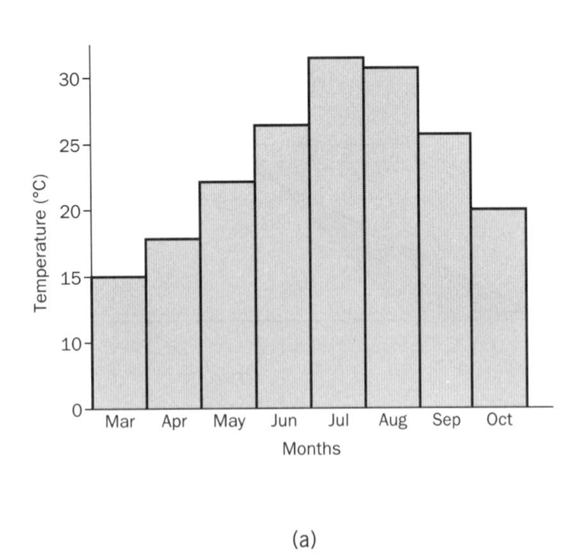

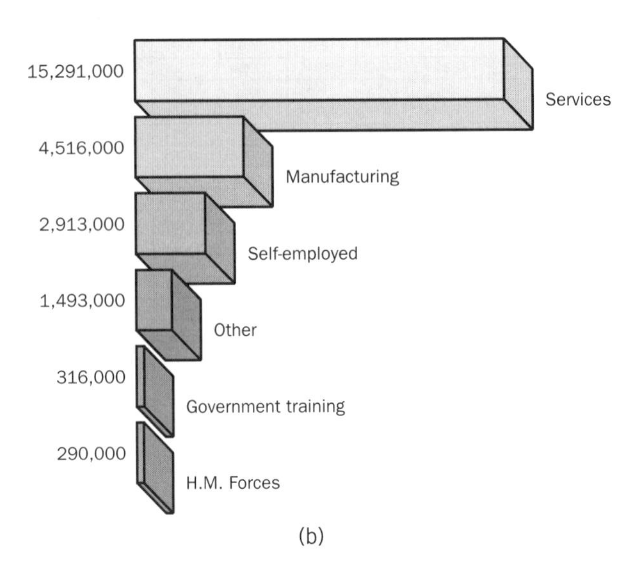

Figure 2.16 Examples of (a) histogram and (b) horizontal bar chart

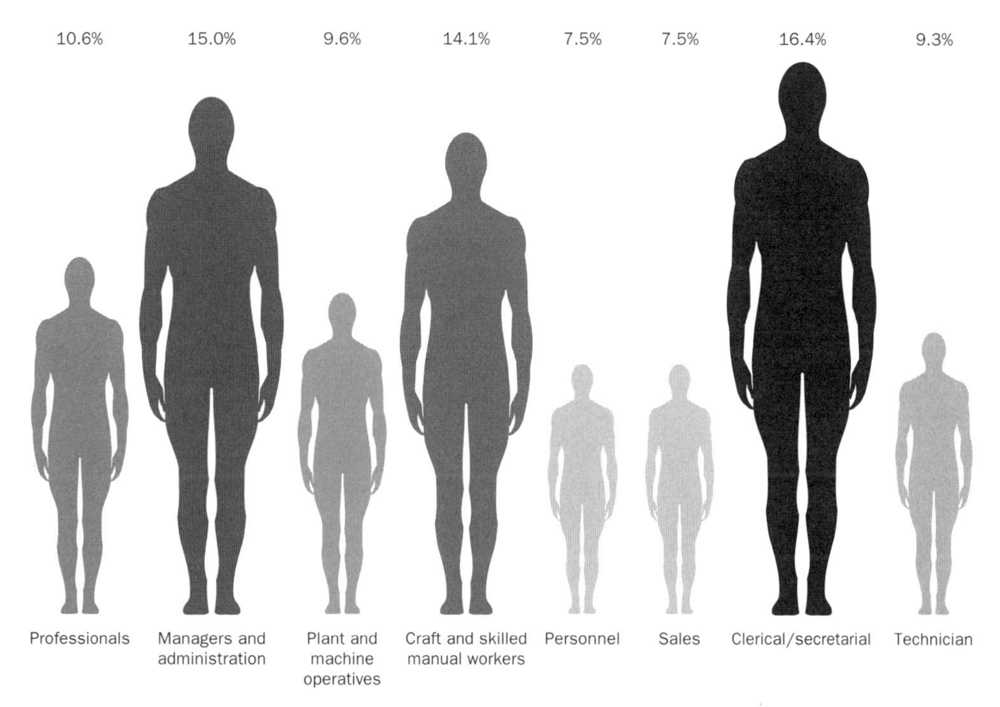

Figure 2.17 Example of a pictogram

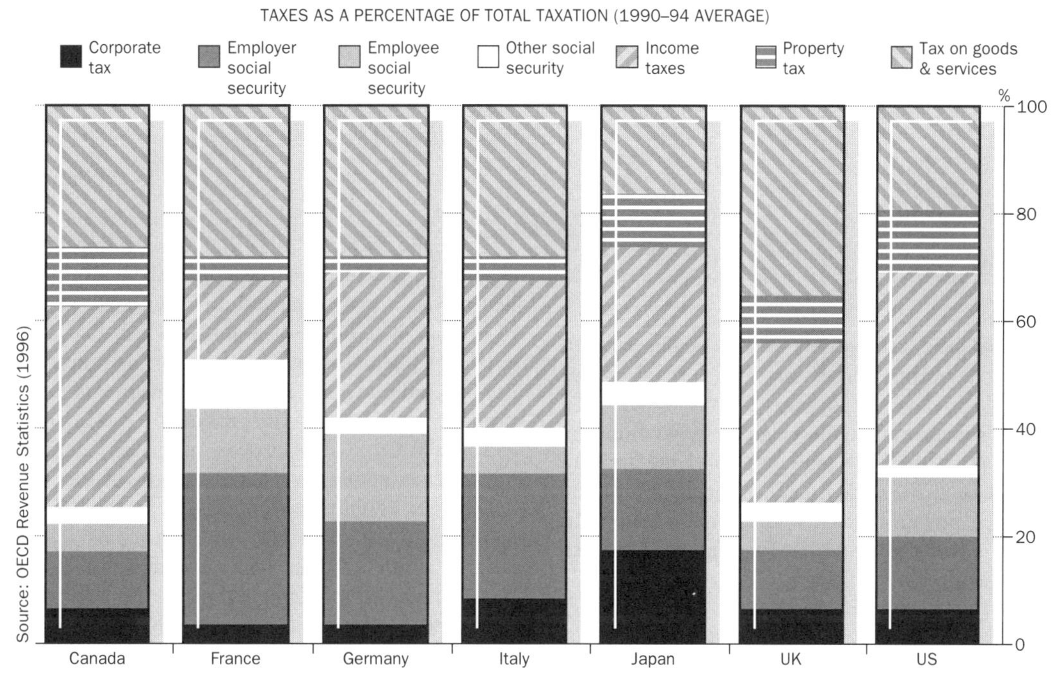

Figure 2.18 Taxation bar chart

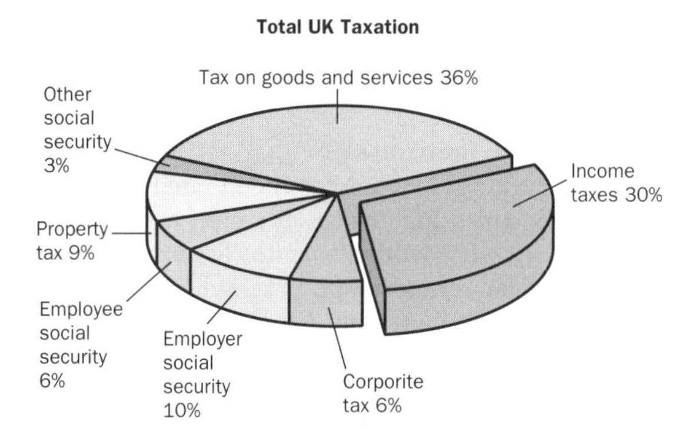

Figure 2.19 UK taxation pie chart

shows graphs and charts generated by Microsoft Excel (a commercially available spreadsheet package). Once data has been entered into the computer it can be organised, manipulated and displayed in a number of ways including:

- spreadsheets
- databases
- tables
- charts
- graphs
- bar charts
- pie charts

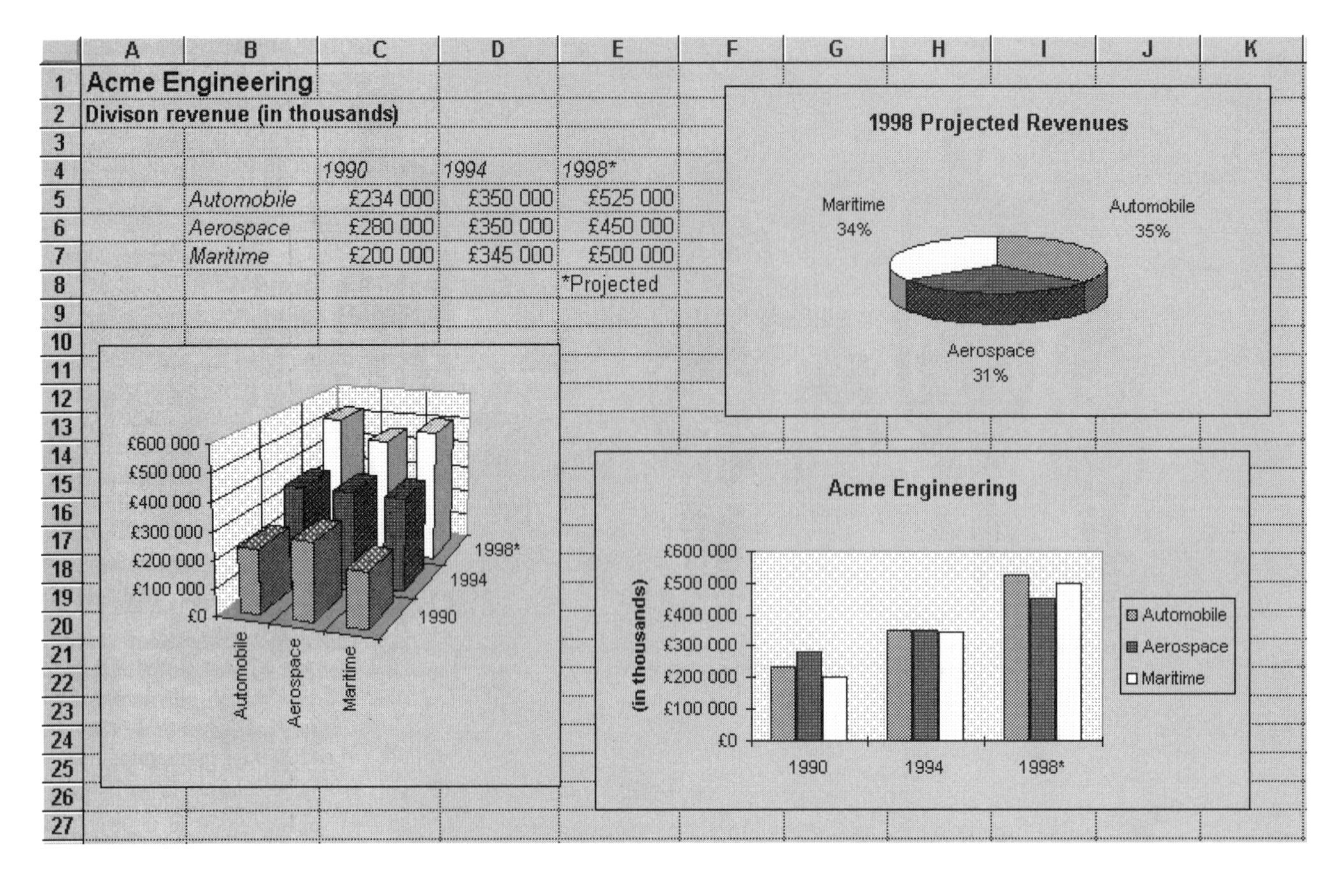

Figure 2.20 Example of using Microsoft Excel to display same data in different graphical formats

Block diagrams

Block diagrams are used very widely within engineering to graphically show concepts, processes and organisational structures. The block diagram in Fig. 2.21 is a graphical representation of 'concurrent engineering'. It shows how the elements overlap along the time axis (left to right) and are therefore taking place at the same time, i.e. concurrently. The block diagram is a powerful means of communication, to describe the process verbally would be long and could lead to confusion.

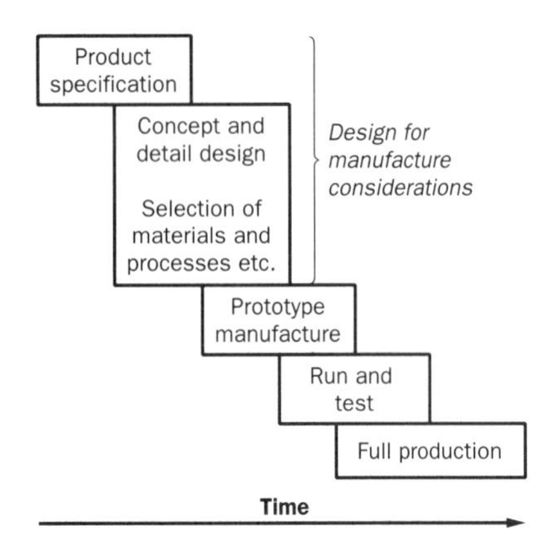

Figure 2.21 Concurrent engineering

There are no strict rules that apply to most block diagrams other than the need for clarity and the convention that if time is involved it progresses either from left to right or from top to bottom. The block diagram shown in Fig. 2.22 describes a product cycle and therefore has no time line. The diagram is used to indicate how and where information technology has been integrated within the production cycle.

Manufacturing processes may be controlled by a 'precedence network'. This is a form of block diagram that adheres to certain conventions as set out in BS 4335 and summarised within PP7307. Precedence networks are used to control processes that have crucial start, finish and delay requirements. For example in Fig. 2.23 'process B' cannot start until '*y*' days after 'process A' has started. This is called a 'lag start'.

Flow charts

Flow charts are a form of block diagram used to organise sequences of events. They are commonly used when writing computer programs and when planning operations that involve decisions of a 'yes/no' nature. Flow charts are also frequently used for diagnostic purposes where a fault-finding procedure must be adhered to. Within PP7307 there is an example of a flow charted procedure for checking a faulty 13 amp electrical plug (Fig. 2.24).

Flow charts should proceed downwards or from left to right. The 'yes' outcome from any decision should follow the

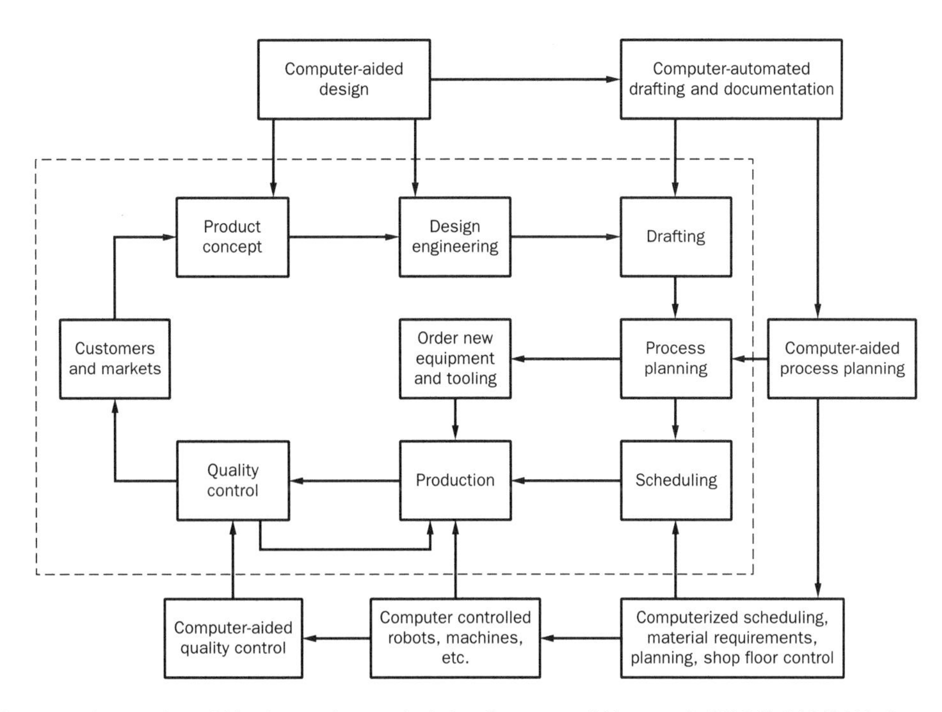

Figure 2.22 Computer integration within the product cycle (*after* Groover and Zimners Jr (1984) *CAD/CAM: Computer-Aided Design and Manufacturing* Prentice-Hall)

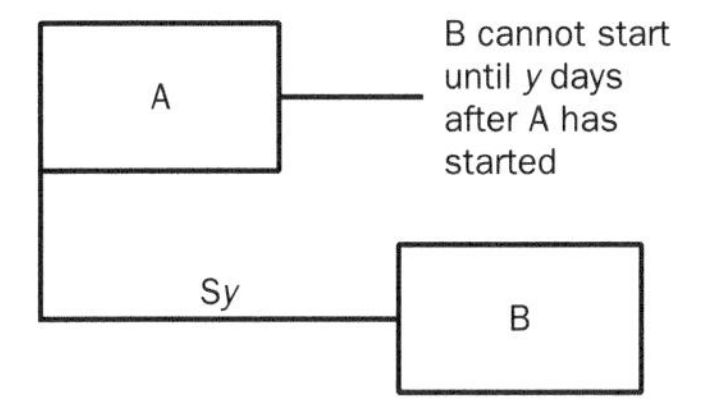

Figure 2.23 Example of a precedence network

principal direction of flow. Arrows may be used to indicate flow direction and increase clarity but these should be kept to a minimum. Flow chart templates that contain all of these standard symbols can be purchased from stationers. The example chart shown here includes most of the commonly used symbols (see Table 2.1).

Table 2.1 Flow chart symbols

Symbol	Name	Meaning
(rounded rectangle)	Terminator	Stop or start, or exit from the process
(rectangle)	Process	Any pre-defined process or operation
(diamond)	Decision	Resulting in 'yes' or 'no' and leading to one of two paths
(parallelogram)	Data (*not used in example*)	An input or output of data to or from the process
(hexagon)	Manual input	Manual or other interruption to the process

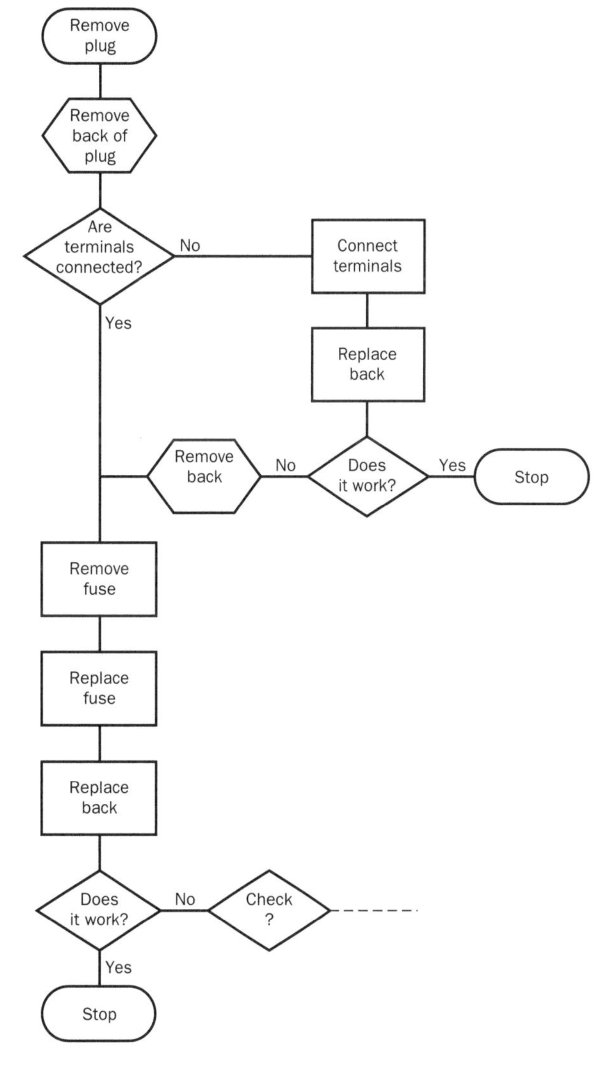

Figure 2.24 Flowchart to analyse a mains plug which is not working

Self-assessment tasks

1. Outline the reasons for using diagrams and charts when communicating information. Give examples from your own experience of occasions when you have referred to a diagram or a chart for information.
2. What criteria should be used to determine the most appropriate type of diagram or chart to use for a particular application.
3. Refer to the information presented in the graph of engine performance in Fig. 2.14. What can you say about the performance of engines A and D?
4. What does the graph in Fig. 2.15 tell you about communication of information?
5. Refer to the information given in the bar chart in Fig. 2.16 and the pictogram in Fig. 2.17. Draw a histogram showing the actual numbers of people employed within the various aspects of manufacturing industry in 1993.
6. Produce a pie chart to show how your typical day is divided up into a range of activities.
7. (a) Enter the following information on bicycle sales into any spreadsheet that you are familiar with, and from it produce a range of graphical interpretations.

	1995	*1996*
1st quarter	1,760	2,140
2nd quarter	3,450	4,025
3rd quarter	1,134	1,560
4th quarter	8,784	10,234

   (b) Comment upon the seasonal nature of these figures in relation to the particular product.
8. Produce a block diagram of any production process that you are familiar with.
   (a) Why should it be crucial within some processes that operations overlap or do not start before a certain point is reached?
   (b) Produce a block diagram in a form that illustrates this using an example of a process that you are familiar with.
9. Produce a flow chart guide to replacing a tap washer or repairing a bicycle puncture.

2.3 Schematic and circuit diagrams

Schematic drawings and circuit diagrams are used to show the arrangement of components within electrical, electronic, hydraulic and pneumatic systems. The various systems use standard symbols that are set down in a large and comprehensive range of British Standards. These are, however, summarised in PP7307 *Graphical Symbols for use in Schools and Colleges* and this document should be made reference to.

Pneumatic and hydraulic systems

It is convenient to group these two types of system together because the symbols used within them are similar. They are described as 'functional' symbols, this means that the symbol describes graphically the function of the component. Valves, whether they control fluid or air, operate in a similar manner, and are therefore represented by the same or a similar symbol.

The simplest valve is a manually operated, two-port, two-way, directional control valve. This means two connections (in and out) and two positions (on and off). It is represented graphically in Fig. 2.25.

The valve is shown connected in a flow line in its 'at rest' position (Fig. 2.25(a)); valves should always be shown at rest. The line is blocked off by two closed ports. The functionality of the symbol is explained by the notion that when the valve is operated the right-hand square is replaced by the left-hand square and so the flow line is joined (Fig. 2.25(b)).

Study the two systems shown in Figs 2.26 and 2.27. The functional nature of the symbols will become clear as you come to understand the function of the systems.

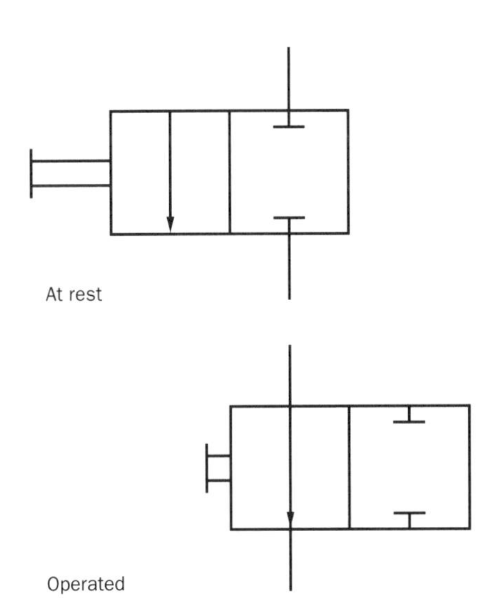

Figure 2.25 Symbol for a simple two-part, two-way, direction control valve

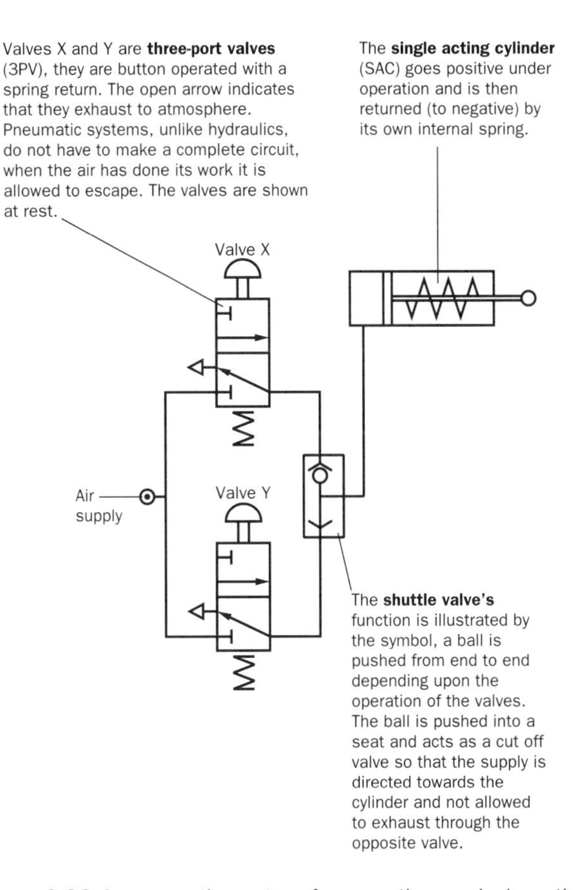

Figure 2.26 A pneumatic system for operating a single acting cylinder from either one of two button operated valves

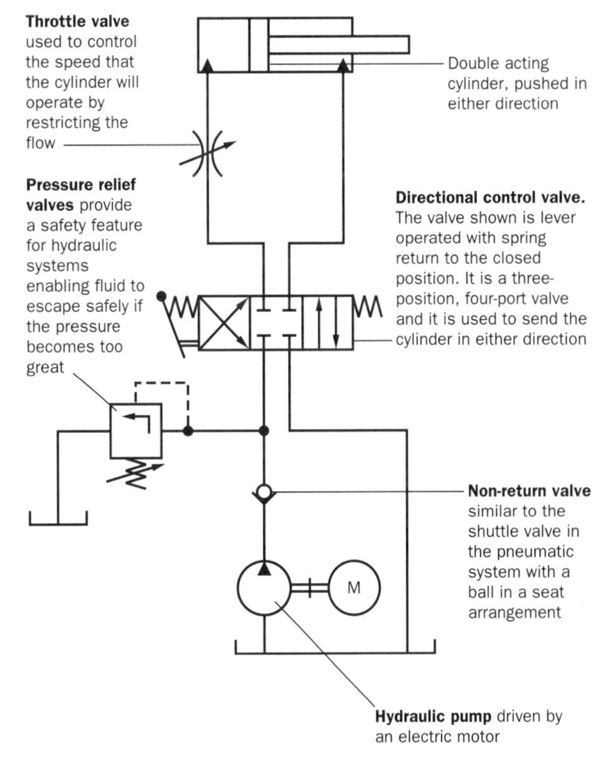

Figure 2.27 A hydraulic system for lever operation of a double acting cylinder

Some other points and a few useful graphical symbols are shown in Fig. 2.28.

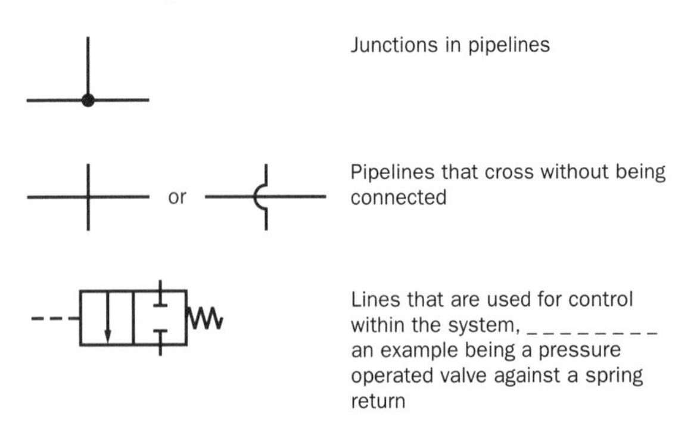

Figure 2.28 Some pneumatic and hydraulic symbols

Electrical and electronic systems

The principal means of communication within electrical and electronic systems is through circuit diagrams. Electronic circuits are often quite complex so it is important that the established international symbols and conventions are used:

- Circuits are 'read' from left to right and top to bottom. Where appropriate they should follow the structure:

  Input – Process – Output
- Positive supply should run along the top of the circuit and the 0V or negative along the bottom.
- Connecting lines should be either horizontal or vertical and should cross each other as little as possible.
- Connections between lines should be made with a clear 'dot'.

Figure 2.29 shows some of the most commonly used component symbols for electronic circuits.

The electronic circuits shown in Fig. 2.30 and 2.31 are based around discrete components like those shown in Fig. 2.29. One is drawn by hand using symbol templates and the other has been drawn using a CAD software package called 'Crocodile Clips'. This software has the advantage that circuits can be both designed and tested on screen before being printed. The electronic symbols that are used are selected from the menu bar and they can be supplemented by animated icons such as the switch example shown here. As you can see the software does not use 'dot' connections uses a 'bridge' where lines cross similar to that used for pneumatic and hydraulic systems. The British Standard recognises this as an acceptable alternative.

Integrated circuits

More commonly integrated circuits (ICs) have replaced many discrete components. An IC within a circuit can be represented with either the connections in the actual pin location or placed for the convenience of the layout and to avoid lines crossing. Figure 2.32 shows the same 555 timer circuit represented in two ways. In circuit (a) the IC is shown as it actually looks with pins numbered 1 to 8. This means that there has to be several lines crossing and the output device, the loudspeaker, is to the left of the IC. Circuit (b) is clearer and the output is in the preferred position on the right-hand side. Notice also that there is no need to show pin 5 because it is not being used.

Circuit (c) shows an amplifier IC in a circuit; it is shown in a triangular form even though it is actually a normal 8-pin IC. Again, for clarity, only the pins that are used are shown. Notice also the −9V supply.

Battery	Resistor	Switch or make contact, normally open
Alternative symbol	Variable resistor	
Earth or ground	Potentiometer with moving contact	Make contact with spring return
Lamp/light bulb	Capacitor	Break contact with spring return
	Polarized capacitor	
Electric bell	Ammeter	Relay
Electric buzzer	Voltmeter	
Crossing of conductors with no electrical connection	Oscilloscope	Semiconductor diode
Junction of conductors	Motor	Photodiode
	Generator	Light-emitting diode (LED)
Fuse	Microphone	NPN transistor
	Loudspeaker	Amplifier

Figure 2.29 Symbols used in electronic circuits

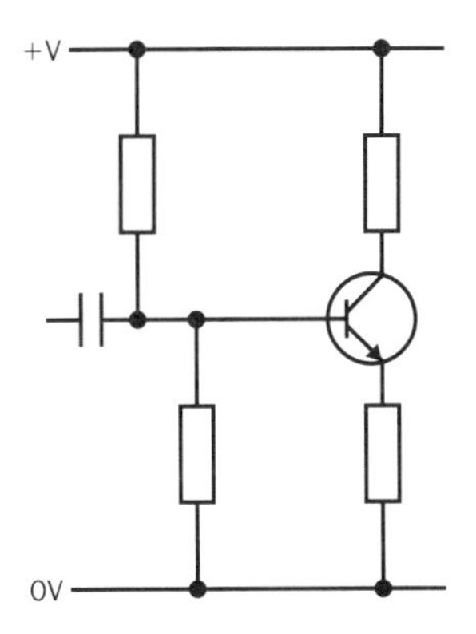

Figure 2.30 Electronic circuit based on discrete components

Communication for manufacture

A circuit diagram is part of the process of designing electronic systems and as such may not be a suitable means of communication to assist the manufacture of printed circuit boards (PCBs) or the location of components. PCB 67284 is a circuit board within a Computer Integrated Training System manufactured by TecQuipment Ltd. The circuit diagram is shown in Fig. 2.33. It is centred around integrated circuit IC1 and two amplifiers that are both located within the integrated circuit IC2. Leading away from the circuit diagram are a number of connections to voltage supplies and to external components.

Figure 2.34 shows the actual layout of the components on the component side of the PCB and the reverse side with the copper tracks making the appropriate connections. Compare the layout with the circuit diagram. The resistors have been grouped together and IC1 takes up only a small amount of space. The dominant features are the connectors CON1 and CON2. The component layout is printed on the component side of the board using silk-screening techniques. The software used to produce these three graphics is part of an integrated computer-aided design package that enables the designer to bring in standard size components and use auto-routing technology. Auto-routing lets the computer make the initial decisions regarding the layout and the routes of the copper track. It is possible then to print out masks for both the copper track and the silk-screening of the component side of the board.

Lastly, the computer system generates a parts list for the PCB and a simple schematic with notes for connection and fitting (Fig. 2.35).

Because these drawings are interrelated and stored electronically modifications and reissues can be monitored by the integrated computer system. This reduces the risk of errors caused by people working to and ordering from out-of-date drawings.

Schematic wiring diagram

Figure 2.36 is a schematic wiring diagram used for the installation of electronic components within a test rig. It is different from the circuit diagram because it is intended for a different audience and purpose. The various meters and

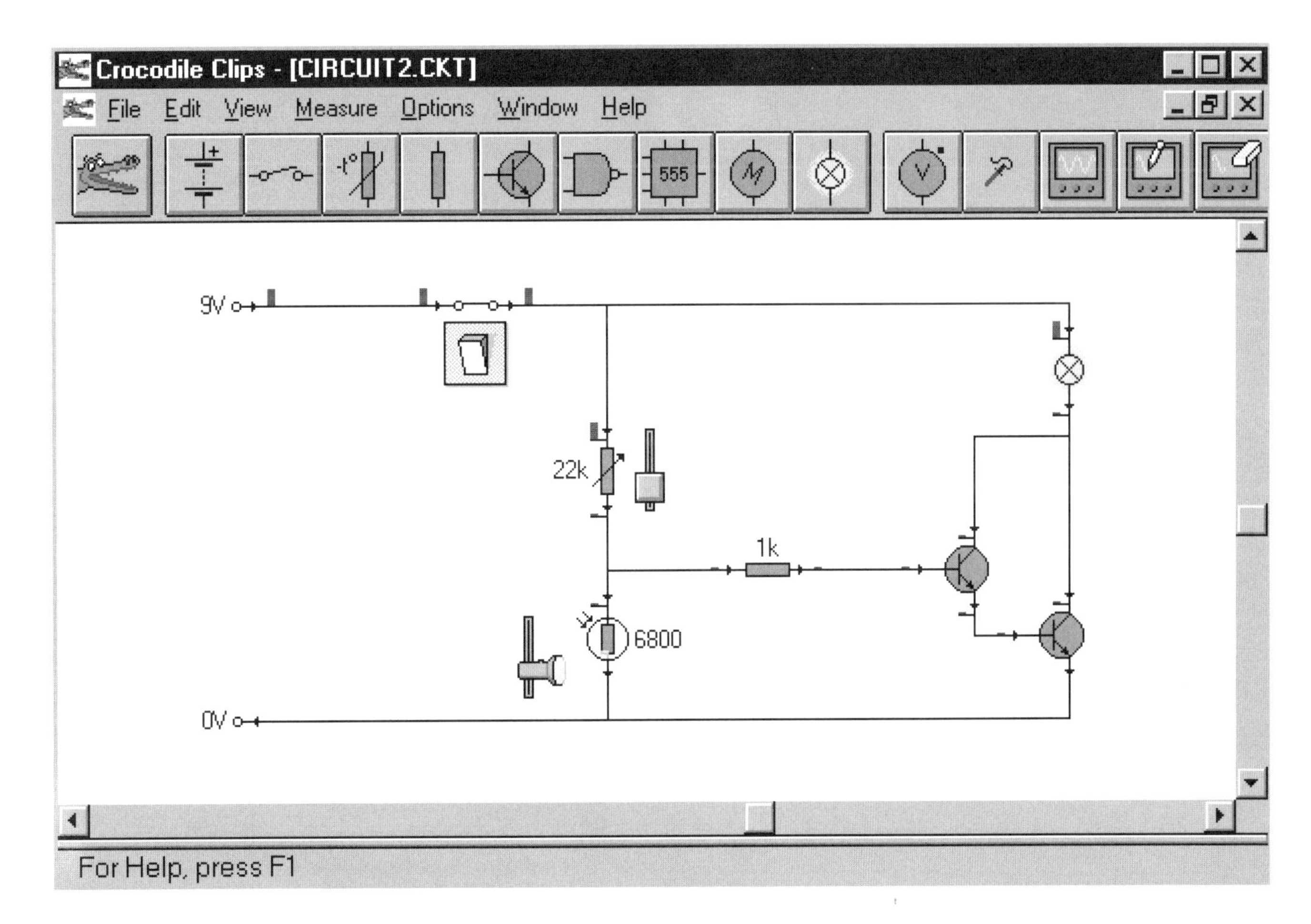

Figure 2.31 Electronic circuit drawn using a proprietary package (*reproduced courtesy of* Crocodile Clips Limited)

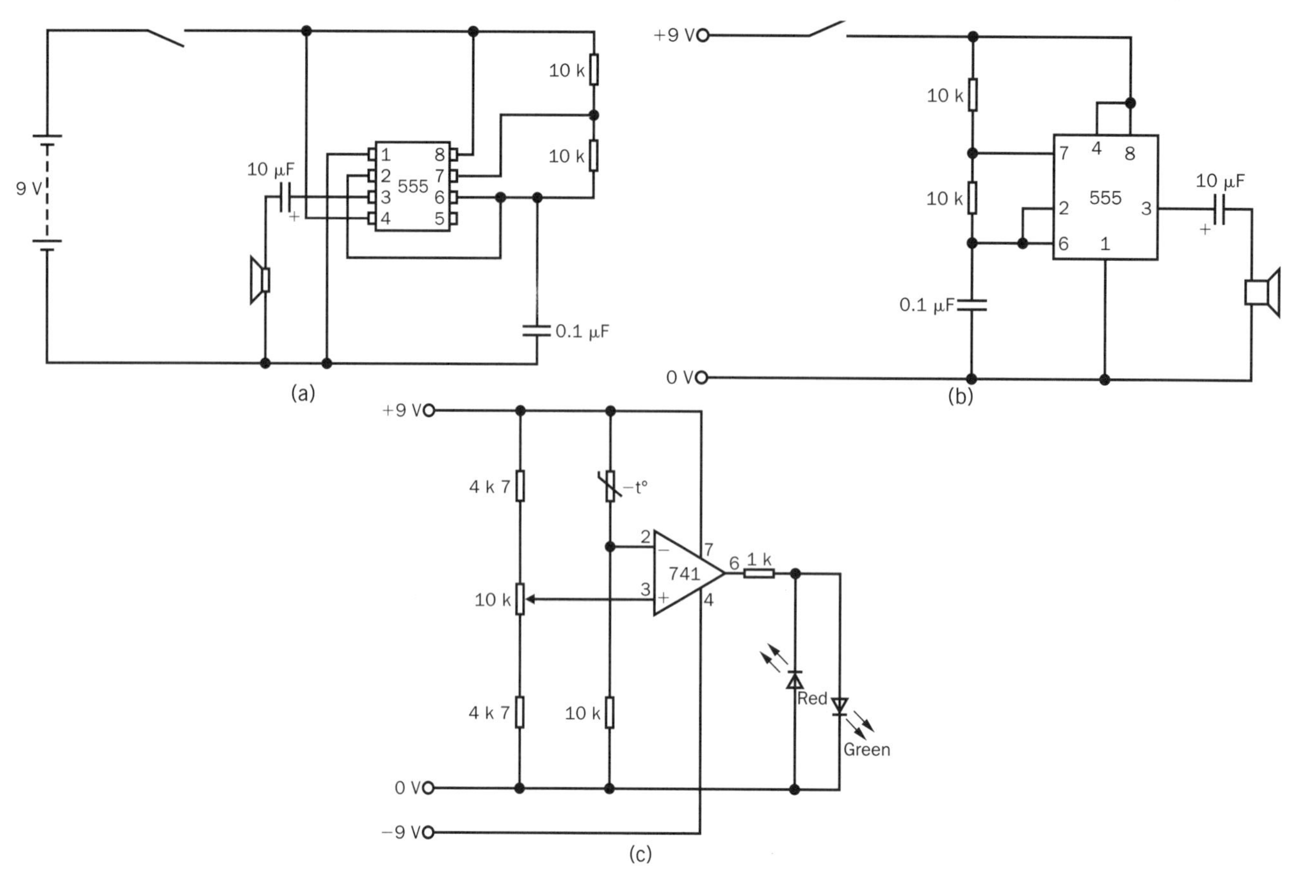

Figure 2.32 IC representation in three circuits

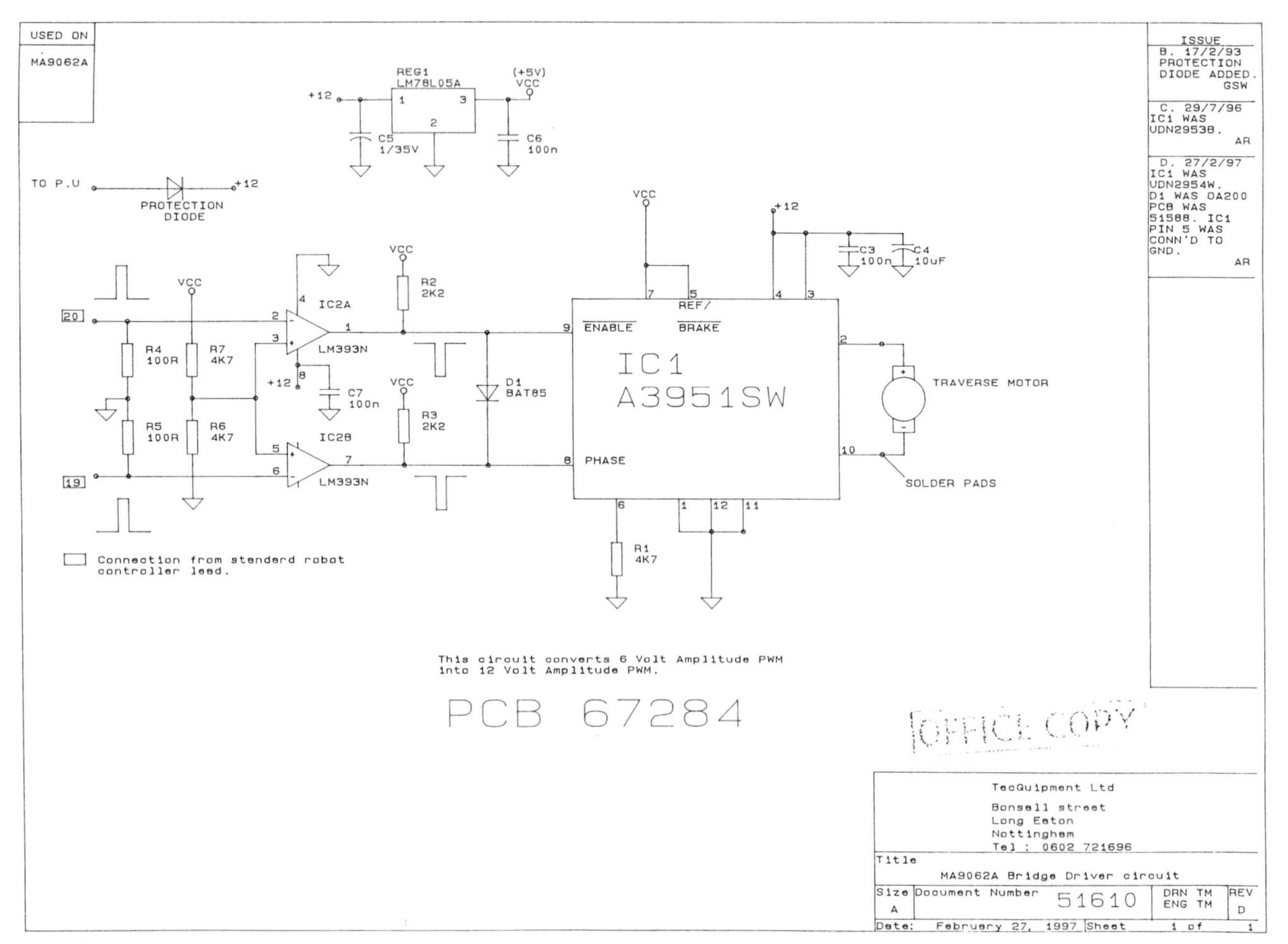

Figure 2.33 Circuit diagram from a computer integrated training system *(reproduced courtesy of* TecQuipment Limited)

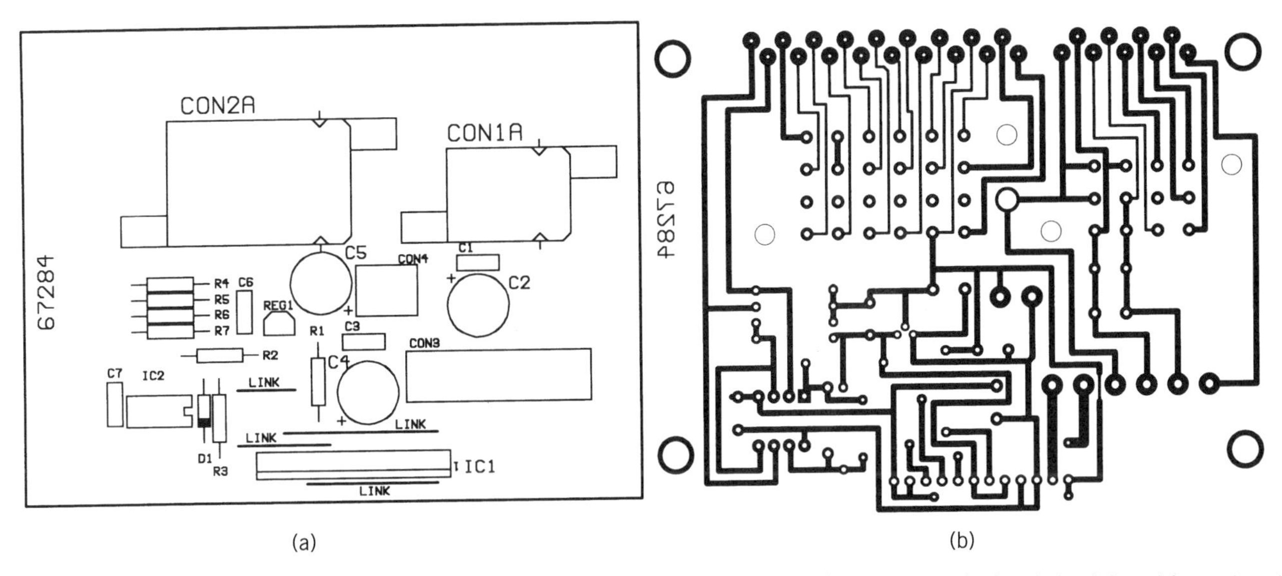

Figure 2.34 Layout of components (a) for circuit given in Fig. 2.32 and (b) copper tracks on reverse of printed circuit board (*reproduced courtesy of* TecQuipment Limited)

components are shown and these may be PCBs or housed components with terminal connectors. The schematic's purpose is to indicate the connections that have to be made and so all of the wire links are numbered at each end to avoid errors in tracking them through the diagram. For example, a connection has to be made between terminal 5 on the PF meter and terminal 3 on the amp SW. This link is shown being made by the line numbered 825. Alternatively, schematic diagrams of this type may show the connections being made using colour-coded wiring.

Measurement systems

The principal development in computer systems and integrated circuit technology is based within American companies and is predominantly located on the West coast of America. The conventional sizes that have evolved, therefore, are based upon imperial measurements. For example the standard spacing for pins on integrated circuits is actually 0.1 inches although you may see this specified as 0.245 mm. It is unlikely that the electronics industry will convert to metric standards.

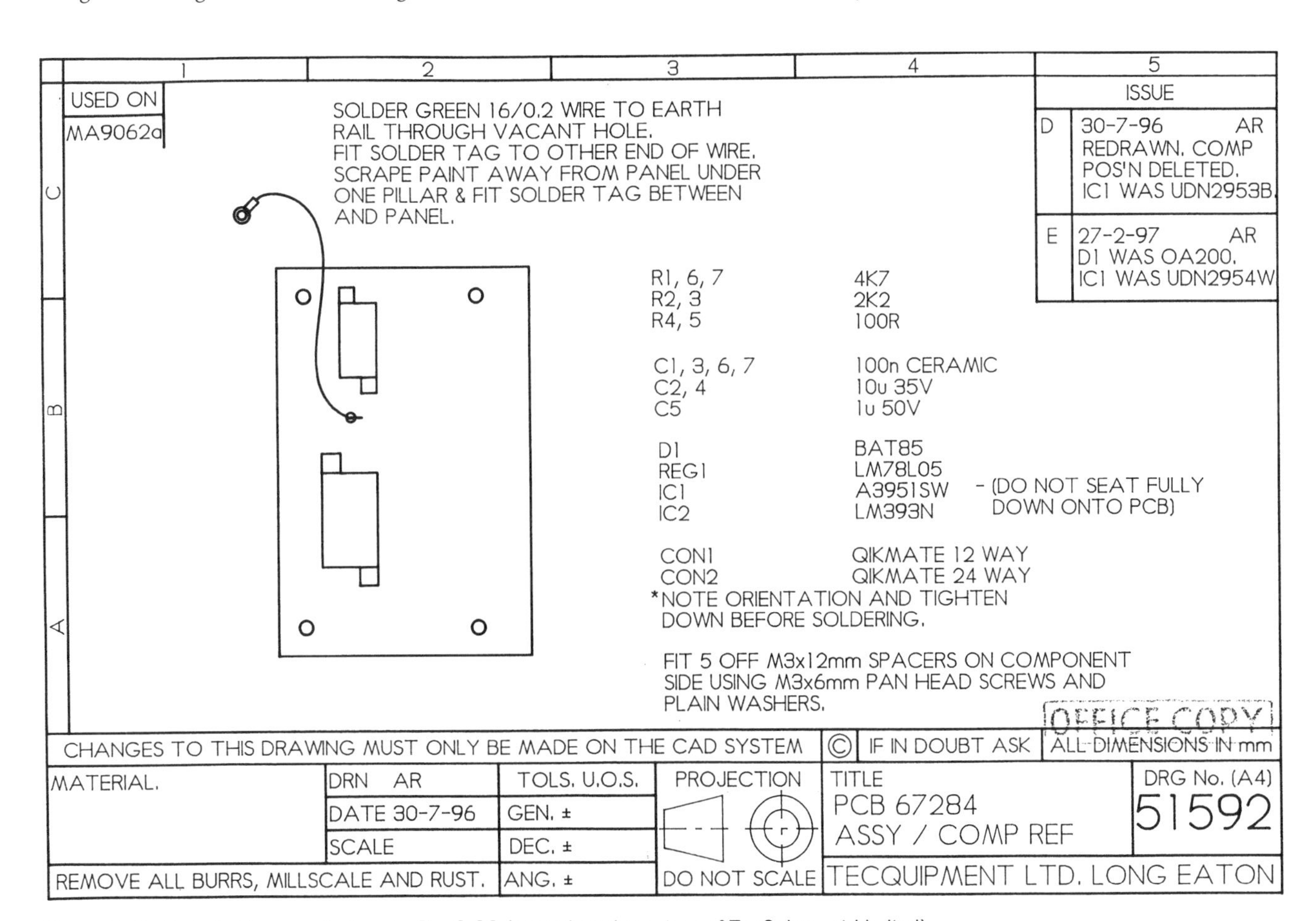

Figure 2.35 Parts list for circuit given in Fig. 2.32 (*reproduced courtesy of* TecQuipment Limited)

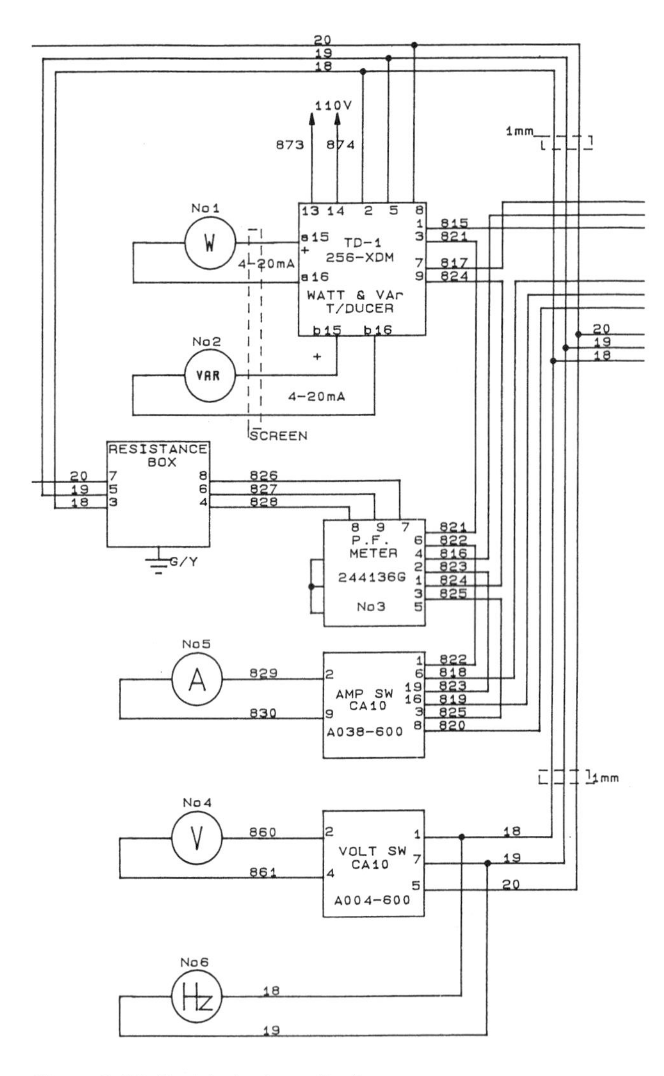

Figure 2.36 Electrical schematic diagram

Self-assessment tasks

1. Use a short sequence of illustrations to show how a button-operated, three-port pneumatic valve operates.
2. Draw a pneumatic circuit that can be used to open and close a door from two positions, i.e. from either the inside or outside.
3. Using standard components draw a circuit that will enable any one of three switches to operate a battery-powered light bulb.
4. Outline the advantages and disadvantages of the two types of representations of an integrated circuit shown in Fig. 2.32.
5. Produce a chart to show the input, process and output components commonly used in electronics.
6. Produce a schematic wiring diagram of a dynamo-generated lighting circuit for a bicycle.

2.4 Engineering drawing

The ability to ensure that manufactured products are produced to a consistently high quality is one of the main priorities within the manufacturing industry. Clear, unambiguous communication underpins the ability to provide customers with quality assurance. The need for formality within much of engineering drawing as a means of communication cannot be avoided or its importance understated. Engineering drawing as a means of graphical communication is a language and is therefore bound by rules and conventions like any spoken language: to be understood it is important for both parties to communicate using the *same* language.

Currently the formal language of engineering is set down within the three volumes of the British Standard BS 308. This publication along with many other BSI (British Standards Institute) publications is being revised as part of the development of European and International Standards. BS 308 is therefore current but will eventually be superseded. The abridged versions, PP7307 and PP7308, are intended for use within schools and colleges.

GNVQ Engineering is concerned with producing and interpreting engineering drawings. Within most engineering companies drawings are created using computer-aided draughting systems, which are vector-based software packages. This means that the software has a mathematical basis for locating points at the ends of lines and within curves. This system of drawing, unlike pixel-based or bitmapped screen graphics, can be correctly scaled and manipulated mathematically by the computer.

Computer-aided design (CAD) offers many advantages to a draughtsperson:

- Drawings can be carried out up to three times faster by using library files of standard components and copy and paste functions.
- Instant access to drawings for re-issue, modification and updating.
- Increased quality and accuracy with consistent text, lines, hatching and shading irrespective of draughtsperson.
- Access to increased range of functions not available conventionally such as re-scaling, auto-dimensioning, rotation of parts, extraction of details for removal or analysis and the combining of drawings.
- Storage is reduced in size, increased in quality and has greater security.
- Allows integration with other computer-based functions such as computer-aided manufacture (CAM), stock control, sales and marketing.

The standards and conventions of engineering drawing apply equally to CAD as they do to conventional technical drawing using a board and pencil.

Lines, text and dimensioning

The drawing shown in Fig. 2.37 is taken from PP7308 and is used to show the application of the various types of lines used within engineering drawings. The line types and how they should be used are given in Table 2.2.

Table 2.2 British Standard line types and their use

Line type		Application
A	Continuous thick	Visible outlines and edges
B	Continuous thin	B1 Dimensions B2 Projection and leader lines B3 Hatching B4 Outline of revolved sections B5 Short centre lines B6 Imaginary intersections
C	Continuous thin	Limit of partial or interrupted view irregular
D	Continuous thin	Sections and parts of drawings if the limit straight with zig-zags of the section is not on an axis
E	Dashed thin	Hidden details, outlines and edges
F	Chain thin	Centre lines, pitch lines and pitch circles
G	Chain thin, thick at ends	Cutting planes for sectional views and changes of direction
H	Chain thin	Outlines of adjacent parts and extreme double dashed positions of moving parts

Lines

- Pencil or black ink should be used for all formal engineering drawing.
- Only two thicknesses of line should be used, thick lines should be twice as thick as thin lines.
- Centre lines should extend just past the outline of the drawing or relevant feature and can be extended to form leader lines for dimensions. They should cross one another and terminate with a long dash.
- Dashed lines and centre lines should meet and cross any other lines on a dash rather than a space.

Text

On many occasions text needs to be added to engineering drawings and the important consideration is clarity.

- Use only capital letters without any embellishments.
- Letters should not be less than 3 mm in height; numbers should be larger.
- Notes should be grouped together and be placed close to the relevant feature.
- Underlining should not be used.

Dimensioning

Dimensions are a vital aspect of component drawings which are intended to be used for manufacture. Mistakes, as simple as an error with a decimal point, can have disastrously expensive consequences.

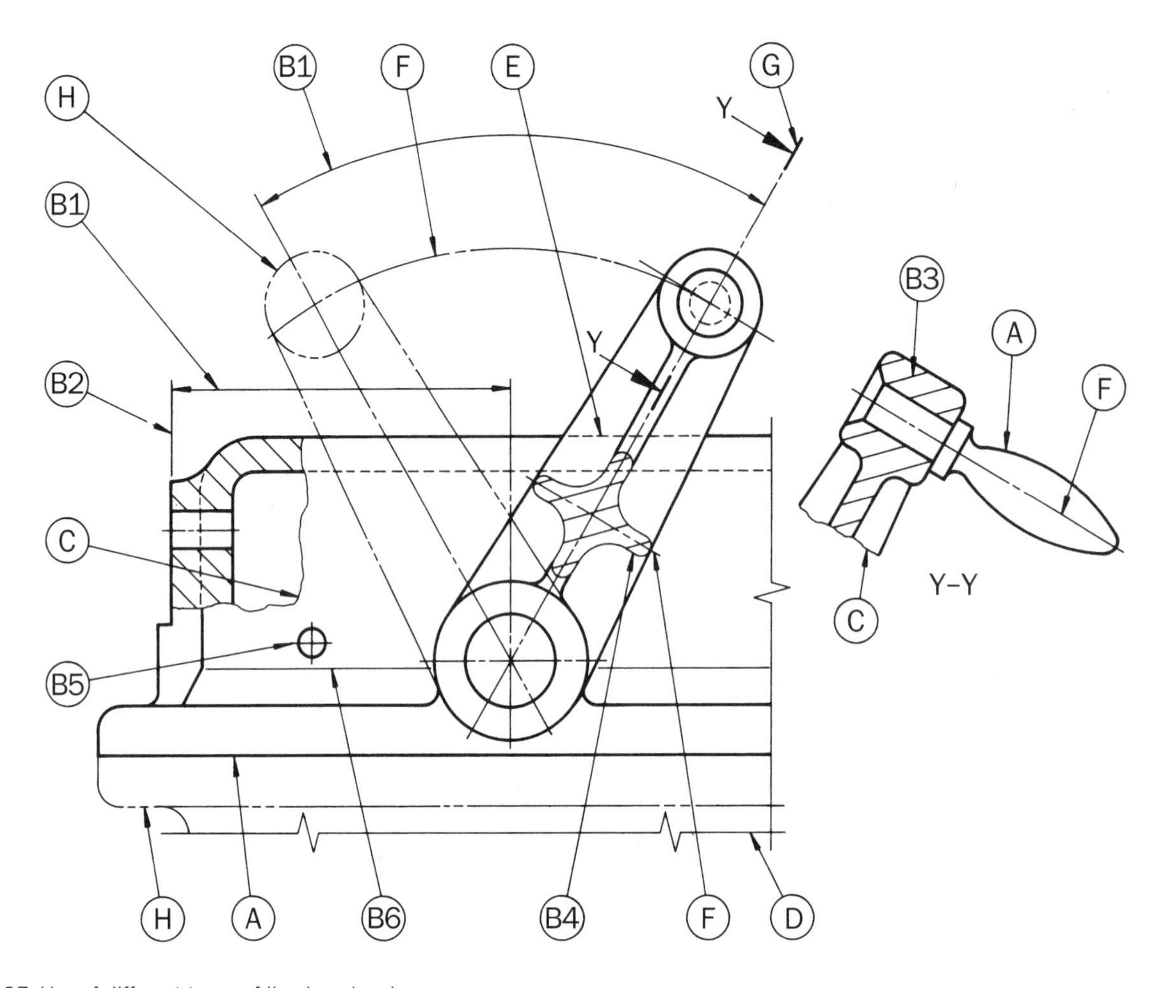

Figure 2.37 Use of different types of line in a drawing

- It should never be necessary for a dimension to be calculated, assumed or scaled from other dimensions on the drawing.
- There should be no more dimensions on a drawing than are necessary.
- Dimensions for a particular feature should, where possible, all be placed on the view that shows the feature most clearly.
- Linear dimensions should be in millimetres with the unit symbol 'mm'. This can be placed as a note, i.e. 'all dimensions in mm unless otherwise stated'.
- Examples of correct methods of expressing numbers:
  - whole numbers: **45** *not* **45.0**
  - less than 1: **0.5** *not* **.5** *or* **1/2**
  - greater than 999: **12500** *not* **12,500** *or* **12 500**
- Projection and dimension lines should be placed outside of the outline wherever possible.
- Projection and dimension lines should not cross over each other unless this is unavoidable.
- Projection lines should start just clear of the outline and extend just beyond the dimension line.
- Dimensions should be placed above the line as viewed from the bottom or the right-hand side of the drawing.
- Arrowheads should be triangular with the length approximately three times the width.

An example of a dimensioned drawing is shown in Fig. 2.38.

Orthographic projection

Orthographic projection provides the engineer with a means of accurately communicating three-dimensional forms on a two-dimensional piece of paper. Pictorial drawing – isometric and oblique (referred to earlier in this unit) – is not accurate enough. Right-angle corners are not shown true and sizes are adjusted to make the drawing 'look' right. For manufacturing purposes this can be confusing and is therefore not acceptable.

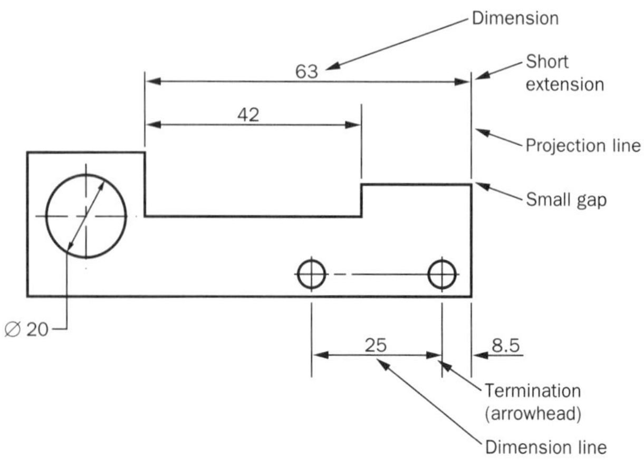

Figure 2.38 Dimensioned drawing

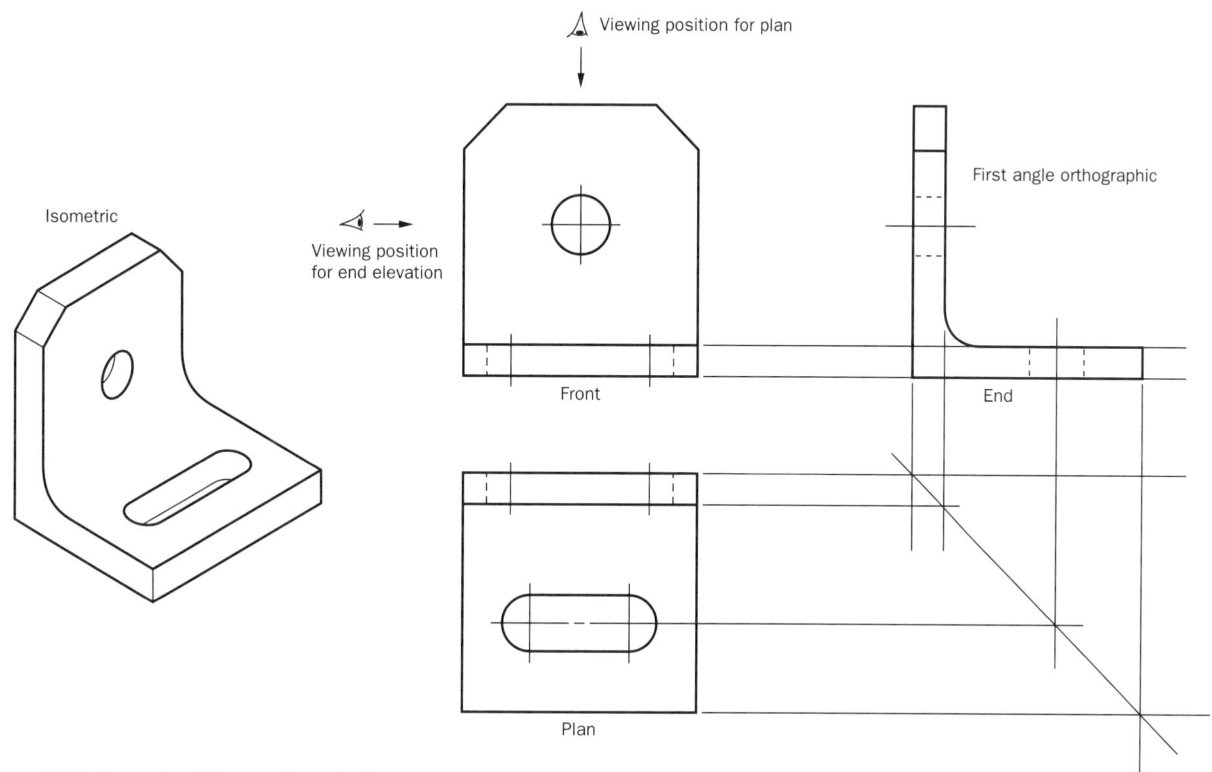

Figure 2.39 Example of first angle projection

Consider the bracket shown as an isometric drawing in Fig. 2.39 and the orthographic representation of the same object. The front view, end view and plan are drawn accurately and placed in such a position that the relationship between them is clear. This method of orthographic projection is called 'first angle projection'. The views are referred to as elevations.

When viewing the front elevation from the left in order to see that end (or side) the view is projected through the bracket and shown on the right-hand side. Viewing from above the bracket in order to see the plan (or top) the view is again projected through the bracket and therefore the plan is shown below.

The notion of projecting is very important and by following the thin projection lines around you can see how the elevations relate to each other and confirm the position of aspects of the drawing.

Figure 2.40 shows the same bracket but this time drawn in 'third angle' orthographic projection. This method of projection is the more popular of the two alternatives. In third angle projection the elevation is placed adjacent to the viewing point rather than projecting it through to the opposite side.

In practice the type of projection used must be indicated on the drawing in writing or by using one of the British Standard symbols shown in Fig. 2.41.

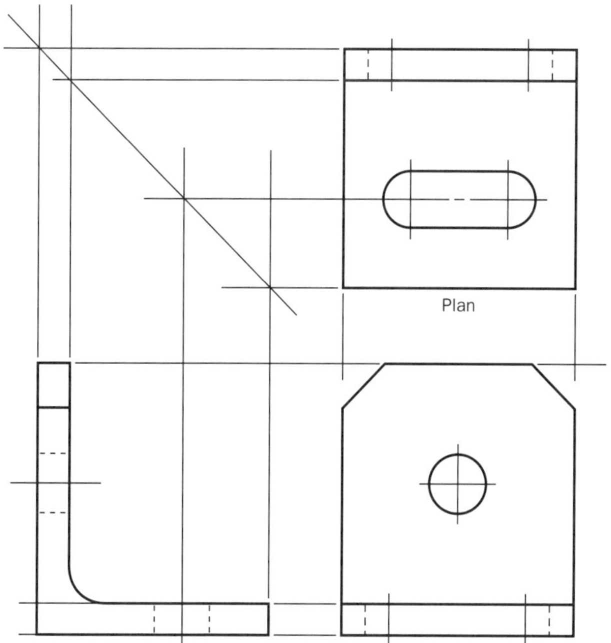

Figure 2.40 Example of Fig. 2.39 in third angle orthographic projection

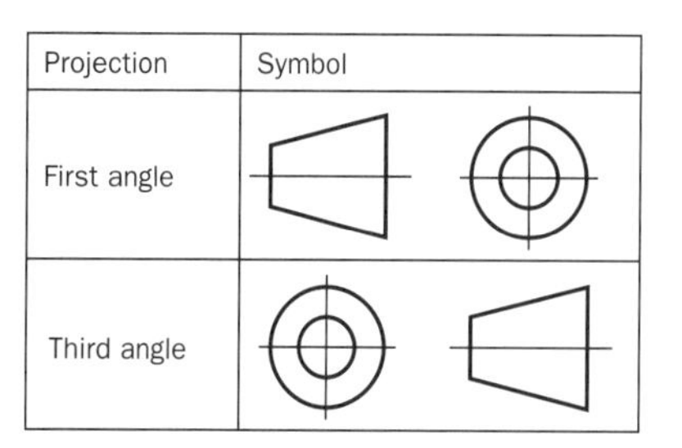

Projection	Symbol
First angle	
Third angle	

Figure 2.41 First and third angle symbols

Self-assessment task

A firm grasp of orthographic projection is a vital aspect of developing the ability to understand and interpret engineering drawings. The diagrams below are an exercise in understanding orthographic projection.

Copy onto graph paper the six drawings shown. All of the drawings are incomplete and require lines and hidden detail adding. Complete the drawings and say which type of orthographic projection has been used.

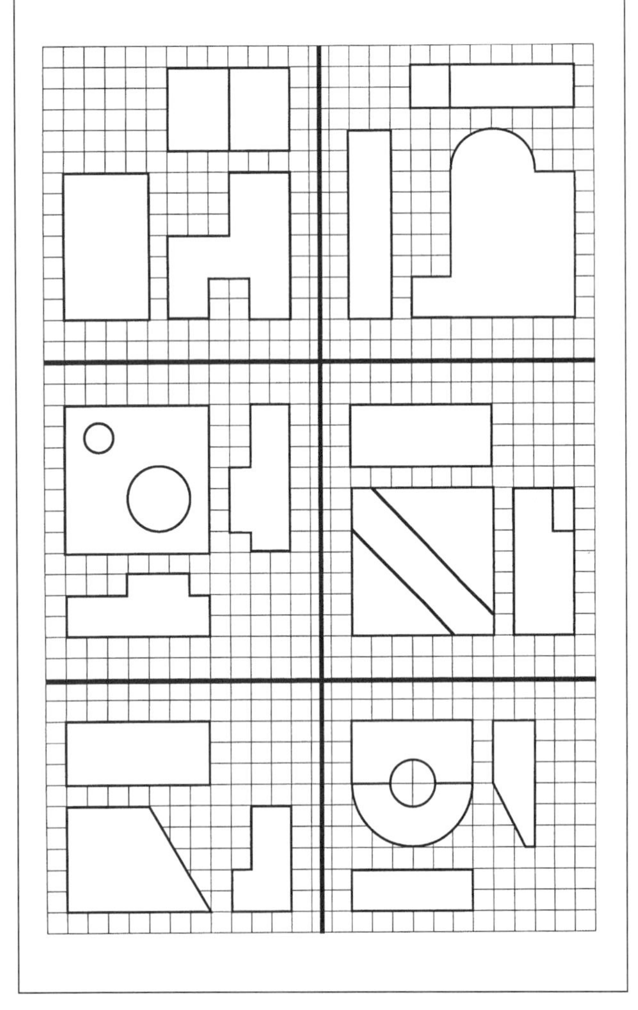

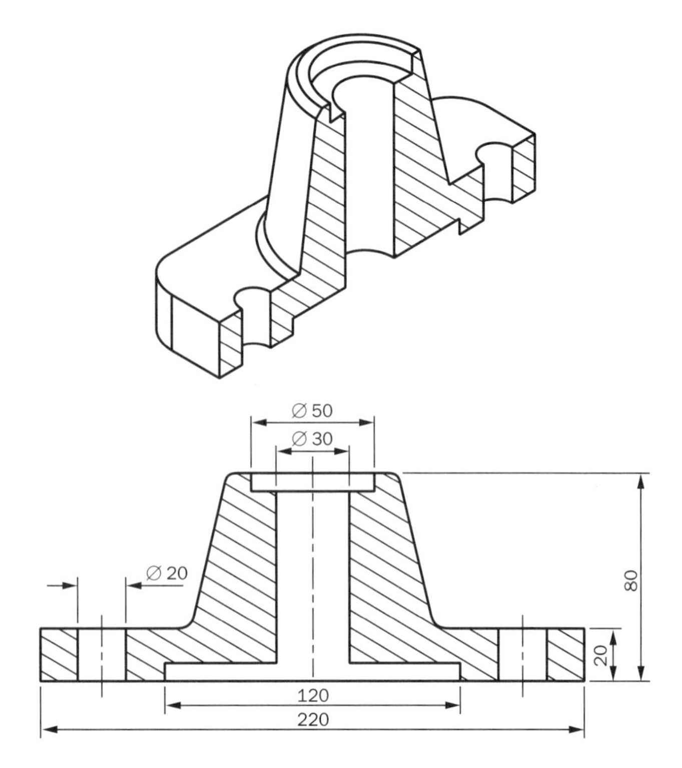

Figure 2.42 Sectional elevations

Sections and auxiliary views

It is often helpful for drawings to show views other than those mentioned above. Sectional views enable the reader to see deeper within what is drawn to locate the internal shapes of features or internal detail that is otherwise hidden from view. Three examples of sectioning are shown in Fig. 2.37 (p. 68): an interrupted view to show the inside of the main casting, a rotated sectional view to show the structure of the lever, and a section designated 'Y–Y' which shows how the handle locates within the lever. You will notice that when a part is sectioned the cutting plane is hatched with lines at 45°.

Figures 2.42 and 2.43 show further examples of sectional elevations. The 3D model and sectional elevation of the foot step bearing have been generated using AutoCAD.

Points about sectional elevations

- The cutting plane should be clearly indicated and the sectional elevation projected in accordance with the method of orthographic projection being used.
- The cutting plane should be hatched at 45° in the sectional elevation, separated areas of the same component should be hatched with the same direction and spacing.
- Where different sectioned parts meet in a drawing the direction and spacing of the hatching should be varied (although not the 45° angle).
- Large areas need only be hatched around the edges and where they connect to adjacent parts.
- Sectional elevations can be made in more than one plane where this clarifies an aspect of the drawing (Fig. 2.43).
- By convention the following details, when they appear on a cutting plane, are not shown hatched when the cutting plane passes through them longitudinally:
  - fasteners, such as nuts, bolts, screws, washers, pins, etc.
  - shafts
  - ribs and webs
  - spokes of wheels

Auxiliary views (Fig. 2.44), like sections, are an opportunity to show aspects of a drawing that would otherwise be unclear or not seen. They are used particularly where a component has angular faces. The rules regarding projection of elevations should be applied to auxiliary views wherever possible although it is often the case, particularly with first angle projection, that clarity is best achieved by placing the auxiliary view adjacent to the face in question rather than projecting it through.

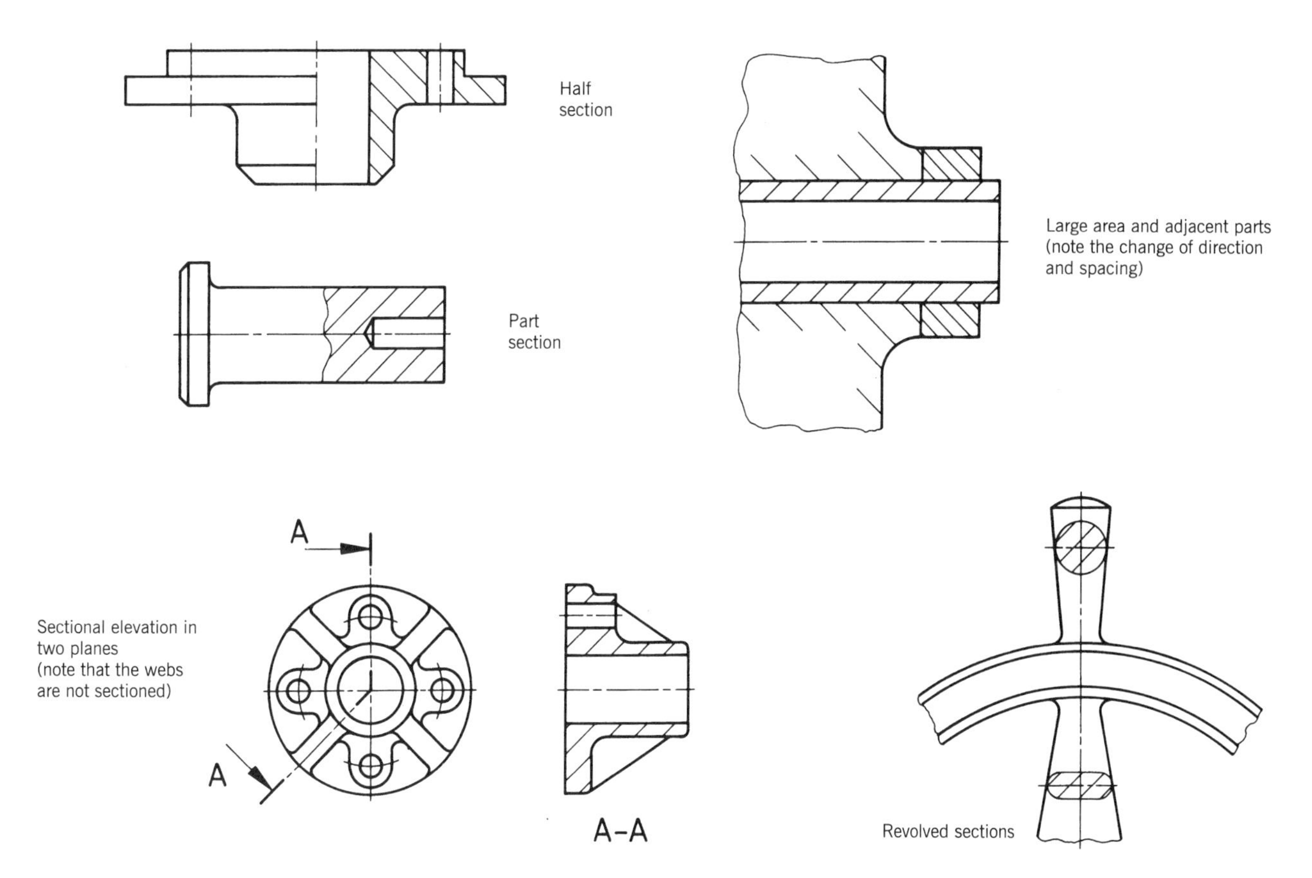

Figure 2.43 Sectional elevations in more than one plane

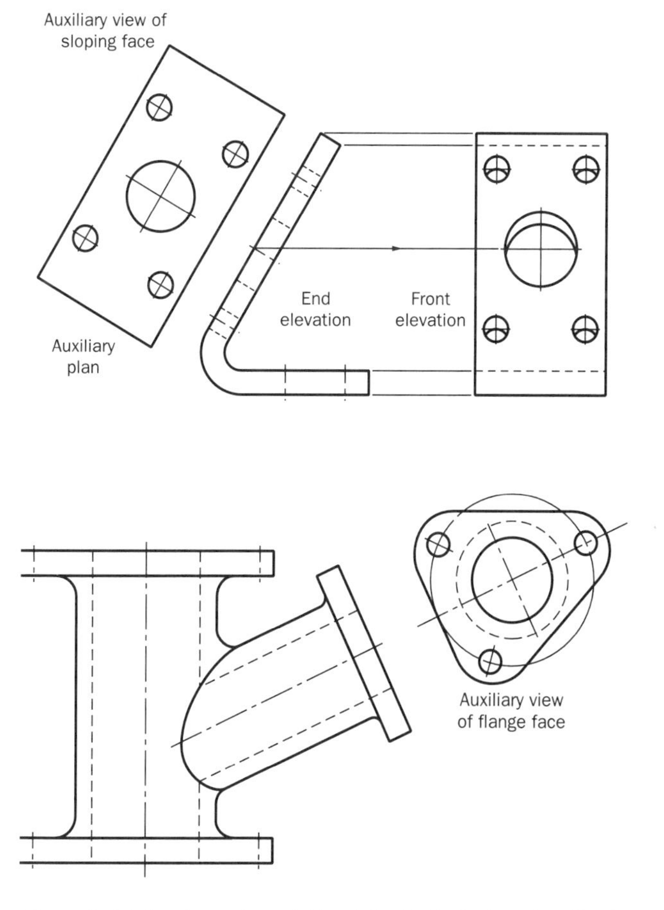

Figure 2.44 Auxiliary views (after L.C. Mott *Engineering Design and Construction* Oxford University Press)

Self-assessment tasks

The diagram below shows a drawing of a block that has been created using AutoCAD LT.

1. Copy the front and end elevations from the drawing using either CAD software or conventional drawing.
2. Add a horizontal cutting plane at mid height and from it project a sectional plan view.
3. From your drawing project an auxiliary elevation at right angles to the inclined face of the block.

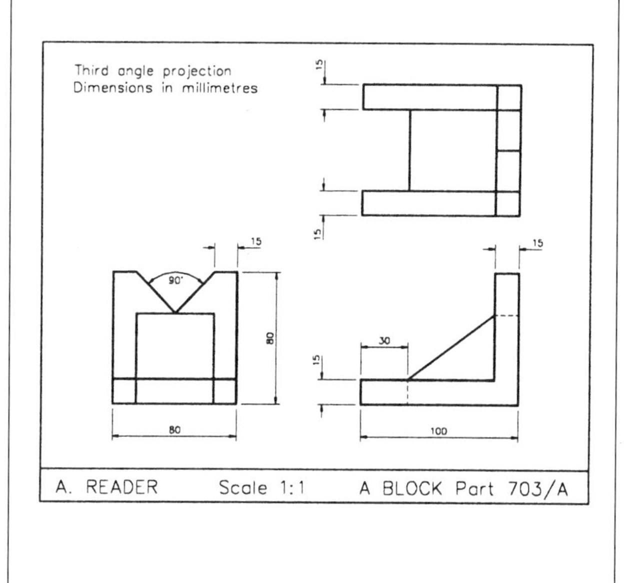

Engineering features

There are many components and features within engineering drawing that should be represented according to conventions set down in British Standards. This use of common conventions avoids the need for explanation and reduces confusion.

Screw threads

Whenever screw threads are shown, irrespective of the thread type, the conventions shown in Fig. 2.45 should be used. This convention enables drawings to be created more quickly and appear less cluttered. Two views of a hexagonal headed bolt are shown, a plan view of a threaded hole and a sectional elevation through a bolt in a hole. Notice the differences between the thread (the broken line) in the end elevation of the bolt and in the thread in the plan. Look also at the detail in the sectional elevation: the drilled hole, the threaded portion of the hole and the bolt within it.

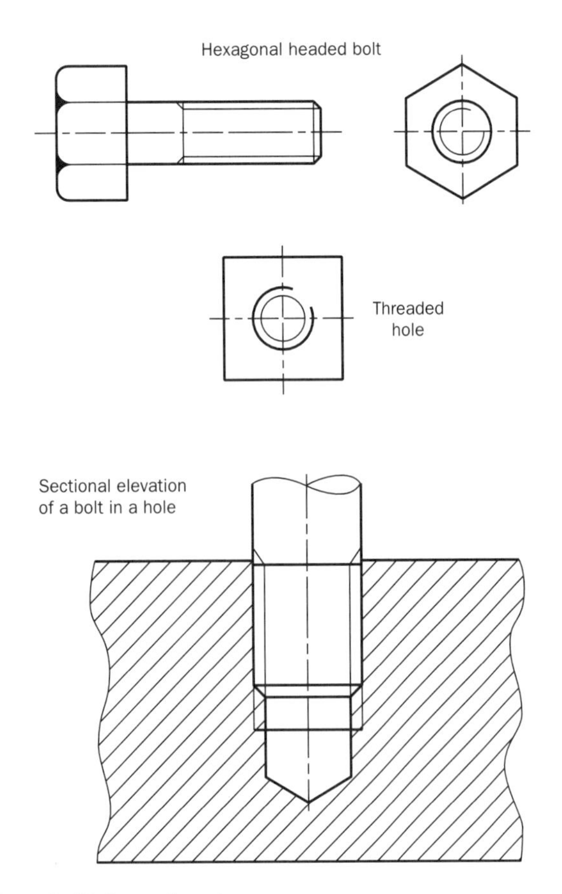

Figure 2.45 Screw threads

Welded joints

A comprehensive range of welding symbols is contained within PP7307. This includes: butt welds; flat, single bevel and double bevel welds; welds with flat, convex or concave surfaces; spot welds and a range of fillet welds. Figure 2.46 shows examples of the symbols used in engineering drawings for welds within fabricated structures. The two examples show an illustration followed by a graphical representation, as it should appear within a drawing, and a symbolic representation using the recommended symbol for that weld type. Symbolic representations are more likely to be used rather than graphical representations when there are a large number of welded joints on a particular drawing.

Gears and bearings

Like screw threads, gears and bearings are simplified and there is no requirement to show the actual detail. Gears are shown as solids without teeth and the pitch circle, the effective diameter of a gear is shown by a chain line. Gear trains are therefore shown with the meshed gears' pitch circles touching. Figure 2.47 shows examples of spur gears and bevel gears.

Bearings, other than plain bearings, are shown as in Fig. 2.48 irrespective of whether they are ball or roller types of bearings.

Common symbols and abbreviations

Only those symbols and abbreviations specified within BS 308 and PP7308 should be used on engineering drawings. Any other necessary notes and information should be clearly printed in full. Table 2.3 includes the most commonly used abbreviations.

Table 2.3 Symbols and abbreviations

Term	Abbreviation or symbol
Across flats	AF
Centres	CRS
Chamfered	CHAM
Countersunk	CSK
Counterbore	CBORE
Diameter	DIA *or* ∅
Drawing	DRG
Equally spaced	EQUI SP
Figure	FIG
Hexagon	HEX
Left hand	LH
Long	LG
Material	MATL
Maximum	MAX
Minimum	MIN
Number	NO
Pitch circle diameter	PCD
Radius (in a note)	RAD
Radius (preceding a dimension)	R
Right hand	RH
Round head	RD HD
Specification	SPEC
Spotface	SFACE
Square	SQ
Standard	STD
Undercut	UCUT

Machining symbols

It is often necessary, particularly with cast components, to indicate those faces that have a critical function or a flatness requirement sufficient for them to require some form of machine finishing. It is the engineering designer's responsibility to make these decisions. Such faces or surfaces that require finishing are indicated by a machining symbol,

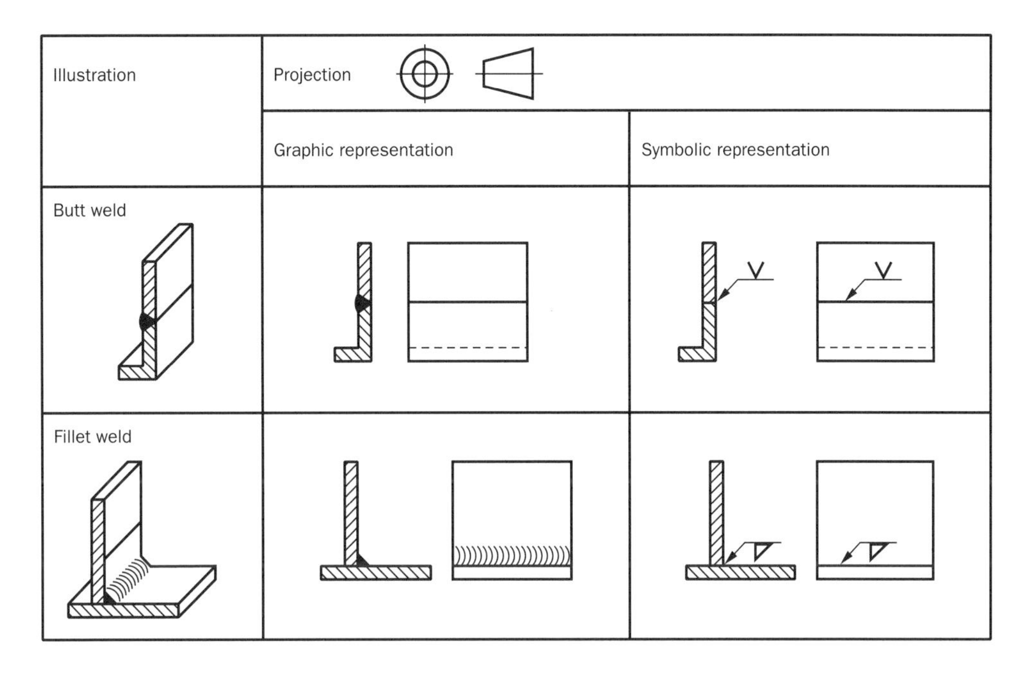

Figure 2.46 Examples of welds

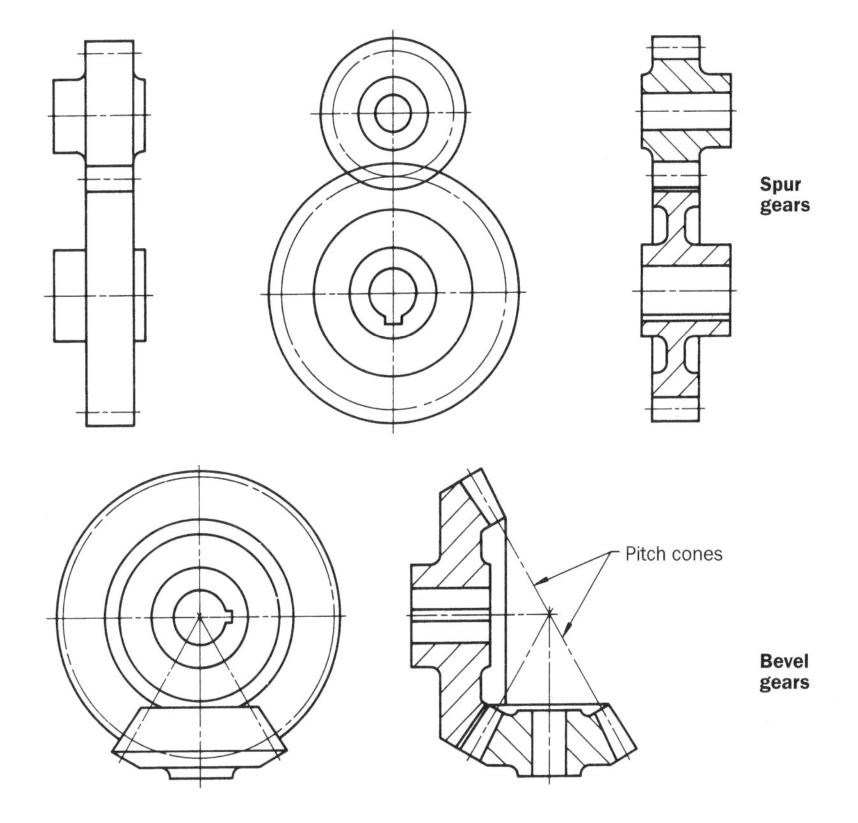

Figure 2.47 Gears

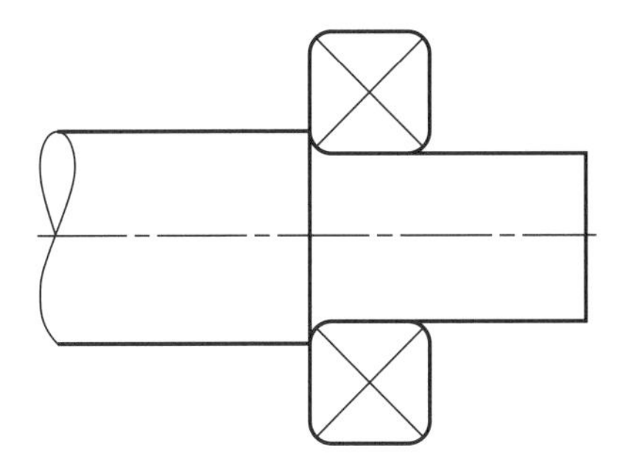

Figure 2.48 A bearing

often including an indication of the acceptable surface finish. This texture or roughness value is a measurement in micrometres (μm) and it is added to the symbol as shown in Fig. 2.49. This value will indicate whether a milled or turned finish is acceptable or if the surface requires further machining such as surface grinding or honing. Where it is necessary to machine finish all over then a general note may be added.

Spotfacing, countersinking and counterboring are other examples of localised surface machining for a particular function (Fig. 2.50). In these instances it is the location or seating of the heads of bolts and screws, and of washers and nuts.

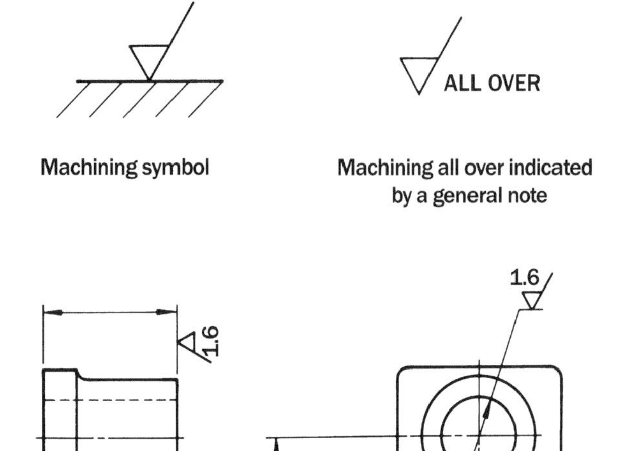

Figure 2.49 Machining symbols

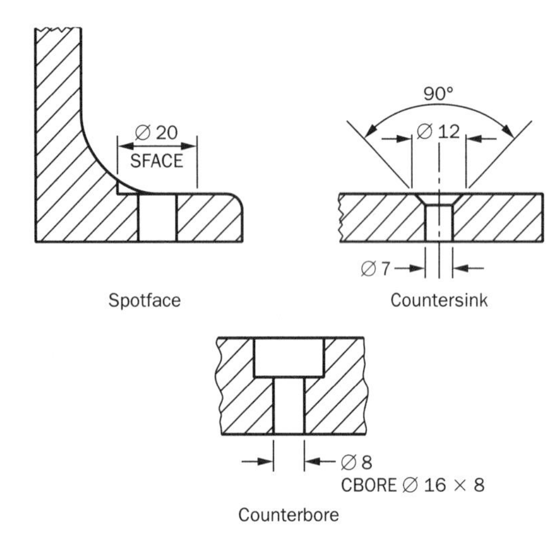

Figure 2.50 Spotfaces, countersinks and counterbores

Scale drawings

The application of scale to engineering drawings is common practice. This is principally to ensure a correct fit regardless of the size of paper used for the drawing, but is also used for clarity of detail. When using CAD drawings are normally made full size and the plot or print size is then scaled if a 'hardcopy' is required. When drawing using conventional instruments scale rules should be used in preference to carrying out calculations which may be prone to error. In the drawings shown in Fig. 2.54 (p. 76) you can see how scale can also be applied to just a part of a drawing, in this instance enabling the dimensioning of a small detail.

Scale is expressed in the form of a ratio and there are a number of British Standard recommended scales to use. For drawings that are smaller than full size there are the following recommended reduction scales:

1:2 1:5 1:10
1:20 1:50 1:100
1:200 1:500 1:1000

For drawings larger than full size there are the following recommended enlargement scales:

2:1 5:1 10:1
20:1 50:1

Scale, including full size, should never be used with the expectation that during manufacture the drawing can be measured and dimensions extracted in this way. As drawings get copied and reproduced they are liable to distortion, for this reason the words 'DO NOT SCALE' are often added to drawings to prompt the user to read rather than measure the drawing.

Tolerancing

It is not actually possible to repeatedly manufacture components to a precisely identical size, nor is it often cost effective to try. There will always therefore be some small variation. When component parts of an assembly are brought together it is important to know that they will fit as required without selection having to take place. Tolerancing dimensions is concerned with what is acceptable in respect of the ability of components to fit together or to function as the design intends. It regards the permitted variation from the nominal or basic size.

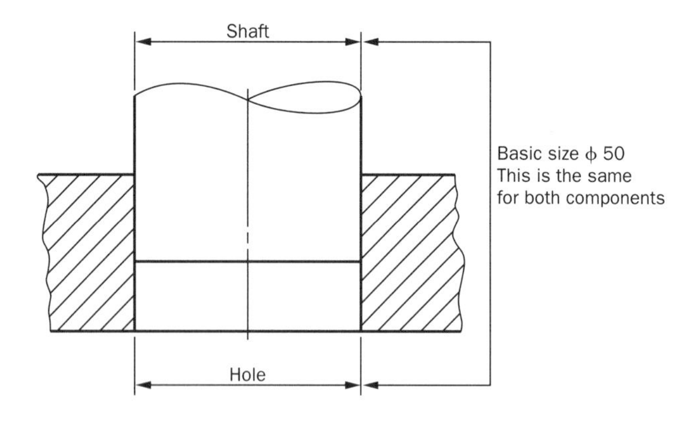

Figure 2.51 Shaft in a hole

Figure 2.51 shows a shaft in a hole. Consider the following in relation to the function of the components and dimensioning them for manufacture (see Fig. 2.52):

- Let us assume that their basic size is 50 mm (diameter) and it is a requirement of the function of these components that clearance exists between them. The hole, therefore, must always be larger than the shaft.
- The design engineer must determine what will be an acceptable functional clearance; this could typically be a minimum of 0.025 mm and a maximum of 0.095 mm: a tolerance band of 0.070 mm.

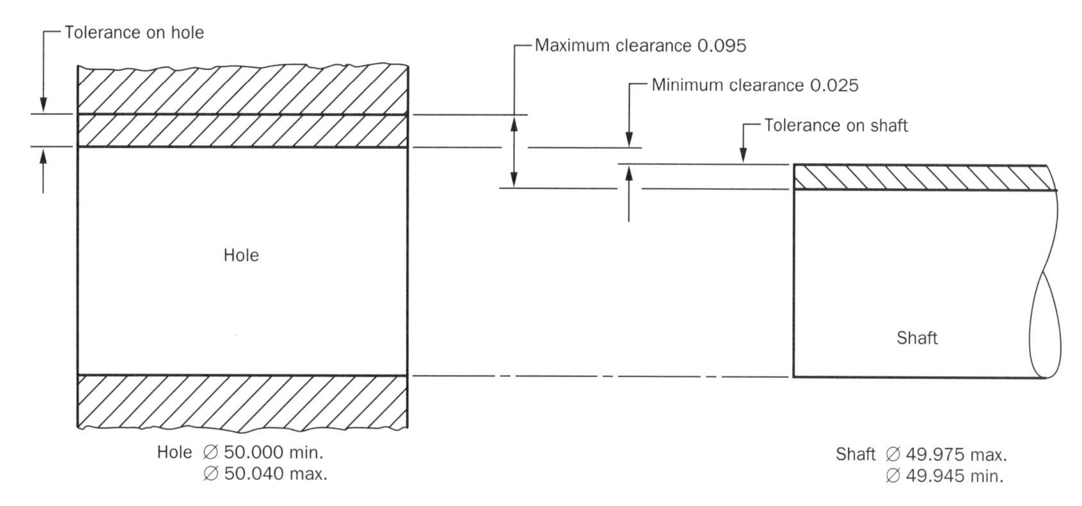

Figure 2.52 Toleranced dimensions

- The task now is to determine the manufacturing tolerances for the two components, i.e. the dimensions for both the shaft and the hole that will ensure that the clearance achieved, with any possible assembly combination, will not go beyond the acceptable tolerance limits.
- By attaching slightly more than half of the tolerance band to the hole (it is always easier to be accurate with a shaft than with a hole), the hole size can be fixed at say ∅ 50.000 min. and ∅ 50.040 max. This means that a ∅ 50.000 hole would be acceptable, as would any size up to a maximum of ∅ 50.040.
- The largest acceptable size of shaft can be determined by considering the smallest possible hole and the minimum clearance, i.e. ∅ 50.000 − 0.025 = ∅ 49.975. The smallest acceptable shaft is therefore the largest hole minus the largest clearance, i.e. ∅ 50.040 − 0.095 = ∅ 49.945.

To assist the design engineer in making decisions regarding the amount of tolerance required to achieve specific functions, ranging from large clearance to heavy interference, British Standards publish tables of limits and fits in BS 4500. The BS guide to tolerancing for schools and colleges, PP7309, does not contain limits and fits specifications.

Communicating tolerance

There are two principal methods of expressing toleranced sizes on engineering drawings. Either with the upper and lower limits of size (Fig. 2.53(a)) or with the tolerances expressed as a variation from the nominal size (Fig. 2.53(b)).

Self-assessment tasks

1. The diagram below shows the front elevation of a pressed steel component. Make a fully dimensioned 2:1 scale copy of this drawing. The hole should be dimensioned with a +/− 0.050 mm tolerance.

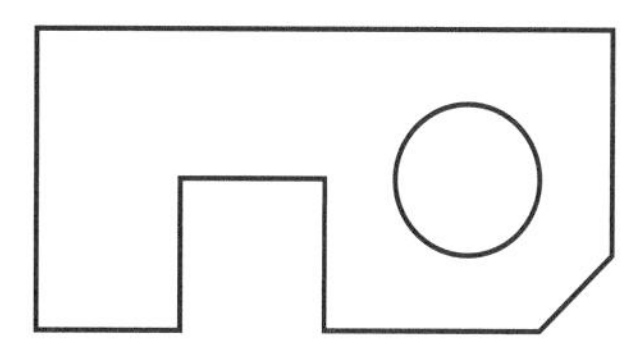

2. The two pieces of 40 × 15 mm mild steel bar shown below are fastened together using a hexagonal headed bolt, a nut and two washers. Draw a full-size sectional elevation through the centre of the hole with all of the components assembled.

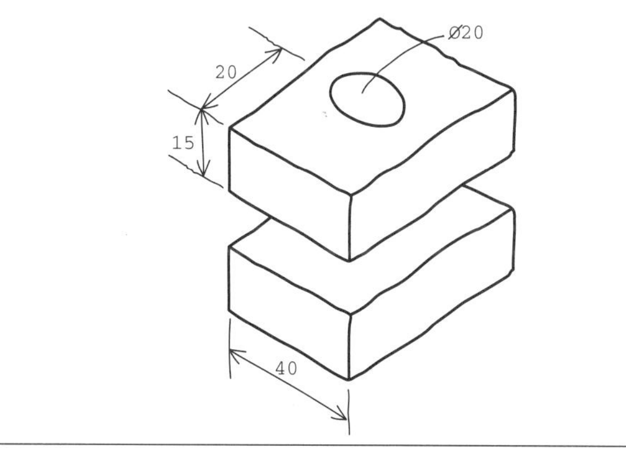

∅ 50.000
∅ 50.040
∅ 49.975
∅ 49.945

(a)

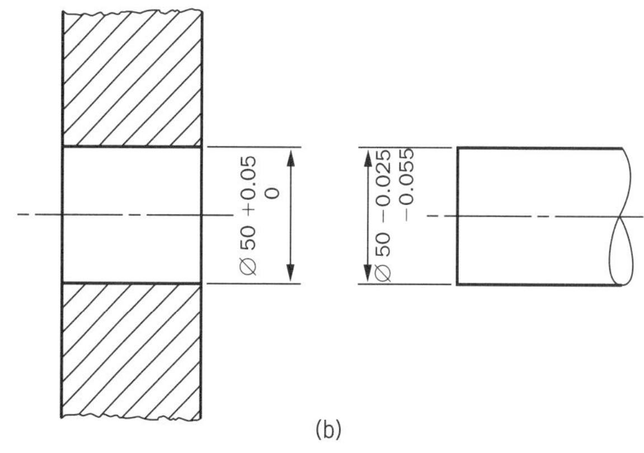

(b)

Figure 2.53 Communicating tolerances

Engineering drawing in practice

Title block

All engineering drawings should have a title block that contains the following basic information (Fig. 2.54):

- name of the draughtsperson
- date
- scale
- title
- drawing and issue number
- method of projection used in words or symbol (this may be located elsewhere on the drawing)

The drawing in Fig. 2.54 of a connector is taken from PP7308 as an example of a typical engineering drawing. Study the drawing to locate and note the following:

- From the front elevation is projected, in third angle projection, a part plan and a part sectional elevation. There is also an enlarged scale detail drawing.
- There is tolerance attached to particular dimensions and as a general note within the title block.
- The 165 mm dimension is shown in brackets to show that this is an overall dimension for guidance only. It is possible that this dimension cannot be achieved within the general tolerance of the drawing because it must depend upon the two dimensions (40 mm and 125 mm) that make it up.
- The drawing also shows examples of:
  - chamfers at 45°
  - knurling (MEDIUM KNURL)
  - a spherical radii (SR)
  - holes equally spaced around a 65 mm pitch circle diameter (EQUI SP ON 65 PCD)
  - screw threads
  - angular dimensioning
  - flat faces on circular parts, indicated by cross lines on the flat
  - machining symbols and surface texture values

Assembly drawings

Figure 2.55 shows an example of an assembly drawing. It shows a ball bearing race located within a split bearing housing bolted down to a cast bedway. Assembly drawings are often shown sectioned to reveal the inner details. This drawing is part sectioned with the vertical centre line acting as the cutting plane. Dimensions are not normally added to assembly drawings other than a few overall dimensions to indicate the assembled size for packaging, storage, transportation, etc.

The numbers around the drawing are used as reference for the parts list that may be included on the assembly drawing. This will be similar to the electronics parts list example shown earlier in this unit (Fig. 2.35). Part lists should be numbered from the bottom of the drawing going upwards, see Fig. 2.56 of from the top going down. This enables additional parts to be added with subsequent modifications and re-issues.

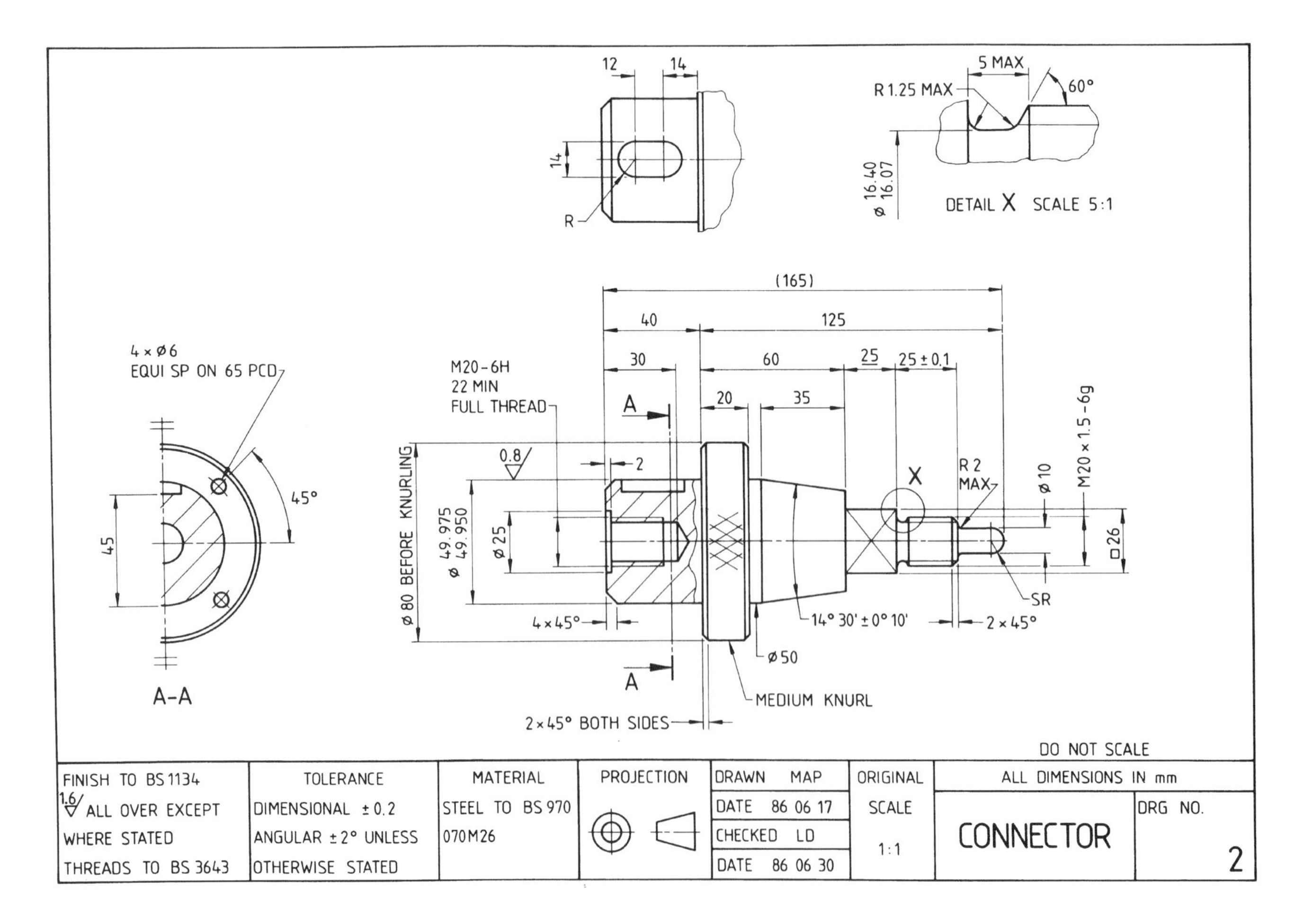

Figure 2.54 Detail drawing

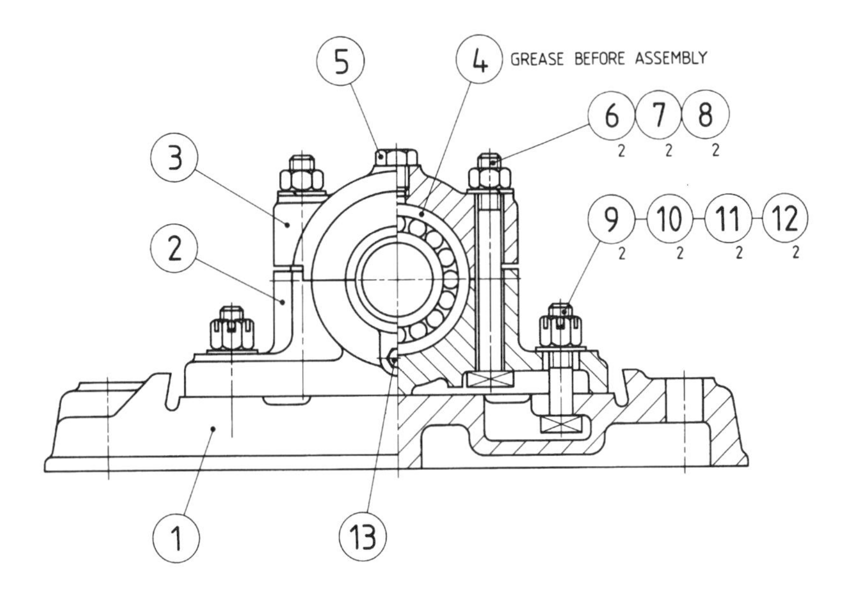

Figure 2.55 Assembly drawing

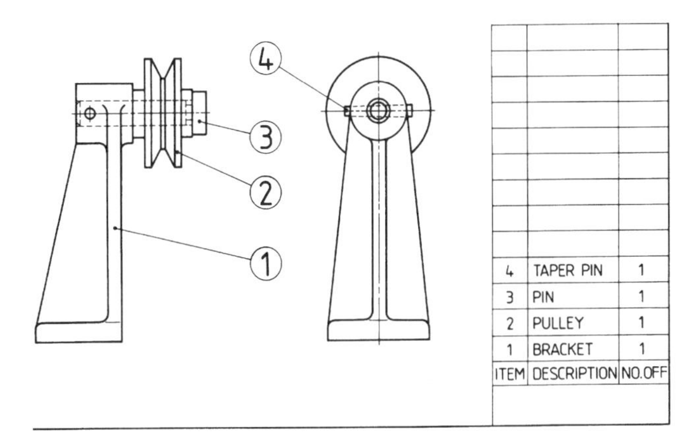

ITEM	DESCRIPTION	NO.OFF
4	TAPER PIN	1
3	PIN	1
2	PULLEY	1
1	BRACKET	1

Figure 2.56 Parts list

Self-assessment tasks

1. Study the drawing of the swivel bracket in Fig. 2.57 and then answer the following questions.
   (i) State the method of projection used.
   (ii) What material and manufacturing process are to be used to make the bracket.
   (iii) With reference to the process by which the bracket has been manufactured explain the note 'FILLET RADII R3 MIN'.
   (iv) With reference to the tolerances and the machining symbols state the maximum and minimum acceptable sizes for the width and the height of the bracket.
   (v) The drawing has two principal views, a plan and a sectional front elevation, from the sectional front elevation draw or sketch two end elevations.
2. From the assembly drawing in Fig 2.55 produce a parts list containing the name of the part, the material from which it is made and the quantity required.
3. Produce detail and assembly drawings of a cassette or CD case.

Engineering drawing case study

This short case study focuses upon an airbox unit that is part of a gas turbine training rig manufactured by TecQuipment Ltd and used by military and civil training establishments. All TecQuipment's drawings are currently carried out using CAD systems but existing conventional drawings of components that are still in use are only converted when a modification or new design takes place. The move from conventional drawing to CAD is expensive but the advantages (outlined on p. 67) soon make the transition worthwhile.

Another advantage of computer integration within engineering is the use of digitised photographic images. Fig. 2.58 shows a digital image of the airbox which is stored along with the drawings as a computer image file. This format enables transmission of the photographs and drawings via land-line or satellite links. This means that discussions can take place with clients and repairs can be undertaken at any distance, including overseas, by the engineer from their desk without the delay involved in waiting for information to be sent and received by post.

The airbox assembly drawing Fig. 2.59 shows what is essentially a sheet metal fabrication. Sheet materials are drawn as thick lines and you can see that this drawing requires a lot of notes regarding adhesives and sealants.

The part numbered 4 on the assembly drawing, the 'intake diffuser', is shown as a detail drawing in Fig. 2.60. Compare this CAD-generated drawing with Fig. 2.61 which is an older conventional drawing of a similar component. The drawing is showing signs of deterioration after some years of storage and handling and you can see in particular the lack of consistency with the textual information on the drawing.

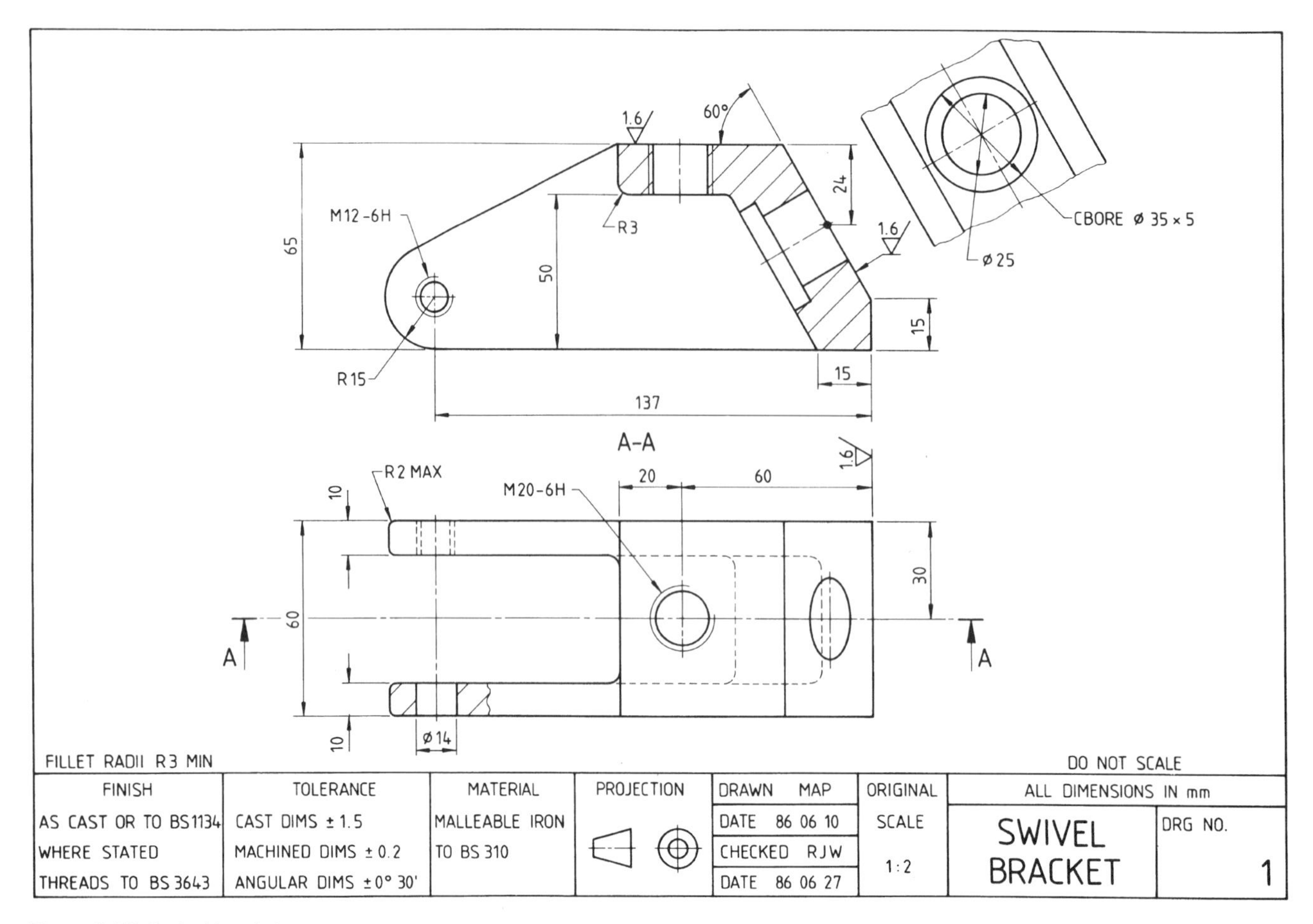

Figure 2.57 Swivel bracket

Figure 2.58 Airbox (*reproduced courtesy of* TecQuipment Limited)

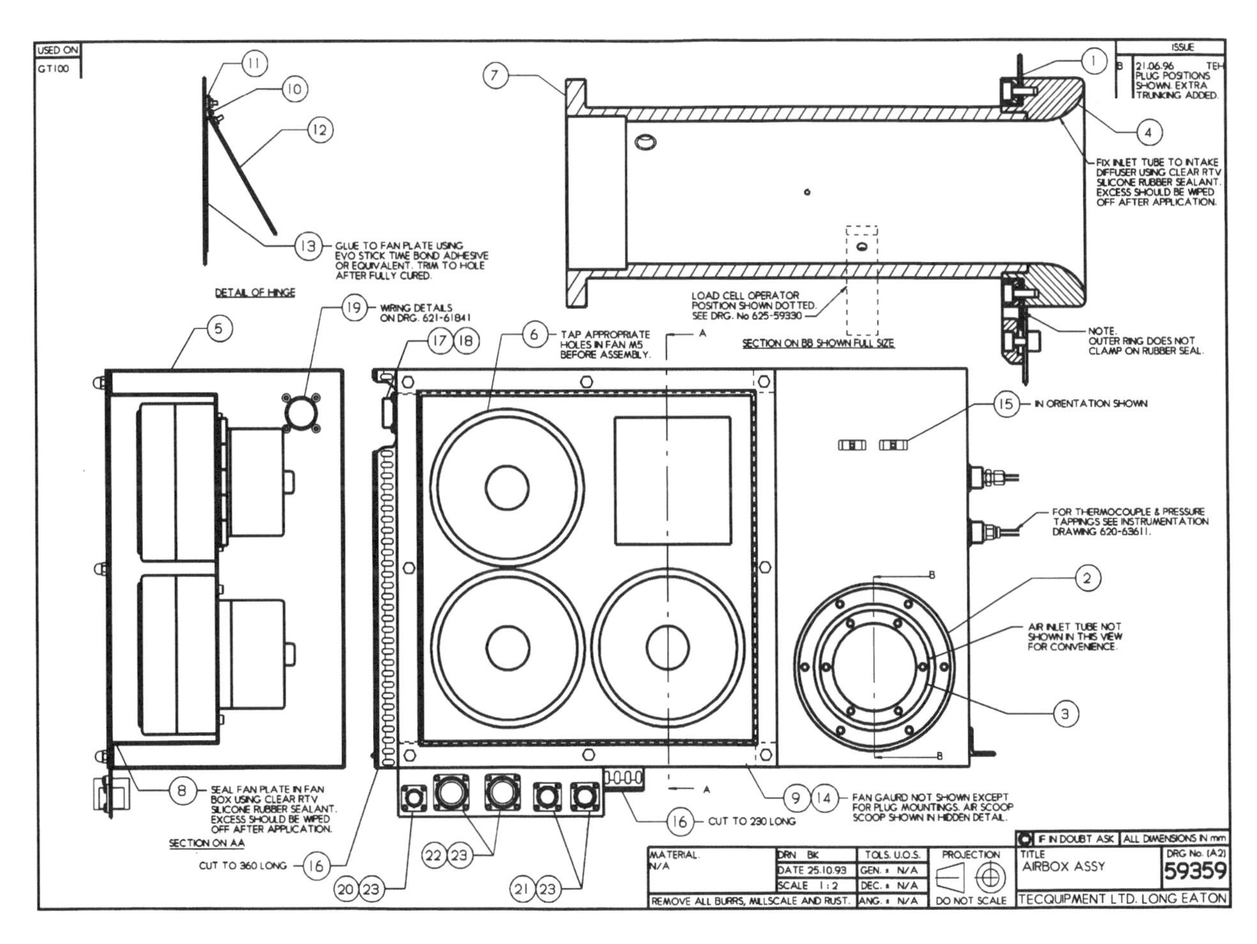

Figure 2.59 Airbox assembly drawing (*reproduced courtesy of* TecQuipment Limited)

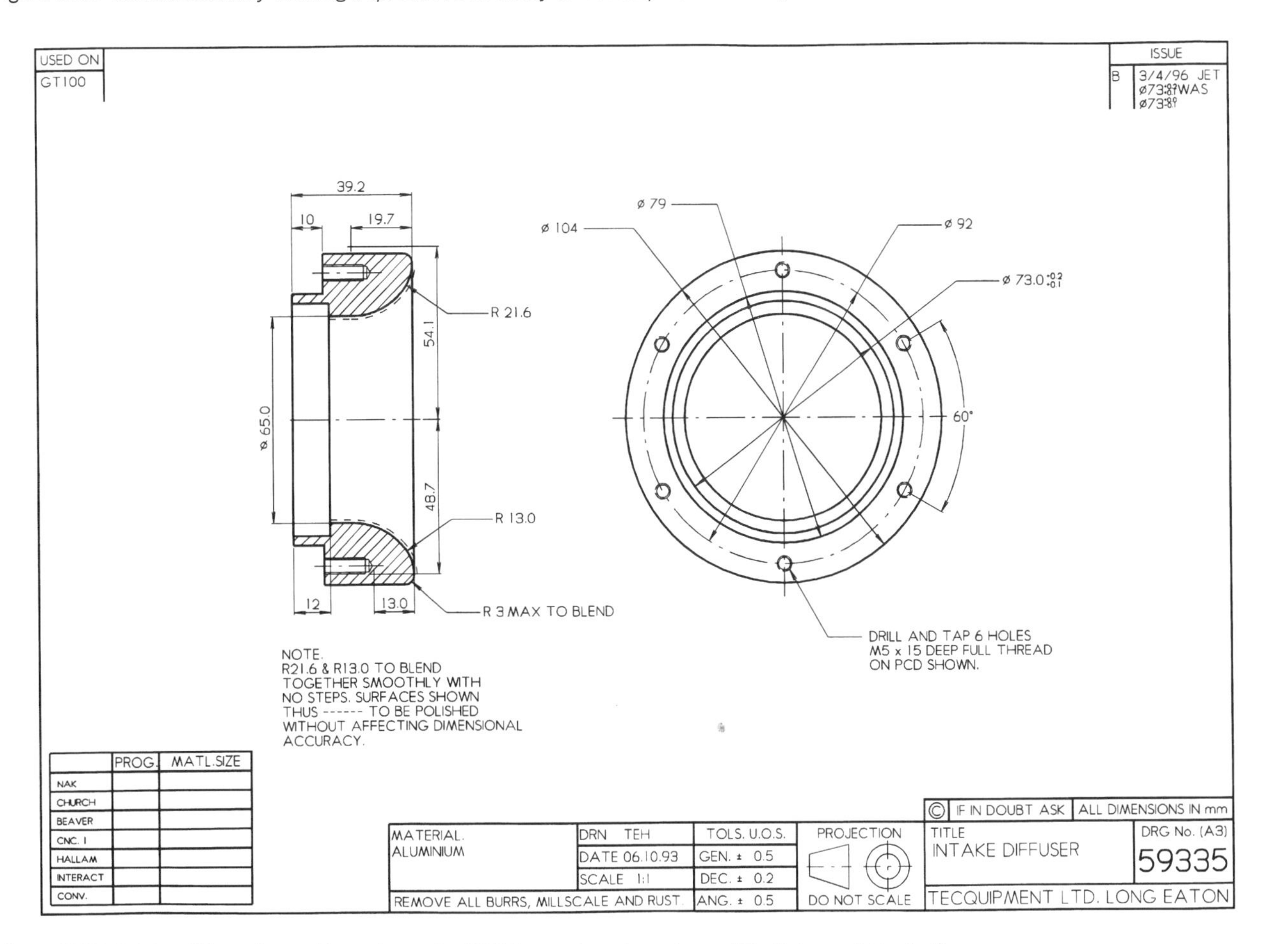

Figure 2.60 Intake diffuser from airbox, drawn by CAD (*reproduced courtesy of* TecQuipment Limited)

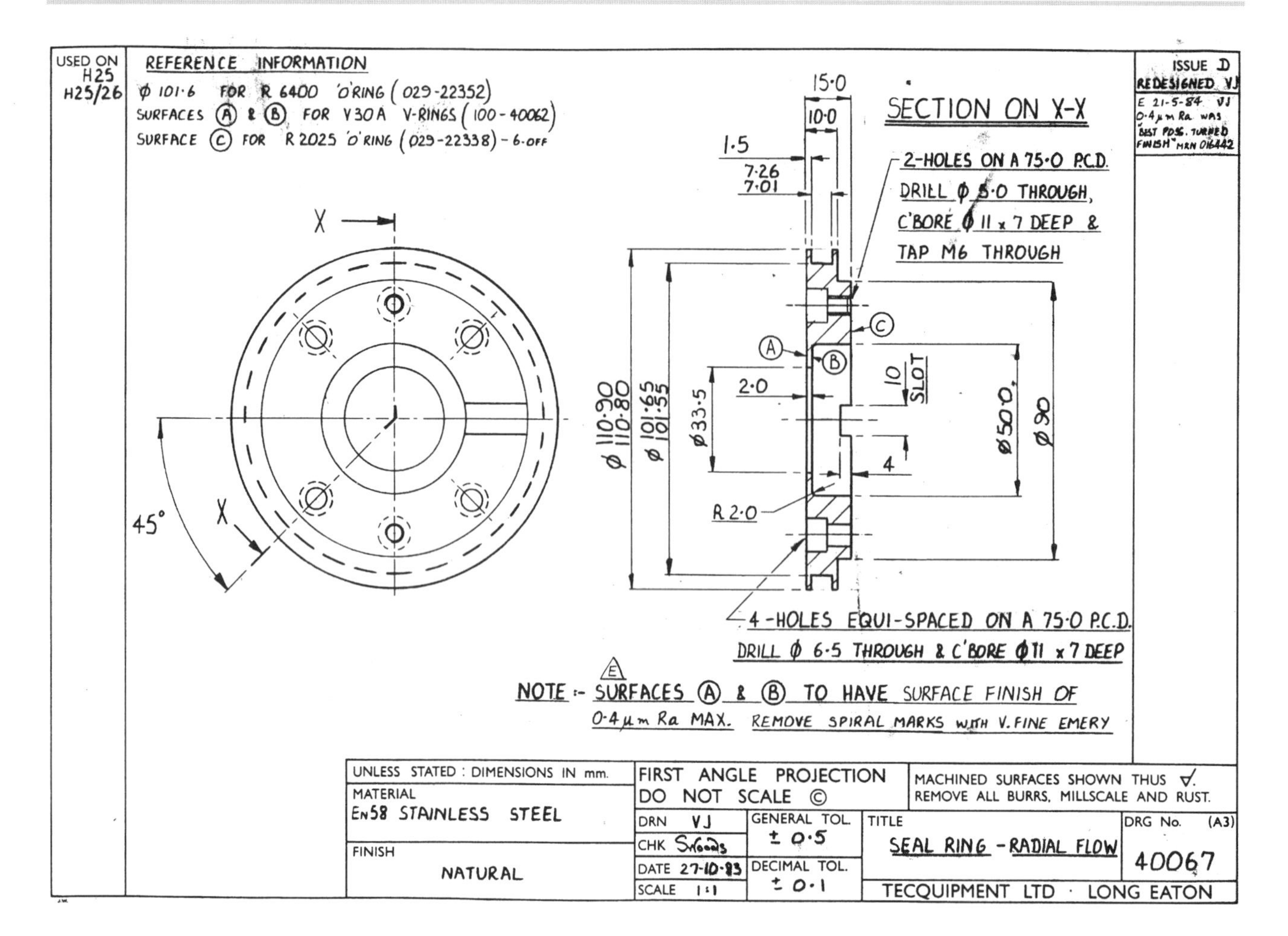

Figure 2.61 Intake diffuser from airbox, drawn conventionally (*reproduced courtesy of* TecQuipment Limited)

2.5 Unit test

Test yourself on this unit with these sample multiple-choice questions.

1. The object shown in Fig. 2.UT1 is a cube that has been drawn in

    (a) isometric
    (b) oblique
    (c) perspective
    (d) orthographic

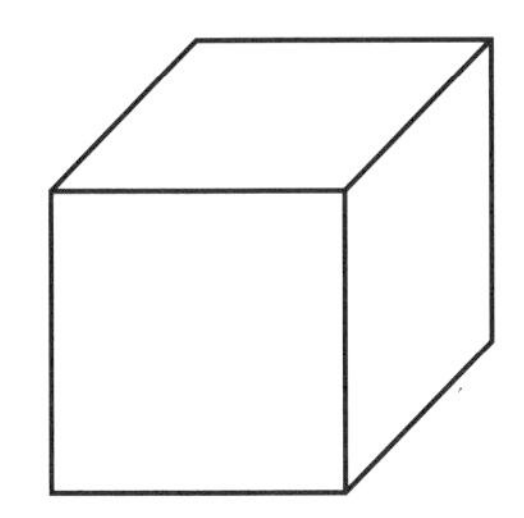

Figure 2.T1

2. The depth of the cube in Fig. 2.T1 is drawn shortened in order to

    (a) make the right angles correct
    (b) give the drawing a sense of perspective
    (c) make the cube look right
    (d) show clearly which is the front face

3. The initials CAD stand for

    (a) computer-assisted drawing
    (b) control and design
    (c) computer-assisted designing
    (d) computer-aided design

4. Parametric design software enables the designer to

    (a) plot the path of moving parts
    (b) produce sectional elevations
    (c) render CAD drawings
    (d) draw to scale

5. Wireframe modelling is

    (a) a wiring diagram
    (b) an electronic circuit diagram
    (c) an aid to visualising 3D forms
    (d) a means of trialing robot production lines

6. AutoCAD is

    (a) a procedure for designing cars
    (b) a means of evaluating robots on car production lines
    (c) a design software package
    (d) automatic dimensioning for drawings

7. 'Time' on a graph is usually displayed on the

    (a) horizontal x axis
    (b) horizontal y axis
    (c) vertical x axis
    (d) vertical y axis

8. Data charts are used to

    (a) plan operations
    (b) organise dates
    (c) display information
    (d) produce computer generated calendars

9. A pictogram is a form of

    (a) software
    (b) photograph
    (c) bar chart
    (d) perspective drawing

10. Pie charts are used to show

    (a) how things are divided up
    (b) ingredients
    (c) taxation
    (d) trends

11. A flow chart is used to

    (a) control pneumatic systems
    (b) evaluate software
    (c) predict current flow in electronic circuits
    (d) sequence events

12. The standards to which all UK formal engineering drawings should conform are

    (a) EC standards
    (b) BSI standards
    (c) ISO standards
    (d) CE standards

13. The cylinder in the pneumatic circuit shown in Fig. 2.T2 is a

    (a) single acting cylinder
    (b) double acting cylinder
    (c) non-return cylinder
    (d) free flow cylinder

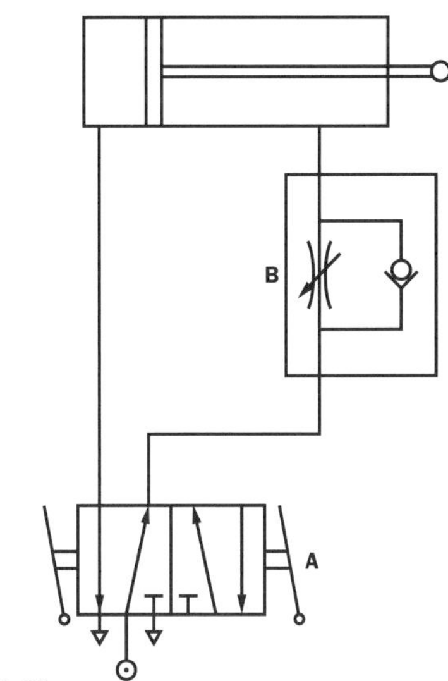

Figure 2.T2

14. Valve 'A' shown in Fig. 2.T2 is a

(a) spring return five-port valve
(b) signal operated five-port valve
(c) double operated five-port valve
(d) lever operated five-port valve

15. Valve 'B' shown in Fig. 2.T2 is a

(a) flow control valve
(b) non-return valve
(c) shuttle valve
(d) reservoir valve

16. Component 'A' in the electronic circuit shown in Fig. 2.T3 is

(a) a circuit board
(b) an amplifier
(c) an integrated circuit
(d) a network

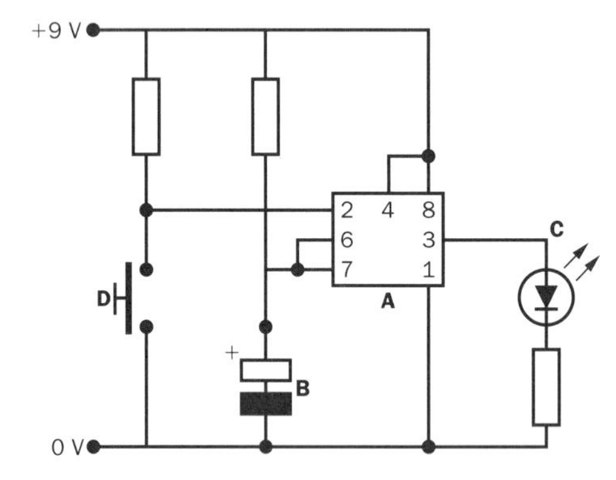

Figure 2.T3

17. Component 'B' shown in Fig. 2.T3 is

(a) a resistor
(b) an LDR
(c) a thermistor
(d) a capacitor

18. Component 'C' shown in Fig. 2.T3 is

(a) a resistor
(b) a light-emitting diode
(c) a transistor
(d) a timer

19. Component 'D' shown in Fig. 2.T3 is

(a) a trigger
(b) an input
(c) a push switch
(d) a capacitor

20. The engineering drawing shown in Fig. 2.T4 is drawn in

(a) first angle projection
(b) second angle projection
(c) third angle projection
(d) fourth angle projection

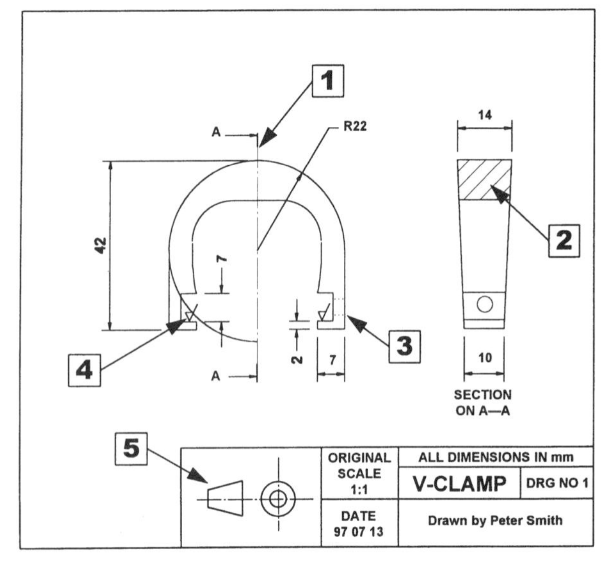

Figure 2.T4

21. The line indicated at Point 1 in Fig. 2.T4 is the

(a) orthographic line
(b) projection line
(c) cutting plane
(d) datum line

22. The lines at Point 2 in Fig. 2.T4 are

(a) sloping
(b) scaling lines
(c) development lines
(d) hatching

23. The lines at Point 3 in Fig. 2.T4 represent

(a) a screw thread
(b) a flat surface
(c) hidden detail
(d) knurling

24. The symbol shown at Point 4 in Fig. 2.T4 is a

(a) machining symbol
(b) cleaning symbol
(c) material symbol
(d) texture symbol

25. The symbol shown at Point 5 in Fig. 2.T4 is the

(a) ISO symbol
(b) projection symbol
(c) BSI symbol
(d) standards symbol

Science and Mathematics for Engineering

Alan Darbyshire

Engineers and technicians may be found working in many different areas of engineering activity. These range from the design and manufacture of engineered products through servicing, and maintenance activities to power generation, and information and data handling.

Engineering technology is the term used for the accumulated knowledge of these different, but often overlapping activities. It is constantly being refined and extended as old methods and procedures are improved and new ones are developed.

Engineering technology is largely science based. Advances and discoveries in science are applied by engineers to solve practical problems and to develop new products and processes. A knowledge of the scientific principles which explain the behaviour of matter and enable energy sources to be efficiently used is essential to engineers as is a knowledge of mathematics which is the language of science.

The key areas covered by this chapter are:

- The physical quantities in engineering systems and the units in which they are measured.
- The mathematical techniques used to manipulate and evaluate data.
- The scientific laws and principles which apply to engineering systems.
- The measurement of physical quantities in engineering systems.

After reading this chapter the student will be able to:

- Identify fixed and variable physical quantities in electro-mechanical engineering systems.
- Describe the relationships between physical quantities in terms of scientific laws and principles.
- Correctly measure and record physical quantities.
- Use mathematical techniques to manipulate and evaluate recorded data.

3.1 Physical quantities and their units

Topics covered in this section are:

- Identification of fundamental, derived and supplementary SI units.
- Use of unit prefixes, multiples and sub-multiples.
- Identification of scalar and vector quantities.

Identification of fundamental and derived SI units

It is essential for all scientists and engineers to measure physical quantities in the same way. The system which is the most widely used throughout the world is the SI system of units. The SI system gets its name from its French title: *Systeme International D'Unite*. It is based on the metric system which was devised in France two hundred years ago following the French revolution.

In the SI system, the units in which physical quantities are measured are classified into:

- Fundamental units
- Derived units
- Supplementary units

Fundamental units

The SI system has seven fundamental, or basic, physical quantities; these are listed in Table 3.1.

Table 3.1

Physical quantity	Fundamental unit	Unit symbol
Mass	Kilogram	kg
Length	Metre	m
Time	Second	s
Electric current	Ampere	A
Temperature	Kelvin	K
Luminous intensity	Candela	cd
Amount of substance	Mole	mol

These units have precise definitions. This ensures that they mean the same to people in different parts of the world.

The standard kilogram is defined by a cylinder of platinum–indium alloy measuring approximately 40 mm in both height and diameter. It is held at the International Bureau of Weights and Measures at Sèvres in France and is known as the International Prototype Kilogram.

The standard metre is defined in terms of the wavelengths of light of a particular colour, and the second is defined in terms of the rate of radioactive decay of a particular radioactive material.

Primary standards such as these have been fixed for all the fundamental SI units against which measuring instruments throughout the world can be precisely calibrated.

Derived units

Other physical quantities have units which are derived from the above fundamental units. For instance, area is measured in square metres (m^2), which is derived from the unit of length. The frequencies of electrical and mechanical oscillations are measured in hertz, equal to the number of oscillations per second. It is derived from the unit of time.

Other common derived units are the newton, which is the unit of force, the watt, which is the unit of power, the volt, which is the unit of electrical potential difference and the ohm, which is the unit of electrical resistance. These units have been given their own names but they are all derived from the fundamental units for mass, length, time and electric current.

Supplementary units

The most commonly used supplementary SI unit is the radian which is used to measure plane angles. For most everyday purposes, angles are measured in degrees, but these are unsuitable for certain engineering and scientific calculations in which radians must be used.

A radian is defined as the plane angle subtended at the centre of a circle by an arc of length equal to the radius (Fig. 3.1).

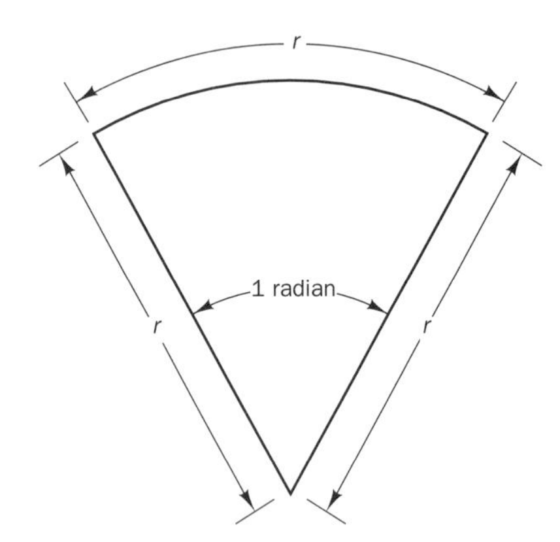

Figure 3.1 Definition of a radian

Use of unit prefixes, multiples and sub-multiples

The basic SI units are not always convenient for measuring very large or very small amounts of a physical quantity. For instance, the metre is too small for measuring distances travelled by motorway and too large for measuring the sizes of screws and rivets. Instead of metres, kilometres and millimetres are used where 'kilo' and 'milli' are called prefixes. They indicate that the basic unit has been multiplied by 1000 to make a bigger unit, or divided by 1000 to make a smaller one. There are other prefixes with bigger and smaller multiplying factors as given in Table 3.2.

Table 3.2

Prefix	Symbol	Multiplying factor
tera	T	10^{12} or 1 000 000 000 000
giga	G	10^{9} or 1 000 000 000
mega	M	10^{6} or 1 000 000
kilo	K	10^{3} or 1 000
milli	m	10^{-3} or 1/1 000
micro	μ	10^{-6} or 1/1 000 000
nano	n	10^{-9} or 1/1 000 000 000
pico	p	10^{-12} or 1/1 000 000 000 000

As can be seen the indices of the multiplying factors go up and down in steps of three. These are termed 'preferred units'. Other prefixes are sometimes used for convenience. The most common is 'centi', as in the centimetres, which is 1/100th or 10^{-2} of a metre.

Examples

1. Express 25.5 km in metres.
   Answer: **25.5×10^3 m**
2. Express 72.5 mm in metres.
   Answer: **72.5×10^{-3} m**
3. Express 45 g in kilograms. (Remember, in the SI system the kilogram is the fundamental unit of mass, not the gram.)
   Answer: **45×10^{-3} kg**
4. Express 650 MN in newtons.
   Answer: **650×10^6 N**

Self-assessment tasks 3.1

1. Express (a) 150 kW in watts and (b) 75 mV in volts.
2. Express (a) 350 μA in amperes and (b) 210 GN in newtons.

Identification of scalar and vector quantities

Scalar quantities are those which have size or magnitude only. Vector quantities are those which have both magnitude and direction. Typical examples of each type are given in Table 3.3.

Table 3.3

Scalar quantities	Vector quantities
Mass	Displacement
Length	Velocity
Time	Acceleration
Density	Force
Energy	Momentum

A vector quantity may be represented on a vector diagram as a line which points in the direction in which it acts and whose length, drawn to some suitable scale, represents the magnitude, or size, of the vector quantity.

Examples

1. Using a scale of 10 mm = 1 N draw the vector of a 5 N force which acts to the right and upwards at an angle of 30°.

   - Draw horizontal and vertical centre lines.
   - Starting where the centre lines cross, draw a construction line at 30° to the horizontal and upwards to the right. This gives the direction of the vector.
   - Measure 5cm from the start of the line. This gives the length of the vector.
   - Draw in the finished vector.

This result is shown in Fig. 3.2.

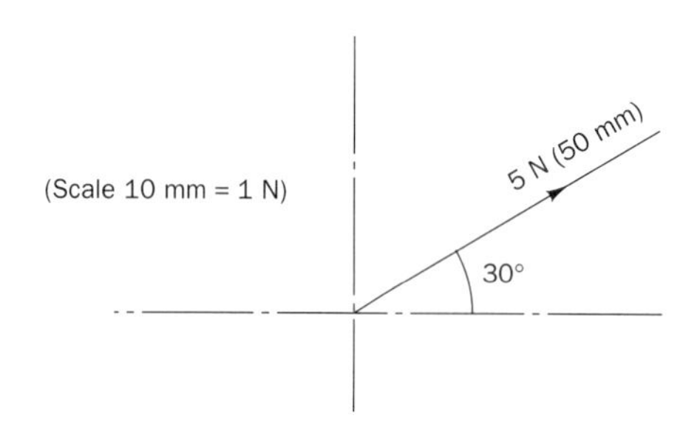

Figure 3.2 Vector diagram – Example 1

2. Using a scale of 1 mm = 1 m/s draw the velocity vector of a vehicle travelling at 40 m/s in a north westerly direction. Use the same procedure (see Fig. 3.3).

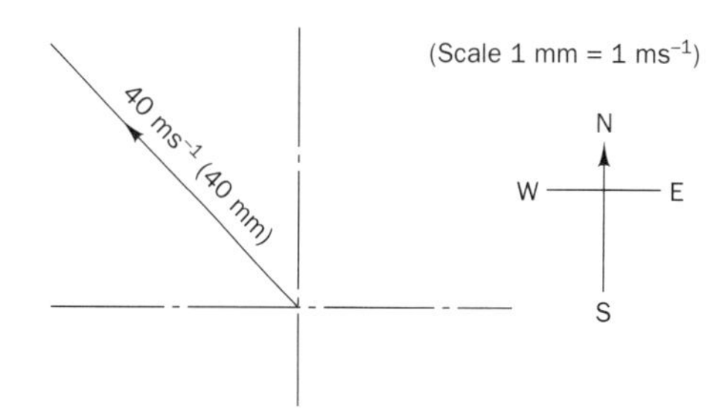

Figure 3.3 Vector diagram – Example 2

Self-assessment tasks 3.2

1. Using a suitable scale, draw the vector diagrams for:
   (a) A 6.5 N force acting downwards and to the left at an angle of 25° to the horizontal.
   (b) A 55 N force acting upwards and to the left at an angle of 40° to the vertical.
2. Using a suitable scale, draw the vector diagrams for:
   (a) An aeroplane flying at 600 kmh^{-1} in a north-easterly direction.
   (b) A motor car travelling at 70 kmh^{-1} in a south-westerly direction.
3. Using a suitable scale, draw the vector diagrams for:
   (a) A 15 kN force acting downwards to the right at an angle of 35°.
   (b) A cutting tool travelling at 0.12 ms^{-1} upwards to the right at an angle of 15°.

3.2 Mathematical techniques for manipulating and evaluating data

As has been stated, mathematics is the language of science and many of the laws and principles used to describe the behaviour of engineering systems are expressed in mathematical terms.

Mathematical techniques are used to investigate and describe the behaviour of engineering systems. The data obtained from tests on systems is manipulated and used to evaluate system performance. Some of the more important mathematical techniques used when investigating engineering systems are:

- Rounding off numerical values.
- Performing arithmetic operations.
- Using multiples and sub-multiples.
- Transposing formulae.
- Plotting graphs.

Rounding off numerical values

Calculated values are often stated correct to a number of significant figures or a number of decimal places.

Three significant figures are sufficient for most calculations. If the fourth figure is 5 or more, the third figure is rounded up, as in the following examples.

Examples

1. State 13.7953 correct to three significant figures.
   Answer: **13.8**
   Reason: The fourth figure is 9, which is greater than 5, the third figure is rounded up from 7 to 8.
2. State 0.025311 correct to three significant figures.
   Answer: **0.0253**
   Reason: The fourth figure is 1, which is less than 5, the third figure stays at 3.

When it is required to give an answer correct to a number of decimal places, the figures before the decimal point stay the same. The number of places is counted after the decimal point and the figure is rounded up if the next one is 5 or more, as in the following examples.

Examples

1. State 21.3469 correct to two decimal places.
   Answer: **21.35**
   Reason: The third figure after the decimal point is 6, which is greater than 5, the second figure after the point is rounded up from 4 to 5.
2. State 0.043401 correct to three decimal places.
   Answer: 0.043
   Reason: The fourth figure after the decimal point is 4, which is less than 5, the third figure after the point stays at 3.

Self-assessment tasks 3.3

1. State the following correct to three significant figures:
   (a) 56.754 (b) 0.0035951 (c) 207959
2. State the following correct to three significant figures:
   (a) 3.14159 (b) 9996 (c) 0.10009453
3. State the following correct to two decimal places:
   (a) 23.68731 (b) 1.02167 (c) 0.89632
4. State the following correct to four decimal places:
   (a) 4.016891 (b) 0.034599 (c) 2.00003

Performing arithmetic operations

Calculations involving the addition, subtraction, multiplication and division of numerical values may be performed speedily and accurately using a scientific electronic calculator. The operations must however be performed in the correct sequence:

1. First of all, work out the value of any bracketed terms.
2. Next, carry out multiplication and division operations.
3. Finally carry out any addition and subtraction operations.

Examples

1. Evaluate $(92 + 135) \times \frac{150}{24}$.

   First evaluate the bracketed quantity as follows.

   [[ ] [92] [+] [135] [ ]]

   Your calculator display should now be showing 227.
   Next perform the multiplication and division operation as follows.

   [×] [150] [÷] [24] [=]

   The answer on your display should be **1418.75**.

2. Evaluate $\frac{(265 - 34)}{58} + 350 - 163$. Give your answer correct to three significant figures.
   First evaluate the bracketed quantity.

   [[ ] [265] [−] [34] [ ]]

   Your calculator display should now be showing 231.
   Next, perform the division operation.

   [÷] [58] [=]

   Your calculator display should now be showing 3.98, when rounded off to three significant figures.
   Finally, perform the addition and subtraction operations.

   [+] [350] [−] [163] [=]

   The answer on your display should be **191** (rounded off to three significant figures).

Self-assessment tasks 3.4

1. Evaluate the following expressions correct to three significant figures:
   (a) $(101 - 33) \times \frac{65}{14}$
   (b) $\frac{(35 + 21)}{22} + 110 - 9$
2. Evaluate the following expressions correct to three significant figures:
   (a) $\frac{5}{(25 - 12)} \times \frac{7}{81}$
   (b) $\frac{(41 - 29)}{(35 + 17)} + 7.5 - 10.8$

Using multiples and sub-multiples

It is often necessary to multiply and divide values expressed as multiples and sub-multiples of 10. Values with multiples and sub-multiples can easily be multiplied by the following method:

1. Multiply the numerical values.
2. Multiply the multiples of 10 by adding together their indices.

Examples

1. Evaluate $(5 \times 10^6) \times (7 \times 10^3)$.

$$5 \times 10^6 \times 7 \times 10^3 = 5 \times 7 \times 10^6 \times 10^3 = 35 \times 10^{(6+3)} = \mathbf{35 \times 10^9}$$

2. Evaluate, $6 \times 10^{-3} \times 12 \times 10^6$.

$$6 \times 10^{-3} \times 12 \times 10^6 = 6 \times 12 \times 10^{-3} \times 10^6 = 72 \times 10^{(-3+6)} = \mathbf{72 \times 10^3}$$

Similarly values with multiples and sub-multiples can easily be divided by the following method:

1. Divide the numerical values of numerator and denominator.
2. Divide the multiples of 10 by subtracting their indices.

Examples

1. Evaluate $\frac{(21 \times 10^6)}{(3 \times 10^3)}$.

$$\frac{(21 \times 10^6)}{(3 \times 10^3)} = \frac{21}{3} \times \frac{10^6}{10^3} = 7 \times 10^{(6-3)} = \mathbf{7 \times 10^3}$$

2. Evaluate $\frac{(125 \times 10^{-3})}{(25 \times 10^6)}$

$$\frac{(125 \times 10^{-3})}{(25 \times 10^6)} = \frac{225}{25} \times \frac{10^{-3}}{10^6} = 5 \times 10^{(-3-6)} = \mathbf{5 \times 10^{-9}}$$

Self-assessment tasks 3.5

1. Evaluate:
   (a) $5 \times 10^3 \times 8 \times 10^9$; (b) $9 \times 10^6 \times 4 \times 10^{-9}$
2. Evaluate:
   (a) $\frac{45 \times 10^8}{5 \times 10^3}$; (b) $\frac{16 \times 10^6}{4 \times 10^{-3}}$

When entering values with multiples and sub-multiples into an electronic calculator, the 'EXP' key is used. To enter 2.5×10^3 the procedure is as follows:

[2.5] [EXP] [3]

Note: It is important not to enter the multiplication sign.

To enter a sub-multiple, which has a negative index such as 1.8×10^{-6}, the procedure is:

[1.8] [EXP] [6] [+/–]

Example

1. Evaluate $\frac{165 \times 10^6}{32}$.

   [165] [EXP] [6] [÷] [32] [=]

   The value shown on the calculator display should now be **5156250.**

2. Evaluate $\frac{54 \times 10^{-3}}{115}$.

   [54] [EXP] [3] [+/–] [÷] [1.15] [=]

   The value shown on the calculator display should now be **0.046956521.**

Very large or very small numbers, such as the above answers, are best expressed with a multiple or sub-multiple. To obtain this on an electric calculator, press the 'ENG' key.

If this is done for the above value of 5156250, it is converted to 5.156250×10^6, which can be rounded off to 5.16×10^6, correct to three significant figures.

If it is done for the above value of 0.046956521, it is converted to 46.956521×10^{-3}, which can be rounded off to 47.0×10^{-3}, correct to three significant figures.

Self-assessment tasks 3.6

1. Evaluate the following expressions and give the answers as multiples or sub-multiples of 10^3:
   (a) $0.025 \times 3 \times 10^6$
   (b) $\frac{546 \times 10^3}{0.15}$
   (c) $\frac{2.6 \times 10^{-3}}{230}$
2. Evaluate the following expressions and give the answers as multiples or sub-multiples of 10^3, correct to three significant figures:
   (a) $37 \times 10^6 \times 51 \times 10^{-3}$
   (b) $\frac{(121 - 39) \times 5 \times 10^6}{91 \times 10^{-3}}$

Transposing formulae

Scientific laws and principles are often expressed as mathematical formulae. The formulae give the relationship between different physical quantities. Depending on which of the physical quantities is required in a calculation, a formula may need to be changed around, or 'transposed', to make that quantity its subject.

In some of the formulae, the quantities may be multiplied together or divided. In others they may be added or subtracted, or there may be a combination of all four operations.

The general rule with formulae in which the terms are **multiplied** or **divided** is that they may be transposed by **dividing** or **multiplying** each side by the term which it is required to move.

Examples

1. Transpose the formula $F = ma$ to make a, the subject.
   Answer: To make a the subject, divide each side by m so that it will cancel on the right-hand side.

$$F = ma$$

Divide by m $\quad \dfrac{F}{m} = \dfrac{\not{m}a}{\not{m}}$

$$\frac{F}{m} = a \quad \text{or} \quad a = \frac{F}{m}$$

2. Transpose the formula $I = \dfrac{V}{R}$ to make R the subject.

   Answer: To make R the subject, multiply each side by R, so that it will cancel on the right-hand side, and divide each side by I, so that it will cancel on the left-hand side.

$$I = \frac{V}{R}$$

Multiply by R $\quad IR = \dfrac{V\not{R}}{\not{R}}$

$$IR = V$$

Divide by I $\quad \dfrac{\not{I}R}{\not{I}} = \dfrac{V}{I}$

$$R = \frac{V}{I}$$

The general rule is that formulae in which the terms are **added** or **subtracted** is that they may be transposed by **subtracting** or **adding** to each side the term which it is required to move.

Examples

1. Transpose the formula $a = b + c$ to make b the subject.
   Answer: To make b the subject, subtract c from each side so that it will cancel on the right-hand side.

$$a = b + c$$

Subtracting c $\quad a - c = b + \not{c} - \not{c}$

$$a - c = b$$

or $\quad b = a - c$

2. Transpose the formula $x = y - z$ to make y the subject.
   Answer: To make y the subject, add z to each side so that it will cancel on the right-hand side.

$$x = y - z$$

Adding z $\quad x + z = y - \not{z} + \not{z}$

$$x + z = y$$

or $\quad y = x + z$

3. Transpose the formula $v = u + at$ to make a the subject.
   Answer: To make a the subject, first subtract u from each side so that it will cancel on the right-hand side.

$$v = u + at$$

Subtracting u $\quad v - u = \not{u} + at - \not{u}$

$$v - u = at$$

Now divide each side by t, so that it will cancel on the right-hand side.

$$\frac{v - u}{t} = \frac{a\not{t}}{\not{t}}$$

$$\frac{v - u}{t} = a$$

or $\quad a = \dfrac{v - u}{t}$

4. Transpose the formula $Q = mc(T_2 - T_1)$ to make T_2 the subject.
   Answer: First of all divide each side by m and by c, so that they will cancel on the right-hand side.

$$Q = mc(T_2 - T_1)$$

Dividing by m and c $\quad \dfrac{Q}{mc} = \dfrac{\not{m}\not{c}(T_2 - T_1)}{\not{m}\not{c}}$

$$\frac{Q}{mc} = T_2 - T_1$$

Now add T_1 to each side, so that it will cancel on the right-hand side.

Adding T_1 $\quad \dfrac{Q}{mc} + T_1 = T_2 - \not{T_1} + \not{T_1}$

$$\frac{Q}{mc} + T_1 = T_2$$

or $\quad T_2 = \dfrac{Q}{mc} + T_1$

Self-assessment tasks 3.7

1. (a) Transpose the formula $F = kx$ to make k the subject.
   (b) Transpose the formula $p = F/A$ to make F the subject.
2. (a) Transpose the formula $E = mgz$ to make g the subject.
   (b) Transpose the formula $P = Fs/t$ to make t the subject.
3. (a) Transpose the formula $v = u + at$ to make t the subject
   (b) Transpose the formula $x = l\alpha(T_2 - T_1)$ to make T_1 the subject.

Plotting graphs

The way in which two variable quantities are related can often be found by plotting their values on a graph.

The quantity whose change can be controlled is called the **independent** variable. The other quantity, which is changing with it, is called the **dependent** variable. The choice of which variable is plotted on the horizontal, or x-axis, and which is plotted on the vertical, or y-axis, depends on the information required from the graph. If one of the variables is time, this is usually plotted horizontally.

Scales should be chosen which make best use of the graph paper so that the points are not clustered together in one corner. The graph should have a title, and the axes should be clearly labelled with the name of the variable and its units.

The points should be carefully plotted and marked with either a small cross or a dot surrounded by a small circle.

When the points do not quite lie on a straight line or a smooth curve, this may be due to errors in the recording equipment, or to human error when taking readings. In such cases a **line of best fit**, or **curve of best fit**, should be drawn.

When a graph is a straight line passing through the origin as shown in Fig. 3.4, the variables are said to have a directly **proportional relationship**. The equation or law connecting the variables is of the form:

$$y = mx$$

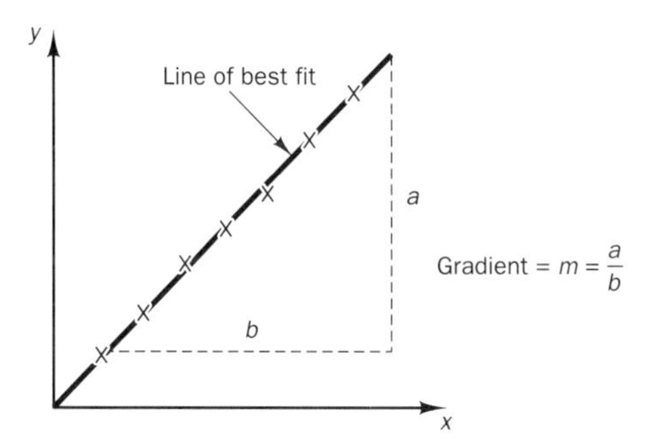

Figure 3.4 Graph of a proportional relationship

The term m is called the constant of proportionality. Its value is given by the slope, or gradient, of the graph. It gives a measure of how much the variable y changes for each unit of change in the variable x.

When a graph is a straight line which does not pass through the origin, the variables are said to have a **linear relationship** (Fig. 3.5). The equation, or law, connecting the variables is now of the form:

$$y = mx + c$$

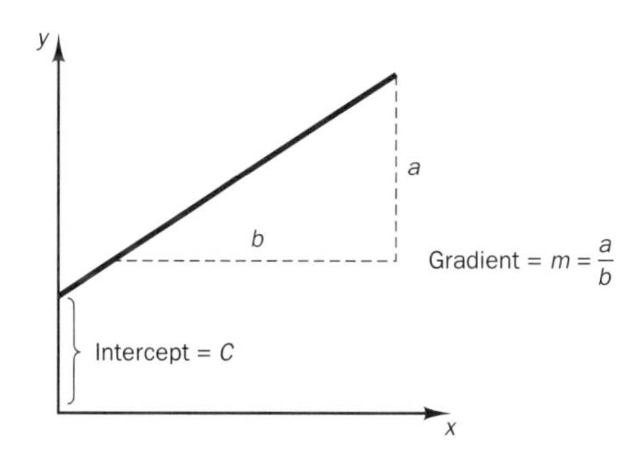

Figure 3.5 Graph of a linear relationship

The term m is again the gradient and c is the value of the intercept where the line crosses the y-axis.

With some graphs, additional information can be obtained by finding the area between the curve and the x-axis. For example, when a graph of velocity against time is plotted for a moving body, the area beneath the curve gives the value of the distance travelled in a given time.

Examples

1. Plot the graph of current against potential difference for the following test results on an electric circuit. Determine the gradient of the graph and state the equation which relates current I and potential difference V. The graph is shown in Fig. 3.6.

Potential difference V (volts)	2	4	6	8	10	12	14	16
Current I (amperes)	0.5	1.0	1.6	2.0	1.5	3.0	3.4	4.0

Answer: The increase in current when the potential difference increases by 1 volt is given by the gradient of the graph:

$$\text{Gradient} = \frac{a}{b} = \frac{3.5}{14} = 0.25 \text{ amperes per volt}$$

Current and potential difference related by the expression:

$$I = 0.25V$$

2. Plot a graph of velocity against time for a vehicle which is accelerating uniformly from the following data. From the graph determine the initial velocity of the vehicle and its acceleration, i.e. the amount by which its velocity is increasing every second. The graph is shown in Fig. 3.7.

Time t (seconds)	5	10	15	20	25	30	35	40
Velocity v (ms^{-1})	6	9	11	15	18	22	24	27

Answer:

Initial velocity of vehicle = y-axis intercept = $4\,ms^{-1}$

The increase in velocity per second is given by the gradient of the graph:

$$\text{Gradient} = \frac{a}{b} = \frac{13}{32.5} = 0.4\,ms^{-1}$$

Velocity and time related by the expression:

$$v = 0.4t + 4$$

Self-assessment tasks 3.8

1. The following shows the results of a test on a spring. Plot a graph of load against extension and from it determine the spring's stiffness, i.e. how much the spring extends for each newton of applied load.

Load (newtons)	0	25	50	75	100	125	150	175	200
Extension (mm)	0	11	21	34	44	55	65	78	88

2. A test was carried out on an electrical resistance element to measure its resistance at different temperatures. Plot a graph of resistance against temperature from the following test results and from it determine: (a) the resistance at 0 °C; (b) the increase in resistance per degree of temperature rise.

Resistance (ohms)	12.0	13.5	15.0	16.4	18.0	19.6	21.0
Temperature (°C)	20	40	60	80	100	120	140

3. The following table shows the total distance travelled by a machine work table with time. Plot the graph of distance travelled against time and from it determine the velocity of the work table in ms^{-1}, i.e. the gradient of the graph.

Distance travelled (mm)	0	15	31	45	60	74	90	105
Time (seconds)	0	3	6	9	12	15	18	21

4. The change in velocity of a car with time is shown below. Plot the graph of velocity against time and from it obtain the expression which relates the two variables.

Velocity (ms^{-1})	18	26	36	45	54	64	72	81
Time (seconds)	5	10	15	20	25	30	35	40

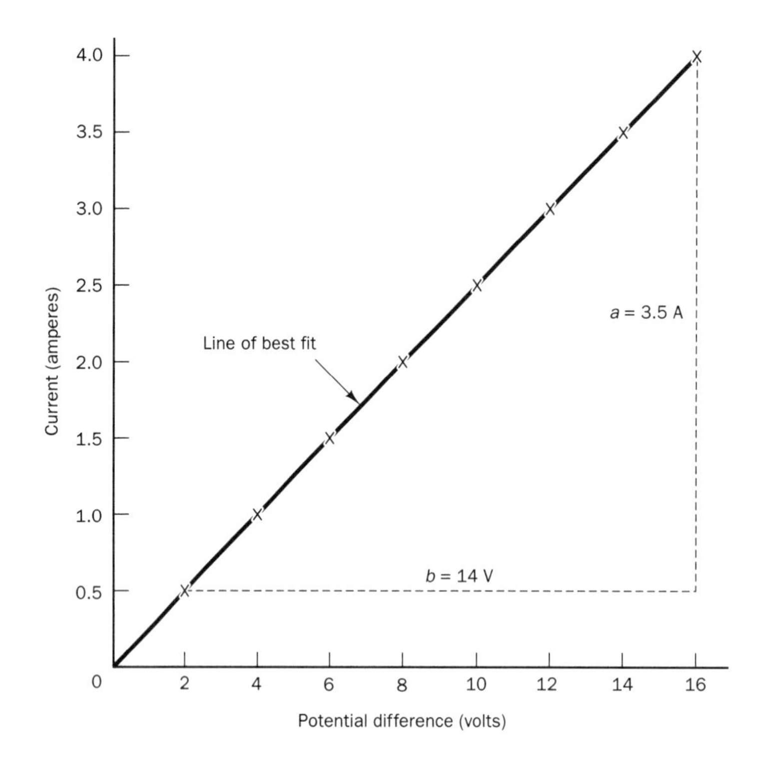

Figure 3.6 Graph of current against potential difference – Example 1

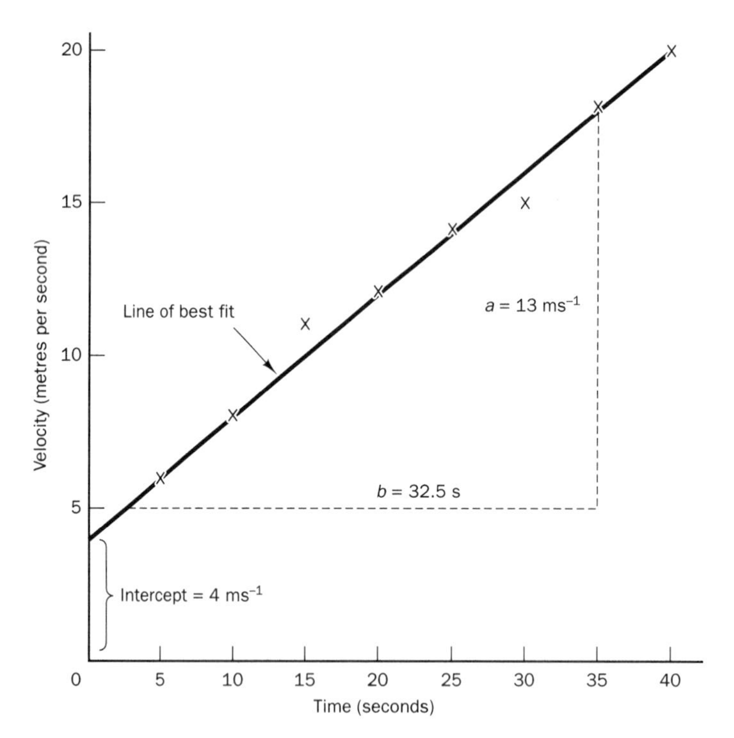

Figure 3.7 Graph of velocity against time – Example 2

3.3 Static systems

Bodies which are stationary are said to be in a state of static equilibrium. Two conditions must apply for this to occur:

1. The vector sum of the forces acting on a body must be zero so that there is no movement in any direction (Fig. 3.8).

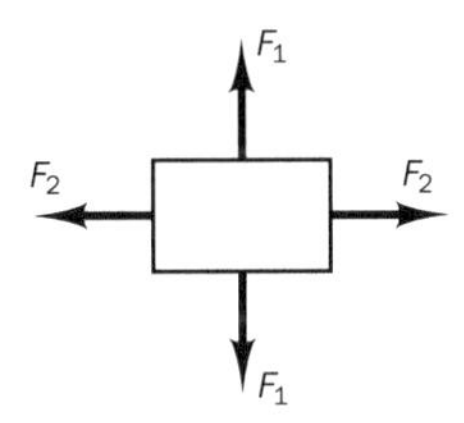

Figure 3.8 Vector sum is zero

2. The turning moments of the forces acting on a body must be equal and opposite so that there is no rotation (Fig. 3.9).

Figure 3.9 Turning moment sum is zero

Before investigating static engineering systems, it is useful to look at what is meant by:

- weight
- pressure
- centre of gravity
- states of equilibrium
- turning moments

Weight

One of the forces which acts on a body is its own weight. Sir Isaac Newton discovered that mass attracts mass and that the larger the masses, the larger the force of attraction. The Earth is a large mass which exerts an attractive force on all of the objects on its surface and on the atmosphere which surrounds it. The force is called the **force of gravity**.

The average pull of the Earth on a mass of 1 kg resting on its surface is 9.81 newtons, which is called the Earth's **gravitational field strength**.

The force of gravity also causes a freely falling body to accelerate, i.e. pick up speed, at the rate of 9.81 metres per second, for every second of fall. This is called the **acceleration due to gravity** which is given the symbol 'g':

i.e. g = 9.81 newtons per kilogram (Nkg^{-1})
or g = 9.81 metres per second for every second of fall (ms^{-2})

The total pull acting on a mass of *m* kg is known as its weight, which is given by the formula:

weight = mass (kg) × 9.81 (Nkg^{-1})
i.e. $W = mg$ (newtons)

Pressure

When a force acts on a surface it may be concentrated at a single point or it may be spread out. Pressure is a measure of the amount of force acting at right angles on each square metre of surface area. Another name for a force which acts at right angles to a surface is a 'normal force'.

The unit of pressure is the **pascal** (Pa). When a force of **1 newton** acts evenly at right angles over an area of **1 square metre** the pressure on the surface is **1 pascal:**

1 Pa = 1 Nm^{-2}

The pressure acting on a surface is calculated using the formula:

$$\text{Pressure} = \frac{\text{Normal force on surface}}{\text{Area of surface}}$$

$$p = \frac{F}{A}$$

In this relationship, the surface area is constant and the normal force and the pressure are variables.

Examples

1. A tank contains 250 kg of water and its base has an area of 1.5 m^2. What is the pressure on its base due to the water?
   Answer:
   Finding weight of water:

   $W = mg$
   $W = 250 \times 9.81$
   $W = 2453$ N

   This is also the force *F* acting on the base of the tank.
   Finding pressure on the base:

   $p = \frac{F}{A}$

   $p = \frac{2453}{1.5}$

   $p = 1635$ **Pa** or **1.64 kPa**

2. The pressure on the piston in an engine cylinder is 1.2 MPa and the piston diameter is 100 mm. Calculate the force on the piston.
   Answer:
   Finding area of piston for diameter $d = 0.1$ m:

   $A = \frac{\pi d^2}{4}$

   $A = \frac{\pi \times 0.1^2}{4}$

   $A = 7.85 \times 10^{-3} m^{-2}$

   Finding force on piston:

   Since $p = \frac{F}{A}$

   transposing gives $F = pA$
   $F = 1.2 \times 10^6 \times 7.85 \times 10^{-3}$
   $F = $ **9425 N** or **9.43 kN**

The force of gravity, pulling downwards on the Earth's atmosphere, gives rise to the atmospheric pressure on the Earth's surface.

Atmospheric pressure is quite large. Its average value is 101.325 kPa. Luckily the human body can adjust to this so that the pressure inside is equal to the pressure outside. Otherwise, the pressure of the atmosphere could quite easily crush us.

The pressure inside boilers, welding cylinders, and hydraulic and pneumatic systems is often measured on pressure gauges. They are usually calibrated to record the amount by which the inside pressure is greater than the outside atmospheric pressure.

This is called the gauge pressure, which is all that is generally needed for everyday monitoring and control purposes. The total pressure inside a pressurised system is called the absolute pressure which can be found from the formula:

Absolute pressure = Gauge pressure + Atmospheric pressure

Self-assessment tasks 3.9

1. The pressure inside a gas cylinder is 400 kPa. What is the force on each end of the cylinder if the inside diameter is 200 mm?
2. A tank contains 750 kg of water. What is the pressure on its base if its dimensions are 2.5 m × 1.5 m?
3. The piston in the hydraulic cylinder of a press must exert a force of 5 kN. If its diameter is 80 mm, what must be the pressure inside the cylinder?
4. The air pressure inside a pneumatic system is 50 kPa. If the piston in a pneumatic cylinder is required to deliver a force of 25 N, what must be the cylinder diameter?

Centre of gravity

Although the force of gravity acts equally on all parts of a body, it is often convenient to assume that it acts at some central point, known as its **centre of gravity**.

The centres of gravity of some commonly shaped objects are as follows. They may be divided into:

- laminae
- three dimensional solids

Laminae

A lamina is a thin, flat shape such as something which has been cut from sheet metal.

1. **Rectangle** (Fig. 3.10) – Centre of gravity is at the intersection of the diagonals at a height of $x = h/2$, above the base.

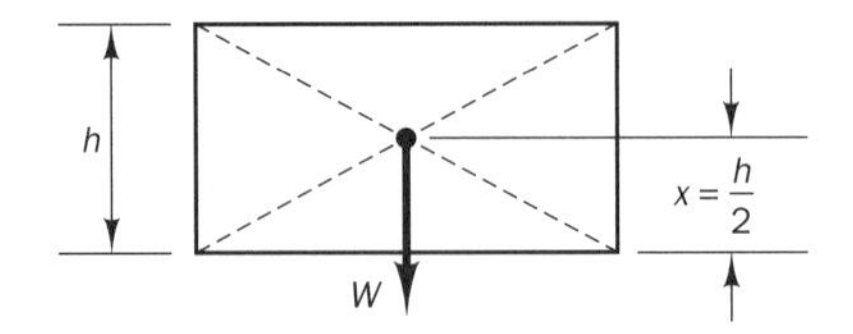

Figure 3.10 Centre of gravity – rectangle

2. **Circle** (Fig. 3.11) – Centre of gravity is at the intersection of the diameters at a height of $x = d/2$, above the base.

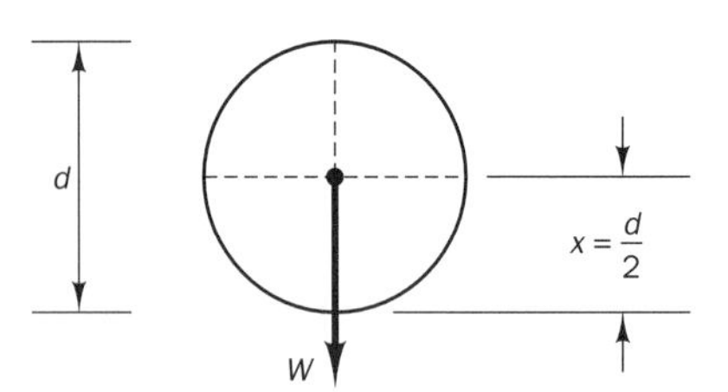

Figure 3.11 Centre of gravity – circle

3. **Triangle** (Fig. 3.12) – Centre of gravity is at the intersection of the medians, i.e. the lines joining the corners to the mid-points of the opposite sides. The height of the centre of gravity is $x = h/3$, above the base.

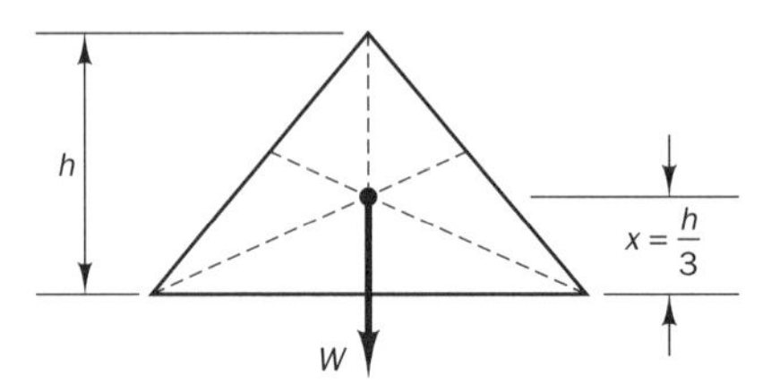

Figure 3.12 Centre of gravity – triangle

4. **Semi-circle** (Fig. 3.13) – Centre of gravity is on the vertical centre line at a height of $x = 4r/3\pi$, above the base.

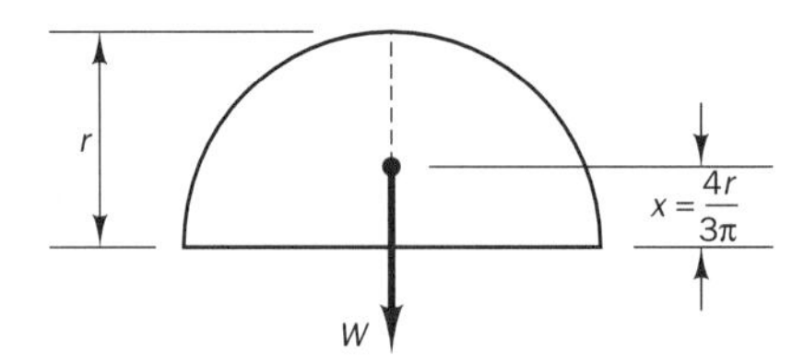

Figure 3.13 Centre of gravity – semicircle

Solid bodies

1. **Rectangular prism** (Fig. 3.14) – Centre of gravity is at the intersection of the diagonals joining the corners of opposite faces, at a height of $x = h/2$, above the base.

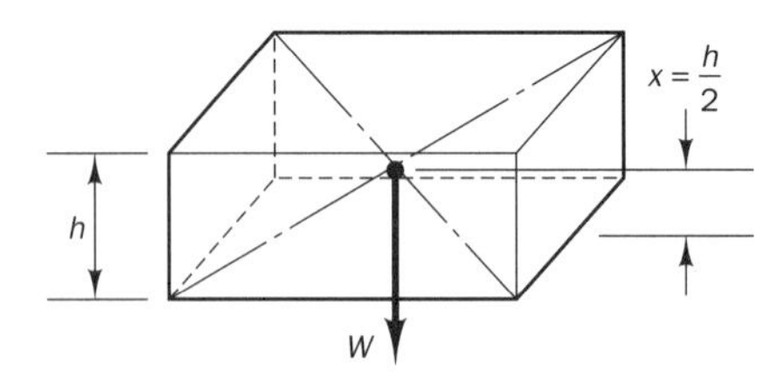

Figure 3.14 Centre of gravity – rectangular prism

2. **Sphere** (Fig. 3.15) – Centre of gravity is at the intersections of the diameters, at a height of $x = d/2$, above the base.

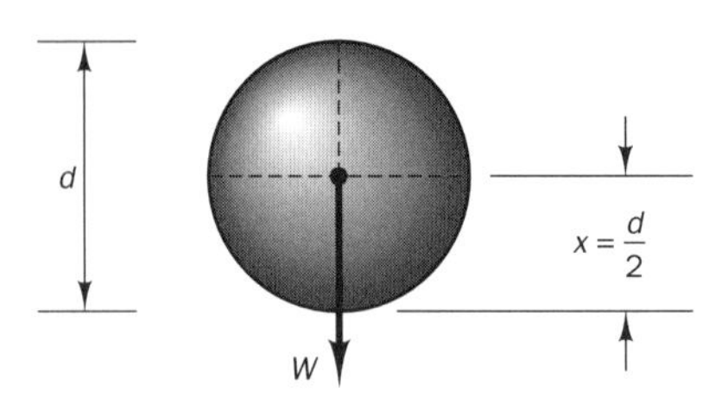

Figure 3.15 Centre of gravity – sphere

3. **Right circular cone** (Fig. 3.16) – Centre of gravity lies on the vertical centre line, at a height of $x = h/4$, above the base.

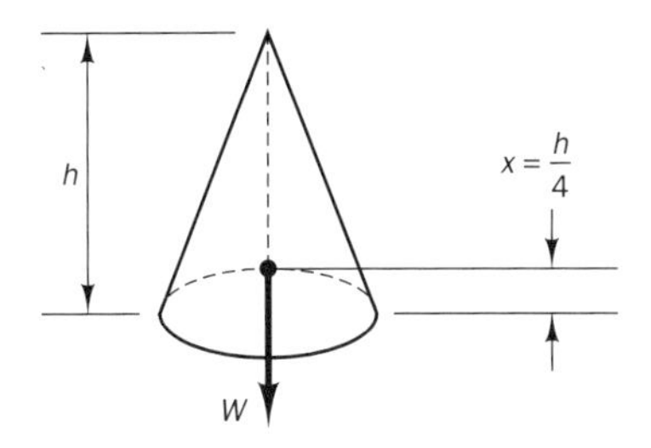

Figure 3.16 Centre of gravity – cone

4. **Hemisphere** (Fig. 3.17) – Centre of gravity lies on the vertical centre line at a height of $x = 3r/8$, above the base.

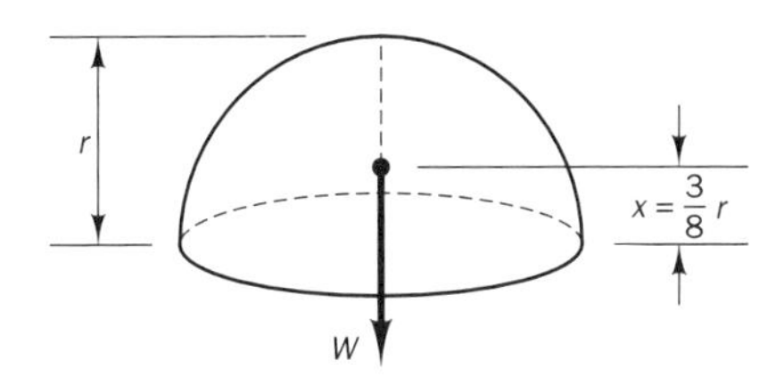

Figure 3.17 Centre of gravity – hemisphere

States of equilibrium

A body or object which is at rest will be in one of three possible states of equilibrium:

- stable equilibrium
- unstable equilibrium
- neutral equilibrium

The three states can be described by considering the different positions in which a cone can be at rest.

Stable equilibrium

A cone is in a state of stable equilibrium when it is resting on its base (Fig. 3.18). If tilted slightly and released it will return to this position. It will only become unstable if it is tilted so that its weight acts outside the base, causing it to topple on to its side.

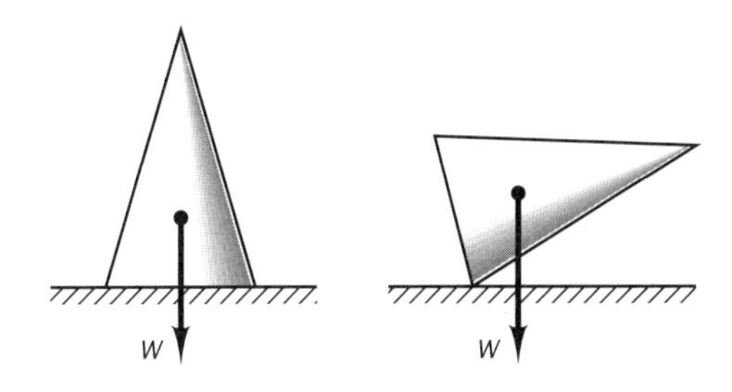

Figure 3.18 Stable equilibrium

Unstable equilibrium

The cone is in a state of unstable equilibrium when balanced on its apex (Fig. 3.19). If displaced slightly, the cone will topple on to its side.

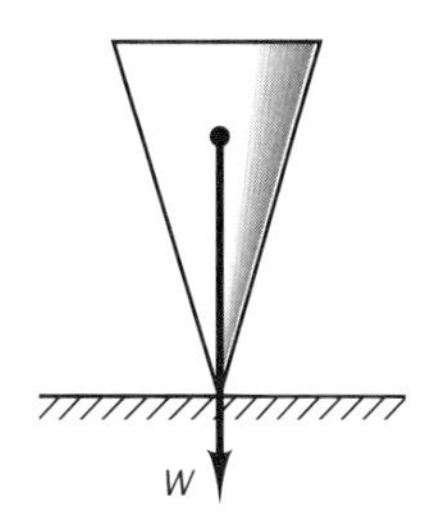

Figure 3.19 Unstable equilibrium

Neutral equilibrium

The cone is in a state of neutral equilibrium when resting on its side (Fig. 3.20). If rolled to a new position, it will stay in that position.

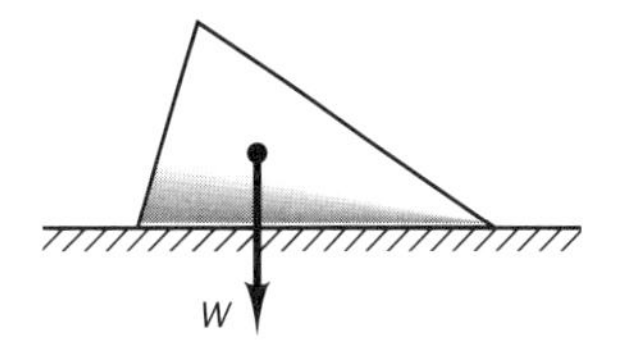

Figure 3.20 Neutral equilibrium

Turning moments

A **turning moment** is a measure of the turning effect which a force has on a body. It is also sometimes called a **torque**. The spanner in Fig. 3.21 has a turning effect on the nut.

The units of turning moments are 'newton-metres' (Nm). The spanner is applying a clockwise turning moment to the nut. Eventually the nut will become tight due to friction between the nut and bolt threads and between the nut face and the component beneath it. Equilibrium will occur when the applied clockwise moment is balanced by the anticlockwise friction moment.

The turning moment, or torque, is found by multiplying the force by the perpendicular distance between its line of action and the point about which it is rotating, i.e. the turning force, measured in newtons, multiplied by its turning radius, measured in metres:

$$\text{Turning moment} = \text{Force} \times \text{Turning radius}$$
$$M = \boldsymbol{Fr} \text{ (Nm)}$$

Engineering systems in static equilibrium

Some engineering systems which are in a state of static equilibrium may now be investigated. Three commonly occurring ones are:

- concurrent coplanar force systems
- balanced moment systems
- elastic spring systems

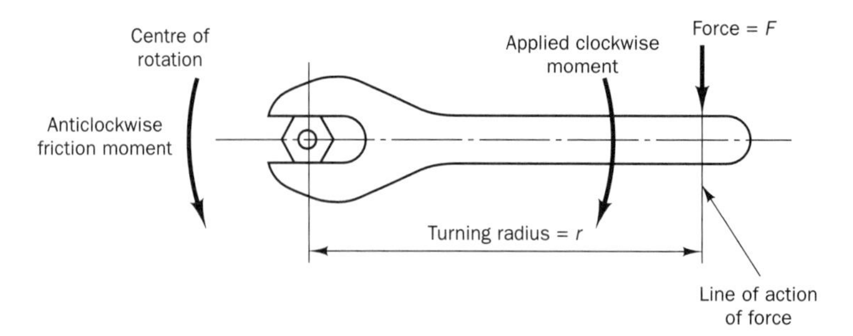

Figure 3.21 Turning moment

Concurrent coplanar force systems

When a number of forces which are all in one plane act on a body, the force system is said to be **coplanar**. If the lines of action of the forces all pass through a single point, the point of concurrence, the system is said to be **coplanar** and **concurrent**. The drawing of the system which shows the values and directions of the forces is called the **space diagram**.

The simplest coplanar force system is when two forces act on a body. For the body to be in equilibrium the forces must be concurrent, equal and opposite (Fig. 3.22).

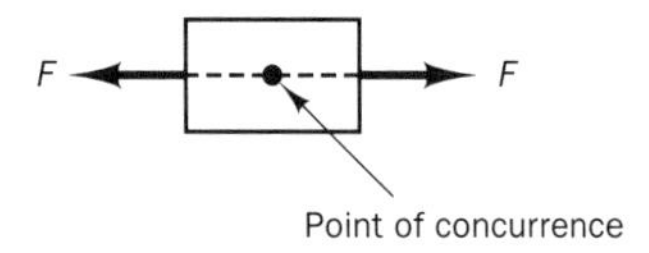

Figure 3.22 Two coplanar forces – body in equilibrium

When three concurrent coplanar forces act on a body, it will be in equilibrium if they are concurrent and if their force vectors form a closed triangle when they are added together. This is done by drawing the force vectors 'end to end' (Fig. 3.23).

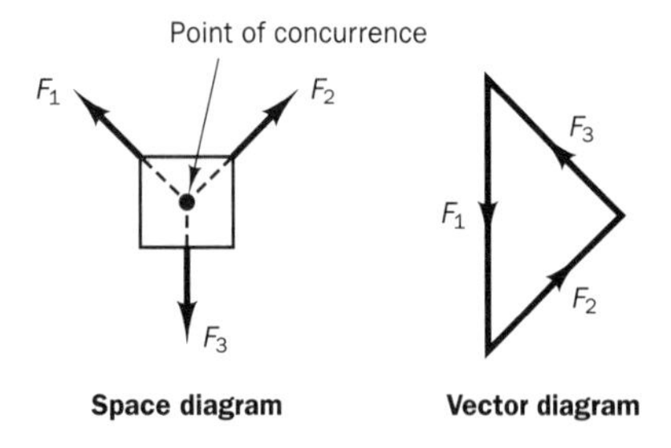

Figure 3.23 Three coplanar forces – body in equilibrium

The **vector diagram** for three concurrent coplanar forces in equilibrium is called the **triangle of forces**. If the magnitude and direction of two of the forces are known, the triangle can be constructed and the magnitude and direction of the remaining force can be found from it.

Alternatively, if the magnitude and direction of one force is known together with the directions of the other two, the triangle can again be constructed and the magnitudes of the unknown forces can be found from it.

Examples

1. Two cables pull on a wall bracket as shown in Fig. 3.24. Find the magnitude and direction of the reaction force exerted by the wall.

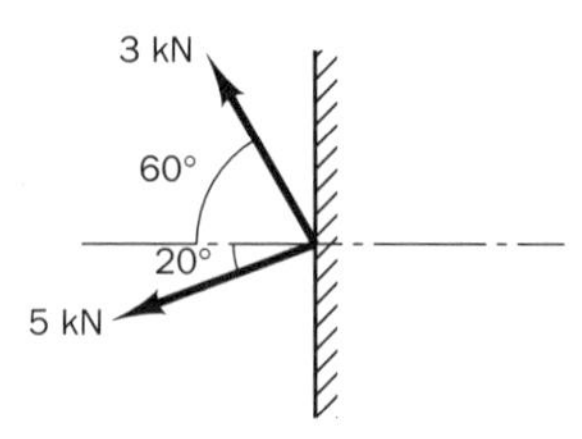

Figure 3.24 Space diagram – Example 1

Answer: Two of the forces are known which enables two sides of the triangle to be drawn. The third side of the triangle gives the magnitude and direction of the reaction of the wall (Fig. 3.25).

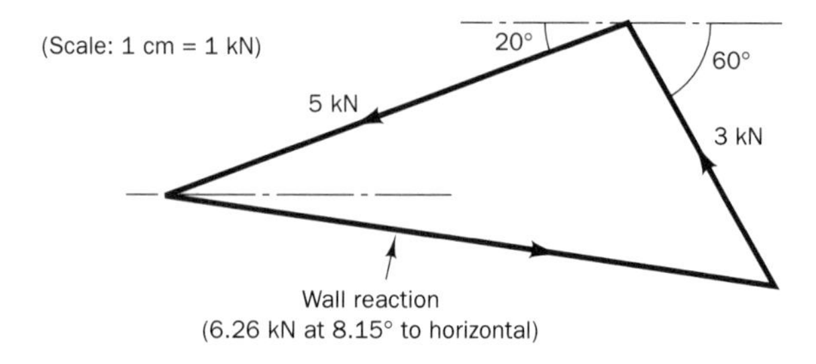

Figure 3.25 Vector diagram – Example 1

2. A crane supports a girder of mass 714 kg on a lifting sling which makes an angle of 45° to the girder as shown in Fig. 3.26. Determine the tension in the sling.

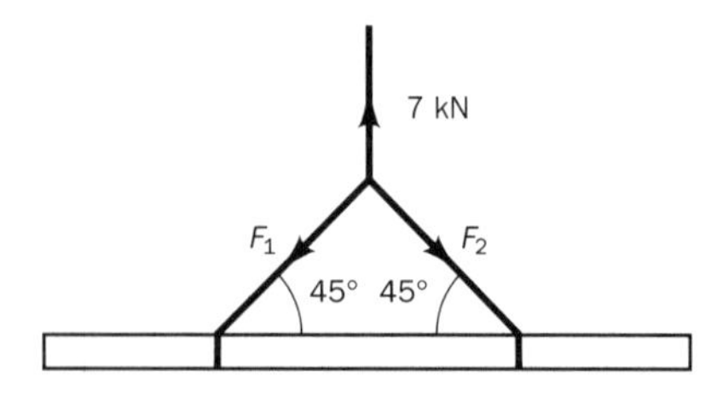

Figure 3.26 Space diagram – Example 2

Answer: Finding the force F_1.

$$F_1 = W = mg$$
$$F_1 = 714 \times 9.81 = 7 \times 10^3\ \text{N}$$
$$F_1 = \mathbf{7\ kN}$$

Here, one force is known and its vector can be drawn. The directions of the other two force vectors can then be drawn from each end of it as construction lines (Fig. 3.27). The intersection fixes the other two sides of the triangle which give the magnitude and direction of the forces in the sling.

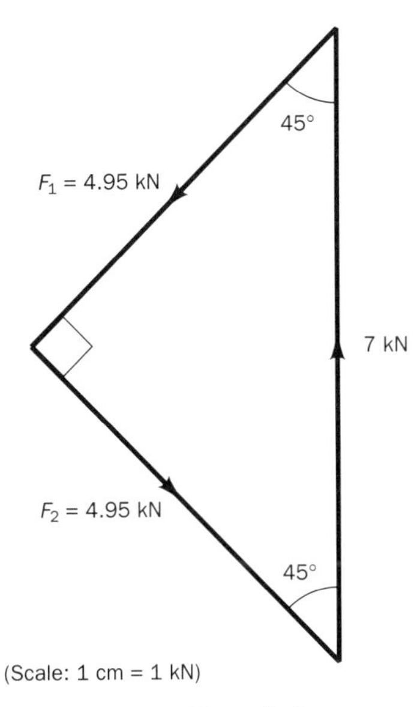

Figure 3.27 Vector diagram – Example 2

Self-assessment tasks 3.10

1. Find the magnitude and direction of the reaction of the wall in the following coplanar force system.

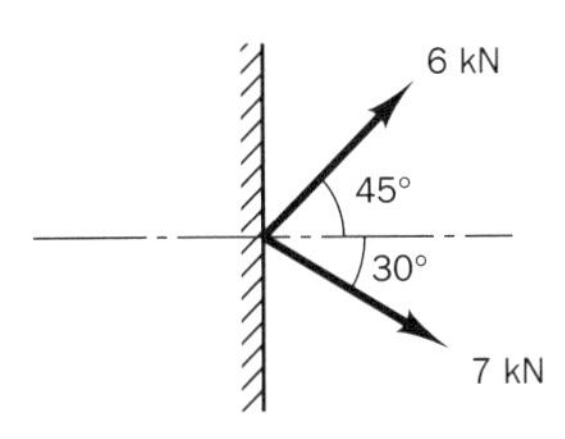

2. A load of mass 714 kg is suspended from a beam by two chains which make an angle of 30° with the beam. Find the tension in the chains.

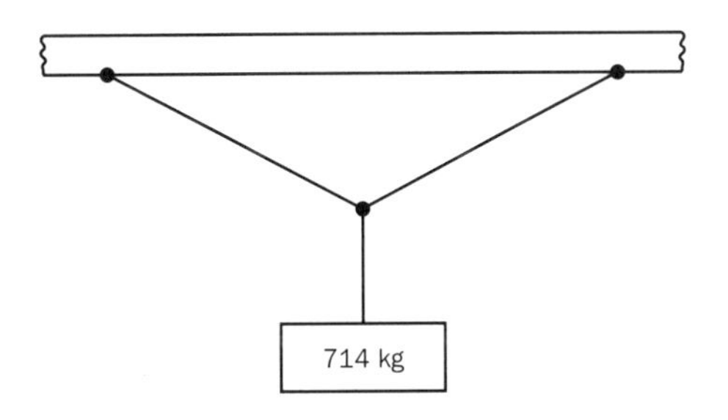

3. A load of 612 kg is hung from a wall mounted jib as shown below. Determine the force carried in the two members of the jib.

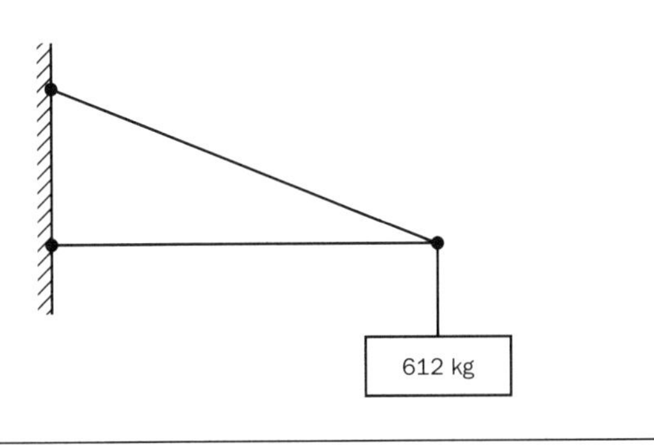

Balanced moment systems

The second of the two conditions for equilibrium is sometimes called the **principle of moments**. It states that 'for a body to be in static equilibrium under the action of a member of forces, the sum of clockwise turning moments taken about any convenient point must equal the sum of the anticlockwise turning moments about the same point.

This principle can be applied to solve problems where it is required to find a turning force which is needed to hold a system in equilibrium.

Examples

1. Find the total force which is required at the handles of the wheelbarrow in Fig. 3.28 which will just raise the load. Find also the load acting on the axle of the wheel.

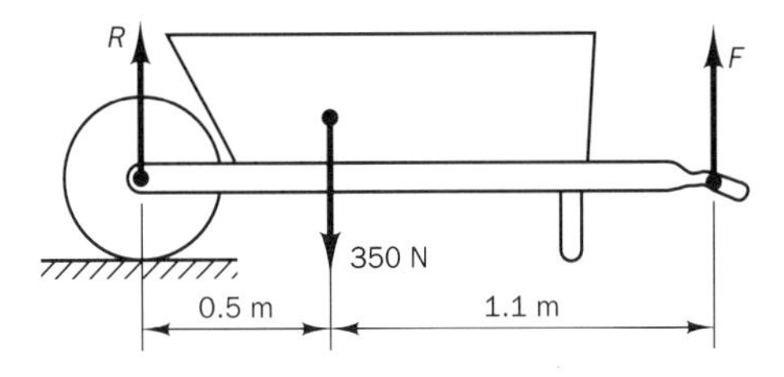

Figure 3.28 Space diagram – Example 1

Answer:
For equilibrium:

$$\text{Clockwise turning moments about centre of wheel} = \text{Anticlockwise turning moments about centre of wheel}$$

$$350 \times 0.5 = F \times 1.6$$
$$175 = F \times 1.6$$
$$\frac{175}{1.6} = F$$
$$\mathbf{F = 109\,N}$$

For equilibrium:

$$\text{Total upward force on barrow} = \text{Total downward force on barrow}$$

$$R + F = W$$
$$R + 109 = 350$$
$$R = 350 - 109$$
$$\mathbf{R = 241\,N}$$

2. Find the missing force F which is needed to hold the following balanced beam in equilibrium in Fig. 3.29. Find also the total force acting on the pivot.

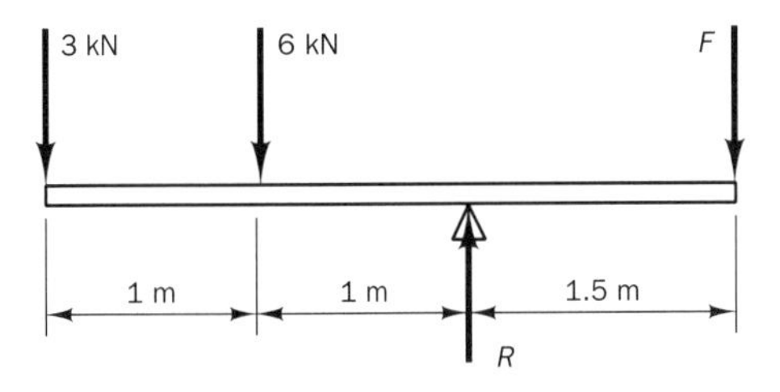

Figure 3.29 Space diagram – Example 2

Answer:
For equilibrium:

$$\text{Clockwise turning moments about pivot} = \text{Anticlockwise turning moments about pivot}$$

$$(F \times 1.5) = (3 \times 2) + (6 \times 1)$$
$$1.5F = 6 + 6$$
$$1.5F = 12$$
$$F = \frac{12}{1.5}$$
$$F = \mathbf{8\,kN}$$

For equilibrium:

$$\text{Total upward force on beam} = \text{Total downward force on beam}$$

$$R = 3 + 6 + 8$$
$$R = \mathbf{17\,kN}$$

Self-assessment tasks 3.11

1. Determine the total horizontal force needed on the handles of the sack truck to raise the load. What is the vertical load on the axle of the truck?

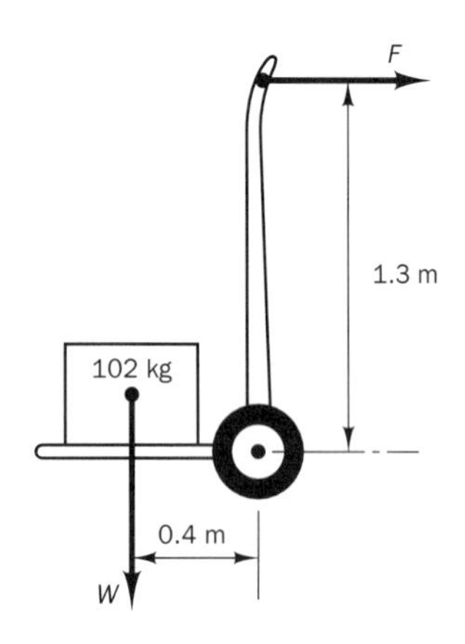

2. What is the force exerted on each side of the bar which is gripped in the tongs, and what is the vertical force exerted on each side of the pivot?

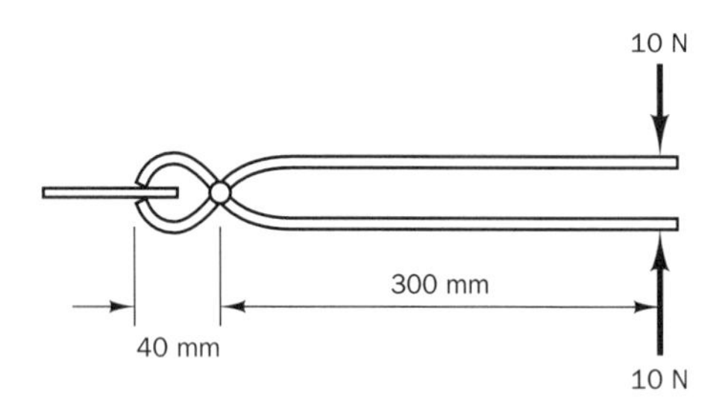

3. Determine the force F which is needed to balance the beam. Find also the total load acting on the pivot.

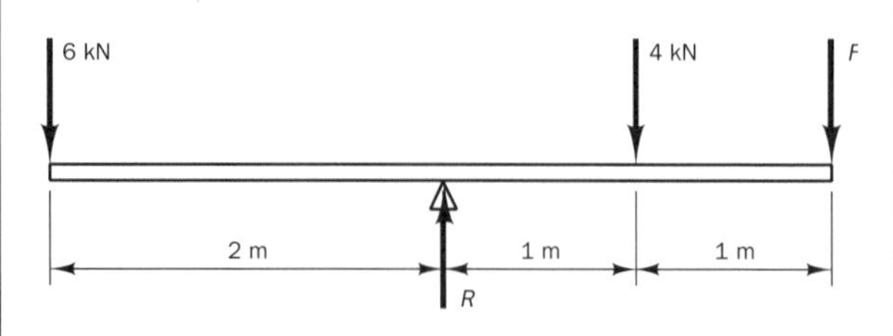

Elastic systems

Elasticity is a physical property which engineering materials possess in varying degrees. The property was first defined by Robert Hooke about two hundred years ago. Hooke's law states that for an elastic material the amount of deformation is proportional to the load applied.

This means that for an elastic material in tension, a doubling of the applied load produces double the extension and three times as much load will produce three times as much extension, etc. A further condition for a material to be elastic is that when the load is removed, the material must return to its original dimensions.

When a graph of load against extension is plotted for a specimen of elastic material, or an elastic component such as a spring, the result is a straight line passing through the origin (Fig. 3.30).

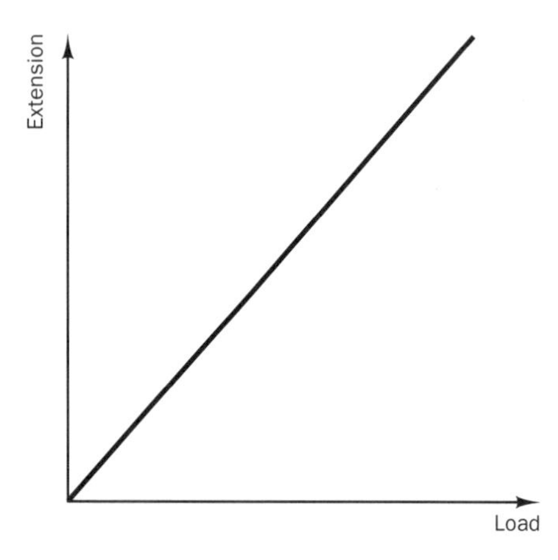

Figure 3.30 Graph of extension against load

The gradient of the graph gives the value of the **stiffness** of the material or spring. This is the force needed to increase the length of that particular piece of material or spring by one unit of length. Stiffness is given the symbol k. The units of stiffness are 'newtons per metre' (Nm^{-1}).

The formula which relates the load, the change in length and the stiffness is:

$$\text{Material stiffness} = \frac{\text{Load applied}}{\text{Change in length}}$$

$$k = \frac{F}{x}\ (\text{Nm}^{-1})$$

The stiffness k is a constant for a given elastic system while the applied load F, and the change in length x, are the variables.

Springs are often listed in catalogues by their stiffness values. For small springs this may be in newtons per mm. The stiffness of a spring is also sometimes called the **spring rate**.

Examples

1. The length of a motor vehicle suspension spring decreases by 3 mm when a load of 500 newtons is applied. Calculate the stiffness of the spring and the extension which a load of 1200 newtons will produce.

Answer:
Finding spring stiffness:

$$k = F = \frac{500}{3 \times 10^{-3}}$$

$$\mathbf{k = 167 \times 10^3\ Nm^{-1}}$$

Finding extension for 1200 N load:

$$x = \frac{F}{k} = \frac{1200}{167 \times 10^3}$$

$$\mathbf{x = 7.19 \times 10^{-3}\ m} \quad \text{or} \quad \mathbf{7.19\ mm}$$

2. An engine valve spring extends by a distance of 2.5 mm when a load of 90 newtons is applied. Calculate its stiffness and the load applied when the compression is 4.0 mm.
Answer:
Finding spring stiffness:

$$k = \frac{F}{x} = \frac{90}{2.5 \times 10^{-3}}$$

$$\mathbf{k = 36 \times 10^3\ Nm^{-1}}$$

Finding load when extension is 4.0 mm:

$$F = kx = 36 \times 10^3 \times 4.0 \times 10^{-3}$$
$$\mathbf{F = 144\ N}$$

Self-assessment tasks 3.12

1. A rubber mounting block is compressed by 1.5 mm when carrying a load of 450 N. Calculate its stiffness and the load which it is carrying when it is compressed by 2.5 mm.
2. In a test on a tension spring, a load of 600 N was seen to produce an extension of 15 mm when hung from its lower end. Calculate the stiffness of the spring and the extension which will be produced when carrying a mass of 85 kg.
3. The following test results were obtained from a tensile test on a component made from an elastic material.

Load (N)	100	150	200	250	300	350	400	450
Extension (mm)	0.80	1.18	1.60	2.00	2.42	2.80	3.19	3.61

Plot a graph of load against extension and from it determine the stiffness of the material specimen. Use this value to calculate the likely extension if a mass of 50 kg is to be carried by the component.

3.4 Dynamic systems

Together with his discovery of the laws of gravity, Sir Isaac Newton put forward three laws which apply to moving bodies. They are known as Newton's Laws of Motion and can be stated as follows.

1. **A body will stay in a state of rest, or travel on with a uniform velocity, until acted upon by an external force.** This means that a body which is at rest tends to stay at rest and a force must be applied to start it moving. Once moving however, a body tends to stay moving and a force has to be applied to stop it. The larger the mass of a body, the larger are the starting and stopping forces needed. This resistance to change is called the **inertia** of a body.
2. **The acceleration of a body is proportional to the applied force.** This law states that if a force is applied to a body it will accelerate, and if the force is doubled, the acceleration will double, i.e. acceleration is directly proportional to the applied force. The formula connecting force, mass and acceleration which newton derived is:

   $$\text{Accelerating force} = \text{Mass} \times \text{Acceleration}$$
   or
   $$F = ma$$

   This is the formula which is used to define the newton as the SI unit of force. It is the force needed to accelerate a mass of 1 kg so that its velocity increases at the rate of 1 metre per second per second. The units of acceleration are normally written as ms^{-2}.

   Except when a body is travelling freely out in space, there is always some resistance to motion caused by friction or air resistance. If this is small it is often neglected but sometimes the additional force needed to overcome friction must be added to the above value of accelerating force:

   $$\textbf{Total applied force} = \textbf{Accelerating force} + \textbf{Force to overcome friction}$$
3. **To every action there is an equal and opposite reaction.** This means that when a force is applied to a body, the body pushes back with an equal and opposite force. The applied force is called an **active force**, while the force of the body pushing back is called a reactive force.

Before investigating dynamic engineering systems it is necessary to define what is meant by:

- work
- power
- energy
- system efficiency

Work

When a force acts on a body and causes it to move, work is done. The work done is defined as the force multiplied by the distance through which it moves along its 'line of action'. The SI unit of work is the joule, where 1 joule is the work done when a force of 1 newton moves a body through a distance of 1 metre along its line of action.

$$\text{Work done} = \text{Force} \times \text{Distance moved}$$
$$W = Fs \text{ (joules)}$$

If the force is constant, the distance moved and work done will be variables in the relationship.

Power

Work can be done slowly or quickly. The amount of work done per second, or the rate of doing work, is the **power** which is developed.

The SI unit of power is the **watt** which is defined as the power developed when work is being done at the rate of 1 joule per second.

$$\text{Power} = \frac{\text{Work done}}{\text{Time taken}} = \frac{w}{t}$$

or

$$\text{Power} = \frac{Fs}{t} \text{ (watts)}$$

In this relationship the force is again constant while distance travelled, the time taken and the power developed are the variables.

Now $s/t = v$, the average velocity of the moving body. The power can thus also be calculated using the formula:

$$\text{Power} = Fv \text{ (watts)}$$

If the force is constant and just sufficient to overcome friction, the body will move with a constant velocity and so the power will also be constant.

If, however, the force is causing the body to accelerate, the velocity will be increasing and the power will be increasing with it, i.e. the velocity and the power will be variables.

Energy

When a force moves and does work **energy** is being changed from one form to another. Energy is stored work, or the capacity of a body or system to do work. In motor vehicles the energy is initially stored in the fuel as chemical energy. This is changed to heat energy as it burns and some of it is then changed into mechanical energy as the hot gases of combustion push the pistons down the engine cylinders. Moving objects are said to have a form of mechanical energy known as **kinetic energy**.

The kinetic energy of a motor vehicle is work which has been done by the accelerating force which has accelerated the vehicle from rest to its final velocity. When the brakes are applied to bring the vehicle to rest, all of the kinetic energy is changed into heat energy and sound given off to the atmosphere by the hot brake drums and discs.

In the case of a hoist which is driven by an electric motor, the electrical energy supplied to the motor is changed to mechanical energy. A force is exerted which raises the load on the hoist against the force of gravity and work is done.

In its raised position, the load is said to have a form of mechanical energy known as gravitational **potential energy**. If the load is allowed to fall freely, this is changed into kinetic energy as it picks up speed. Finally as it hits the ground, the kinetic energy is given up to the ground and the surrounding atmosphere, in the form of heat energy and sound.

System efficiency

Systems which convert energy into a more useful form always waste some of it. It is usually lost in the form of heat. In mechanical systems heat is generated due to friction between the moving parts. In electrical systems, heat is generated as the current flows through the conductors and circuit components.

The **efficiency** of a system is a measure of its effectiveness in converting energy from one form to another which is more useful. It is calculated using the formulae:

$$\text{Efficiency} = \frac{\text{Energy output from system}}{\text{Energy input to system}}$$

$$\text{or Efficiency} = \frac{\text{Work output from system}}{\text{Work input to system}}$$

$$\text{or Efficiency} = \frac{\text{Output power from system}}{\text{Input power to system}}$$

The calculated value of efficiency may be given as a decimal fraction or as a percentage, i.e. 'efficiency = 0.9 or 90 per cent' means that nine tenths of the input power or energy has been usefully converted.

Examples

1. A force of 20 N is required to lift a casting through a distance of 4.5 metres in a time of 3.5 seconds. Find (a) the work done; (b) the power developed.
   Answer:
   (a) Finding work done:

   $$W = Fs = 20 \times 4.5$$
   $$\mathbf{W = 90\,J}$$

   (b) Finding power developed:

   $$\text{Power} = \frac{Fs}{t} = \frac{W}{t}$$
   $$\text{Power} = \frac{90}{3.5}$$
   $$\mathbf{Power = 25.7\,W}$$

2. A cutting tool moves at a steady speed of $0.5\ \text{ms}^{-1}$ with a force of 150 N. What is (a) the power developed; (b) the work done by the tool in three seconds?
   Answer:
   (a) Finding power developed:

   $$\text{Power} = Fv = 150 \times 0.5$$
   $$\mathbf{Power = 75\,W}$$

   (b) Finding work done in three seconds.

   $$\text{Since} \quad \text{Power} = \frac{\text{Work done}}{\text{Time taken}}$$
   $$\text{Work done} = \text{Power} \times \text{Time taken}$$
   $$= 75 \times 3$$
   $$\mathbf{W = 225\,J}$$

3. The input power to a robot is 500 W. If its output arm applies a force of 300 N and moves at a steady speed of $1.5\ \text{ms}^{-1}$, calculate (a) the output power; (b) the efficiency of the robot.
   Answer:
   (a) Finding output power:

   $$\text{Output power} = Fv = 300 \times 1.5$$
   $$\mathbf{Output\ power = 450\,W}$$

   (b) Finding efficiency:

   $$\text{Efficiency} = \frac{\text{Output power}}{\text{Input power}}$$
   $$= \frac{450}{500}$$
   $$\mathbf{Efficiency = 0.9} \quad \text{or} \quad 90\,\%$$

Self-assessment tasks 3.13

1. A force of 75 N is applied by an actuator to move a component through a distance of 0.5 m in a time of 1.25 seconds. Calculate (a) the work done; (b) the power output of the actuator.
2. A hoist raises a load of 2.5 kN at a steady speed of $1.5\ \text{ms}^{-1}$ through a height of 12 metres. Find (a) the work done; (b) the power output of the hoist.
3. A hydraulic jack requires an input force of 40 N which moves through a distance of 0.5 metre to raise a load of 2 kN through a distance of 8 mm. Find (a) the work input; (b) the work output; (c) the efficiency of the jack.
4. The input power to a cutting tool is 750 W. If the tool exerts a cutting force of 5 kN and moves through a distance of 0.5 metre in 8 seconds, determine (a) the work done; (b) the output power; (c) the efficiency of the system.

It may now be appropriate to investigate some engineering examples of dynamic systems. They can be classified into:

- uniform velocity systems
- uniform acceleration systems

Uniform velocity systems

A spacecraft, travelling in the vacuum of space, far away from the Earth's atmosphere or gravity, experiences no resistance to its motion. When its rocket motors are fired, the spacecraft accelerates and when they are shut off, it continues on at a steady speed in a straight line. It obeys Newton's First Law of Motion exactly and will only be slowed down if the rocket motors are fired in the reverse direction.

Objects in motion on the Earth's surface always experience some resistance to motion. This might be due to friction between sliding and rolling surfaces or air resistance. When friction is present a body will move with uniform velocity when the force acting upon it is just sufficient to overcome the frictional resistance. If the active force is less, it will slow down and if it is more, it will accelerate.

Velocity is measured in metres per second (ms^{-1}) or kilometres per hour (kmh^{-1}). To convert the velocity of a body from kilometres per hour to metres per second, multiply by 1000 to change kilometres to metres and divide by 3600 (60^2) to change hours to seconds:

$$\text{Velocity in } ms^{-1} = \text{Velocity in } kmh^{-1} \times \frac{1000}{60^2}$$

or $$\textbf{Velocity in } \mathbf{ms^{-1}} = \textbf{Velocity in } \mathbf{kmh^{-1}} \div 3.6$$

To convert back again use the formula:

$$\textbf{Velocity in } \mathbf{kmh^{-1}} = \textbf{Velocity in } \mathbf{ms^{-1}} \times 3.6$$

If a graph is plotted of distance s against time t for a body travelling with constant velocity, it will be a straight line which passes through the origin (Fig. 3.31).

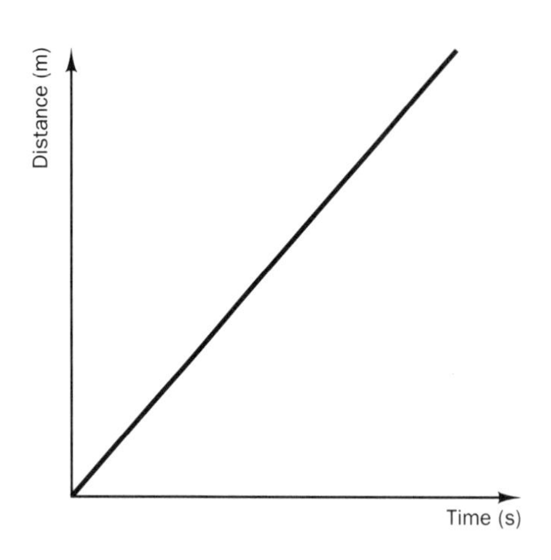

Figure 3.31 Graph of distance against time

The gradient of the graph gives the value of the velocity of the body, which is the distance travelled for each unit of time.

The general formula for calculating average velocity is:

$$v = \frac{s}{t} \quad (ms^{-1} \text{ or } kmh^{-1})$$

In this relationship, the velocity is constant and the distance travelled and the time taken are variables.

Examples

1. A car travels a distance of 200 metres in a time of 12 seconds. What is its average velocity in (a) metres per second; (b) kilometres per hour?
   Answer:
   (a) Finding velocity in ms^{-1}:

$$v = \frac{s}{t}$$
$$v = \frac{200}{12}$$
$$\mathbf{v = 16.7\,ms^{-1}}$$

   (b) Finding velocity in kmh^{-1}:

$$v\,km^{-1} = v\,ms^{-1} \times 3.6$$
$$v = 16.7 \times 3.6$$
$$\mathbf{v = 60.1\,kmh^{-1}}$$

2. A conveyor belt travels at a uniform velocity of 0.25 metres per second. Calculate (a) the time taken to transport a component through a distance of 18 metres; (b) the distance travelled by a component in 1.5 minutes.
   Answer:
   (a) Finding time to move component through 18 m:

$$\text{Since } v = \frac{s}{t}$$
$$t = \frac{s}{v}$$
$$t = \frac{18}{0.25}$$
$$\mathbf{t = 72\,s}$$

   (b) Changing 1.5 minutes to seconds:

$$t = 1.5 \times 60 = 90\,s$$

   Finding distance travelled in 90 s:

$$\text{Since } v = \frac{s}{t}$$
$$s = vt$$
$$s = 0.25 \times 90$$
$$\mathbf{s = 22.5\,m}$$

3. A robot arm exerts a force of 50 newtons to move a component through a horizontal distance of 1.5 metres at a uniform velocity in a time of 3.5 seconds. Calculate (a) the velocity; (b) the work done; (c) the power developed.
   Answer:
   (a) Finding velocity of robot arm:

$$v = \frac{s}{t}$$
$$v = \frac{1.5}{3.5}$$
$$\mathbf{v = 0.429\,ms^{-1}}$$

   (b) Finding the work done:

$$\text{Work done} = Fs$$
$$\text{Work done} = 50 \times 1.5$$
$$\textbf{Work done} = \mathbf{75\,J}$$

   (c) Finding the power developed:

$$\text{Power} = \frac{Fs}{t} = \frac{w}{t}$$
$$\text{Power} = \frac{75}{3.5}$$
$$\textbf{Power} = \mathbf{21.4\,W}$$

4. A hoist, driven by an electric motor, raises a load of 2 kN through a distance of 3 metres with uniform velocity in a time of 15 seconds. If the input power to the motor is 0.5 kW determine (a) the velocity; (b) the work done; (c) the power output; (d) the efficiency of the system.
   Answer:
   (a) Finding the velocity:

$$v = \frac{s}{t}$$
$$v = \frac{3}{15}$$
$$\mathbf{v = 0.2\,ms^{-1}}$$

(b) Finding the work done:

$$\text{Work done} = Fs$$
$$\text{Work done} = 2 \times 10^3 \times 3$$
$$\textbf{Work done} = \mathbf{6 \times 10^3\,J} \quad \text{or} \quad \mathbf{6\,kJ}$$

(c) Finding the power output:

$$\text{Power} = \frac{Fs}{t} = \frac{w}{t}$$
$$\text{Power} = \frac{6 \times 10^3}{15}$$
$$\textbf{Power} = \mathbf{400\,W}$$

(d) Finding the efficiency:

$$\text{Efficiency} = \frac{\text{Output power}}{\text{Input power}}$$
$$\text{Efficiency} = \frac{400}{0.5 \times 10^3}$$
$$\textbf{Efficiency} = \mathbf{0.8} \quad \text{or} \quad \mathbf{80\%}$$

Self-assessment tasks 3.14

1. A lift cage is raised at a steady rate through a height of 18 metres in a time of 7.5 seconds. Calculate (a) its velocity; (b) the distance travelled in 12.5 seconds.
2. A train travels at a steady speed of 110 kilometres per hour. Calculate (a) its velocity in metres per second; (b) the time taken to cover a distance of 250 metres; (c) the distance travelled in a time of 30 seconds.
3. Copper wire is formed by drawing it through a die. The force required is 210 newtons and it travels through the die at a speed of 1.5 metres per second. Calculate (a) the time taken to fill a roll containing 150 metres; (b) the work done; (c) the power developed.
4. A hoist raises a load of 150 kg at a steady speed of $0.75\,ms^{-1}$ through a distance of 3.5 metres. Calculate (a) the time taken; (b) the output power; (c) the efficiency if the input power is 1.5 kW.

Uniform acceleration systems

Newton's First Law of Motion states that a body continues in a state of rest or travels with a uniform velocity unless it is acted upon by some external force. The external force can be frictional resistance, which slows the body down, or an accelerating force which speeds it up. Newton's Second Law gives the relationship between the accelerating force and the acceleration which it produces in the formula,

$$\text{Force} = \text{Mass} \times \text{Acceleration}$$

or

$$F = ma$$

If the mass is constant, the force and the acceleration will be the variables in this relationship.

Examples

1. If there is no frictional resistance, calculate the force needed to accelerate a space vehicle of mass 2400 kg so that it increases its velocity at the rate of $20\,ms^{-2}$. If the mass is reduced to 1200 kg after releasing empty fuel tanks what will be its acceleration if the same force is applied?

   Answer:
   Finding accelerating force:

   $$F = ma$$
   $$F = 2400 \times 20$$
   $$\mathbf{F = 48 \times 10^3\,N}$$

   Finding new acceleration:

   Since $F = ma$

   $$a = \frac{F}{m}$$
   $$a = \frac{48 \times 10^3}{1200}$$
   $$a = 40\,ms^{-2}$$

   That is, if the mass is halved the same force will produce twice as much acceleration.

2. An automatic guided vehicle of mass 500 kg is propelled by a tractive force of 300 newtons. If frictional resistance is neglected, what acceleration will this produce, and what will be the force needed to give the same vehicle twice as much acceleration?

   Answer:
   Finding acceleration produced:

   Since $F = ma$

   $$a = \frac{F}{m}$$
   $$a = \frac{300}{500}$$
   $$a = \mathbf{0.6\,ms^{-2}}$$

   Finding force needed for twice the acceleration, i.e. $2 \times 0.6 = 1.2\,ms^2$:

   $$F = ma$$
   $$F = 500 \times 1.2$$
   $$\mathbf{F = 600\,N}$$

   That is, if the force is doubled, the acceleration is doubled.

Suppose a body travelling at the steady velocity of $u\,ms^{-1}$ is accelerated uniformly for a period of t seconds until its velocity has increased to $v\,ms^{-1}$. Its motion can be plotted on a graph of velocity against time (Fig. 3.32).

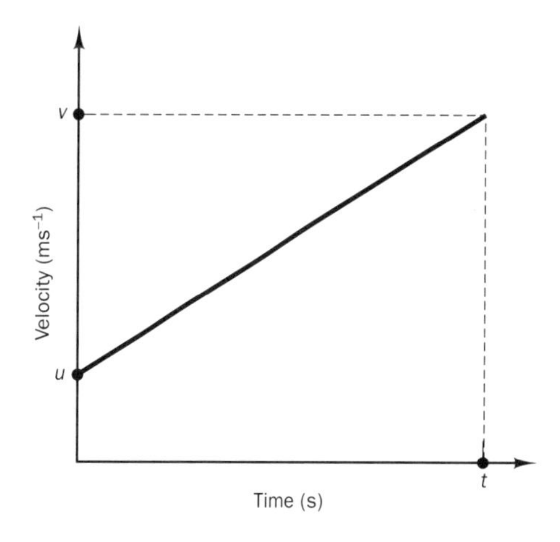

Figure 3.32 Graph of velocity against time

As can be seen, there is a linear relationship between velocity and time. The intercept on the velocity axis is the initial velocity u. The acceleration a which is the increase in velocity each second, is given by the gradient of the graph.

$$a = \frac{v - u}{t} \quad \text{or} \quad v = u + at$$

In this relationship, the initial velocity u and the acceleration a are constants. The time taken t and the final velocity v are variables.

Example

1. A pneumatic actuator exerts a force of 25 N on a component which has a mass of 2 kg and which is initially at rest. If frictional resistance is neglected, and the force acts for 0.8 second, calculate (a) the acceleration produced; (b) the final velocity of the body.
   Answer:
   (a) Finding acceleration:

   $$a = \frac{F}{m}$$
   $$a = \frac{25}{2}$$
   $$a = \mathbf{12.5\ ms^{-2}}$$

   (b) Finding final velocity:

   $$v = u + at$$
   $$v = 0 + (12.5 \times 0.8)$$
   $$v = \mathbf{10\ ms^{-1}}$$

2. A carriage on a fairground ride has a mass of 600 kg and is accelerated from a velocity of 10 ms^{-1} to 30 ms^{-1} in a time of 12 seconds. If the effects of friction are neglected, calculate (a) the acceleration; (b) the accelerating force.
   Answer:
   (a) Finding the acceleration:

   $$a = \frac{v - u}{t}$$
   $$a = \frac{30 - 10}{12}$$
   $$a = \frac{20}{12}$$
   $$a = \mathbf{1.67\ ms^{-2}}$$

   (b) Finding the accelerating force:

   $$F = ma$$
   $$F = 600 \times 1.67$$
   $$F = \mathbf{1000\ N} \quad \text{or} \quad \mathbf{1\ kN}$$

3. A motor vehicle of mass 750 kg and travelling at a velocity of 15 ms^{-1} is retarded uniformly to a velocity of 3 ms^{-1} in a time of 6.5 seconds. Calculate (a) its retardation; (b) the retarding force.
   Answer:
   (a) Finding the retardation:

   $$a = \frac{v - u}{t}$$
   $$a = \frac{3 - 15}{6.5}$$
   $$a = \mathbf{-1.85\ ms^{-2}}$$

   Note: The calculated value of the acceleration is always negative if retardation is taking place.

   (b) Finding retarding force:

   $$F = ma$$
   $$F = 750 \times -1.85$$
   $$F = \mathbf{-1.39 \times 10^3\ N} \quad \text{or} \quad \mathbf{-1.39\ kN}$$

   The negative sign may be omitted if it is clearly stated that the force is a retarding or braking force.
   Note: A velocity against time graph, like that shown above, can also be used to calculate the distance travelled by a body in a given time. It can be shown that for any such graph, the distance travelled is given by the area under the curve. For dynamic systems moving with uniform acceleration or uniform velocity, the area can be divided up into triangles and rectangles whose areas can be calculated separately and then totalled, to give the total distance travelled.

4. A motor vehicle travelling at a velocity of 18 kilometres per hour is accelerated uniformly to 72 kilometres per hour in a time of 12 seconds. Show this on a velocity against time graph and from it calculate (a) the acceleration; (b) the distance travelled while accelerating (see Fig. 3.33).

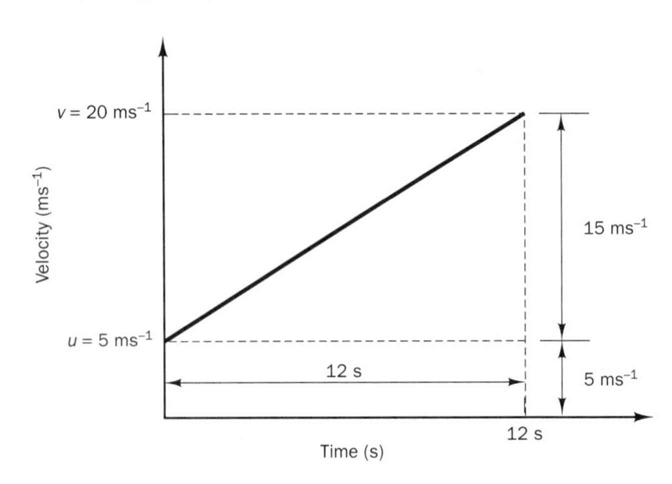

Figure 3.33 Graph of velocity against time

   Answer:
   To change velocities to ms^{-1}, divide by 3.6:

   $$u = \frac{18}{3.6} = 5\ ms^{-1}$$
   $$v = \frac{72}{3.6} = 20\ ms^{-1}$$

   (a) Finding acceleration:

   Acceleration = Gradient of graph

   $$a = \frac{20 - 5}{12}$$
   $$a = \frac{15}{12}$$
   $$a = \mathbf{1.25\ ms^{-2}}$$

   (b) Finding distance travelled:

   Distance travelled = Area under graph

   $$s = (5 \times 12) + (\tfrac{1}{2} \times 12 \times 15)$$
   $$s = \mathbf{150\ m}$$

5. A hoist raises a load with uniform acceleration from rest to a steady velocity of 2.5 ms^{-1} in a time of 4 seconds. It stays at this speed for 3 seconds after which it is retarded uniformly to rest in a further period of 6 seconds. Draw the velocity against time graph and from it calculate (a) the acceleration; (b) the retardation; (c) the total distance travelled (see Fig. 3.34).

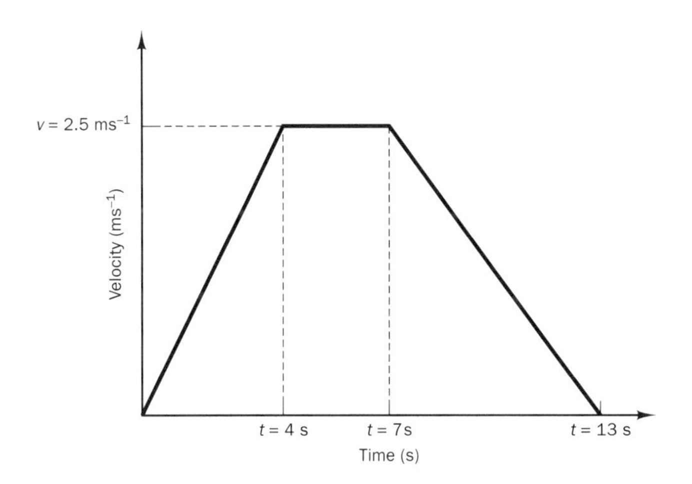

Figure 3.34 Graph of velocity against time

Answer:
Graph of velocity against time:
(a) Finding acceleration (let this be a_1 ms^{-2}):

Acceleration = Gradient of graph for first 4 seconds

$$a_1 = \frac{2.5}{4}$$

$$a_1 = \mathbf{0.625\ ms^{-2}}$$

(b) Finding retardation (let this be a_2 ms^{-2}):

Retardation = Gradient of graph for last 6 seconds

$$a_2 = \frac{-2.5}{6}$$

$$a_2 = \mathbf{-0.412\ ms^{-2}}$$

(c) Finding total distance travelled:

Distance travelled = Area under graph

$$s = (\tfrac{1}{2} \times 4 \times 2.5) + (3 \times 2.5) + (\tfrac{1}{2} \times 6 \times 2.5)$$
$$s = 10 + 7.5 + 7.5$$
$$s = \mathbf{25\ m}$$

Self-assessment tasks 3.15

1. A machine worktable of mass 55 kg is accelerated from rest at the rate of 4 ms^{-2}. If frictional resistance is ignored, calculate (a) the accelerating force; (b) the work done in moving the worktable through a distance of 0.75 metres; (c) the average power developed if this takes 1.5 seconds.
2. A vehicle of mass 750 kg is accelerated uniformly from an initial velocity of 12 ms^{-1} to a final velocity of 35 ms^{-1} in a time of 11 seconds. Neglecting friction, calculate (a) the acceleration; (b) the accelerating force; (c) the maximum power developed.
3. A train travelling at 144 kmh^{-1} has its brakes applied and slows down uniformly to rest in a time of 1 minute. Show this on a velocity against time graph and from it calculate (a) the retardation; (b) the distance travelled in coming to rest.
4. An overhead gantry crane starts from rest and is accelerated uniformly to a velocity of 1.8 ms^{-1} in a time of 6 seconds. It continues at this velocity for a further period of 7 seconds and is then retarded uniformly to rest in a period of 5 seconds. Draw the velocity against time graph and from it calculate (a) the acceleration; (b) the retardation; (c) the total distance travelled.

3.5 Investigate thermal systems in terms of scientific laws and principles

In basic engineering terms, a thermal system can be any application where materials or components receive or lose heat energy.

When heat energy is transferred to or from a substance, one or more of the following changes may occur:

- a change of temperature
- a change of state
- a change of dimension

Other changes such as changes in colour, changes in chemical composition and changes in electrical properties may also occur but will not be described here. Practical thermal systems may be investigated while considering the meaning and definitions of:

- heat energy
- temperature
- specific heat capacity
- specific latent heat
- linear expansivity

Heat energy

Heat and temperature are closely related but should not be confused with each other. Heat energy is measured in **joules**, just like any other form of energy.

The atoms and molecules of all substances are thought to be in a state of motion. In solids they are thought to vibrate about fixed positions while in liquids and gases they are thought to be in a state of random motion. As energy is supplied to a substance, their motion increases and as energy is lost, their motion slows.

Anything which is in motion has kinetic energy. The heat energy content of a substance is the sum of all the separate kinetic energies of its atoms and molecules.

Temperature

Temperature is a measure of the hotness of a substance. It is an indication of the energy level of the individual atoms and molecules. The 'absolute zero' of temperature is when there is no heat energy remaining and all movement of the atoms and molecules has ceased.

Temperature is measured in degrees on the Celsius scale:

$-273\,^{\circ}\mathrm{C}$ = the absolute zero of temperature
$0\,^{\circ}\mathrm{C}$ = the freezing point of water
$100\,^{\circ}\mathrm{C}$ = the boiling point of water at normal atmospheric pressure

It will be remembered however, that the SI unit of temperature is the kelvin, and temperature is also measured on the Kelvin scale. A kelvin is exactly the size as a degree on the Celsius scale. The difference between the two scales is that the kelvin scale starts at the absolute zero of temperature so that:

$0\,\mathrm{K}$ = the absolute zero of temperature
$273\,\mathrm{K}$ = the freezing point of water
$373\,\mathrm{K}$ = the boiling point of water at normal atmospheric pressure

Specific heat capacity

The specific heat capacity c of a substance is the amount of heat energy required to raise the temperature of a mass of 1 kg by 1 kelvin, or 1 °C. Its units are thus joules per kilogram per kelvin ($\mathrm{J kg^{-1} K^{-1}}$).

Typical values for common materials are given in Table 3.4.

Table 3.4

Material	Specific heat capacity ($\mathrm{kJ kg^{-1} K^{-1}}$)
Water	4.187
Ice	2.110
Aluminium	0.896
Steel	0.486
Brass or copper	0.394
Lead	0.130
Gold	0.130

The amount of heat energy Q required to raise the temperature of a mass m kg of a substance from an initial temperature T_1 to a final temperature T_2 is given by the formula:

$$Q = mc(T_2 - T_1) \text{ (joules)}$$

In this relationship the specific heat capacity c is a constant and the other quantities are variables.

If the time taken for the flow of heat energy to take place is t seconds, the rate of energy supply, or the power input, is given by:

$$\textbf{Input power} = \frac{Q}{t} \text{ (watts)}$$

Examples

1. A steel forging of mass 20 kg has its temperature raised from 20 °C to 350 °C. If its specific heat capacity is $486\,\mathrm{J kg^{-1} K^{-1}}$, calculate the heat energy received.
   Answer:

$$Q = mc(T_2 - T_1)$$
$$Q = 20 \times 486 \times (350 - 20)$$
$$Q = \mathbf{3.21 \times 10^6\,J} \quad \text{or} \quad \mathbf{3.21\,MJ}$$

2. A water heater raises the temperature of 12 kg of water at 18 °C to 100 °C in 5 minutes. If the specific heat capacity of the water is 4.187 kJkg^{-1}K^{-1} calculate (a) the amount of heat energy received; (b) the power output.
Answer:
(a) Finding heat energy received:

$$Q = mc(T_2 - T_1)$$
$$Q = 12 \times 4.187 \times 10^3 \times (100 - 18)$$
$$\mathbf{Q = 4.12 \times 10^6\,J \quad or \quad 4.12\,MJ}$$

(b) Finding output power:

$$\text{Output power} = \frac{Q}{t}$$
$$\text{Output power} = \frac{4.12 \times 10^6}{5 \times 60}$$
$$\textbf{Output Power} = \mathbf{13.7 \times 10^3\,W} \quad \text{or} \quad 3.7\,\text{kW}$$

Specific latent heat

When a solid substance such as ice or a metal is heated its temperature rises. It is said to be receiving **sensible heat**, i.e. heat which produces temperature rise.

Eventually a temperature is reached at which the solid cannot absorb any more heat energy without undergoing a change of state, or **phase change**. At this point the material starts to melt, and although heat is still being received, the temperature remains constant until it has become completely molten.

Ice changes to water, although it can change directly into a gas under certain conditions. The process is called **sublimation** but it can only occur with ice under conditions of very low pressure. Other substances however, such as carbon dioxide will change directly from a solid to a gas at normal atmospheric pressure.

The heat received at this constant temperature, which changes the state of a substance, is called **latent heat**, i.e. heat which does not show itself by producing temperature rise.

Latent heat is also being received when a liquid changes into a vapour, i.e. when it evaporates. Figure 3.35 shows a graph of temperature rise against heat energy received as a substance changes from a solid, to a liquid, and then to a gas.

The same amounts of heat energy are given out when a vapour changes back to a liquid, i.e. condenses, and when a liquid solidifies. The temperature again remains constant until the changes are complete.

The latent heat, required to change the state of 1 kg of a substance from solid to liquid at constant temperature, is called its **specific latent heat of fusion**, and that required to change 1 kg from liquid to vapour at constant temperature, is called its **specific latent heat of vaporisation.**

The units of specific latent heat are joules per kilogram (Jkg^{-1}). Some typical values are given in Table 3.5.

Table 3.5

Substance	Specific latent heat of fusion (kJkg^{-1})
Ice	335
Aluminium	387
Copper	180
Cast iron	96
Tin	60
Lead	23
Water	2248
Alcohol	863
Ether	379
Turpentine	309

It should be noted that the latent heat of vaporisation is affected by pressure. The values given are at standard atmospheric pressure.

The total latent heat required to change the state of a mass of m kg of a substance is given by:

$$Q = \boldsymbol{mL} \text{ (joules)}$$

In this relationship the specific latent heat L is constant; the mass m and the heat transferred Q are variables.

In some engineering applications it may be required to calculate the total sensible and latent heat needed to raise the temperature of a substance to its boiling or melting point and to change its state. In such cases the total heat energy Q, is given by:

$$Q = \text{Sensible heat} + \text{Latent heat}$$
$$Q = \boldsymbol{mc}(T_2 T_1) + \boldsymbol{mL} \text{ (joules)}$$

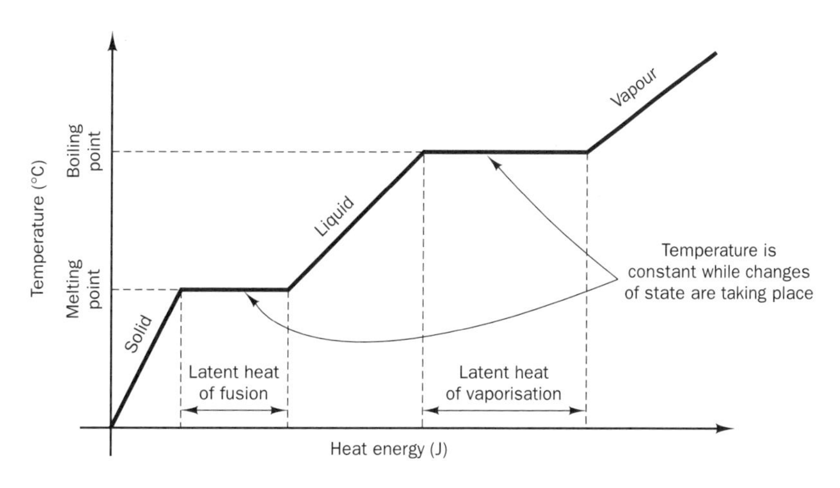

Figure 3.35 Graph of temperature against heat energy received

Examples

1. A boiler contains 150 kg of water at 100 °C. How much heat energy is required to change it completely to steam. The specific latent heat of vaporisation is 2248 kJkg^{-1}.
Answer:
Finding heat energy required.

$$Q = mL$$
$$Q = 150 \times 2248 \times 10^3$$
$$Q = 337 \times 10^6\,\text{J} \quad \text{or} \quad \mathbf{337\,MJ}$$

2. A quantity of aluminium scrap of mass 50 kg and temperature 20 °C is heated to its melting point which is 659 °C. Calculate the heat energy which it has received when melting is complete. The specific heat capacity of aluminium is 896 Jkg^{-1}K^{-1} and its specific latent heat of fusion is 387 kJkg^{-1}.
Answer:
Finding total heat energy received.

$$Q = mc(T_2 - T_1) + mL$$
$$Q = (50 \times 896 \times (659 - 20) + (50 \times 387 \times 10^3)$$
$$Q = 48.0 \times 10^6\,\text{J} \quad \text{or} \quad \mathbf{48.0\,MJ}$$

Self-assessment tasks 3.16

1. A steel ingot of mass 25 kg is heated from a temperature of 180 °C to 9500 °C ready for forging. If the specific heat capacity of the steel is 486 Jkg^{-1}K^{-1}, calculate the amount of heat energy received.
2. A water heater raises the temperature of 15 kg of water from 170 °C to 400 °C in 4 minutes. Calculate (a) the heat energy received; (b) the power output to the water. The specific heat capacity of the water is 4187 kJkg^{-1}K^{-1}.
3. A condenser receives 1.5 kg of steam at 100 °C and standard atmospheric pressure which is completely free of any water droplets. How much heat energy is given out by the steam as it condenses to water at 100 °C. The specific latent heat of vaporisation is 2248 kJkg^{-1}.
4. 6 kg of lead at 210 °C is heated to its melting point of 3270 °C until it is completely molten. Calculate the total heat energy which it receives. The specific heat capacity of the lead is 130 Jkg^{-1}K^{-1} and its specific latent heat of fusion is 23 kJkg^{-1}.

Linear expansivity

Unless they are securely constrained in some way, materials will expand when heated. The amount of free expansion, of a solid material depends on its temperature rise, its original length and the value of its **linear expansivity**. This is a constant for the material which is sometimes also called its **coefficient of linear expansion.**

The linear expansivity (α) of a material is defined as its increase of length per unit of its original length per degree of temperature rise. Its units are simply written as per kelvin, or per degree Celsius, i.e. K^{-1} or °C^{-1}.

The expansion x of a material of length l metres for a temperature change from T_1°C to T_2°C is given by:

$$x = l\alpha(T_2 - T_1)$$

The units of the expansion x will be the same as those used to measure the length, i.e. metres or mm.

In this relationship the original length and the linear expansivity are constants. The temperatures and the change in length are variables.

Some typical values of linear expansivity for different materials are given in Table 3.6.

Table 3.6

Material	Linear expansivity (K^{-1})
Aluminium	24×10^{-6}
Brass	19×10^{-6}
Copper	17×10^{-6}
Mild steel	12×10^{-6}
Cast iron	11×10^{-6}

Examples

1. An aluminium tie bar of length 1.5 metres is heated from an initial temperature of 20 °C to a final temperature of 200 °C. Calculate the change of length if it is allowed to expand freely. The linear expansivity of aluminium is 24×10^{-6} K^{-1}.
Answer:
Finding change in length.

$$x = l\alpha(T_2 - T_1)$$
$$x = 1.5 \times 24 \times 10^{-6} \times (200 - 20)$$
$$x = 6.48 \times 10^{-3}\,\text{m} \quad \text{or} \quad \mathbf{6.48\,mm}$$

2. One end of a steel bridge of length 20 metres is supported on a roller so that it can expand freely. If the bridge is assembled in position when the temperature is 150 °C, how much of a gap must be left for expansion if the maximum expected summer temperature is 350 °C? The linear expansivity of steel is 12×10^{-6} K^{-1}.
Finding required gap.

$$x = l\alpha(T_2 - T_1)$$
$$x = 20 \times 12 \times 10^{-6} \times (35 - 15)$$
$$x = 4.8 \times 10^{-3}\,\text{m} \quad \text{or} \quad \mathbf{4.8\,mm}$$

Self-assessment tasks 3.17

1. A copper pipe is 4.5 metres long at a temperature of 19 °C. By how much will it expand when hot water at a temperature of 80 °C flows through it? The linear expansivity of copper is 17×10^{-6} K^{-1}.
2. A precision length bar is made from steel and has a length of exactly 500 mm at 20 °C. What will be its length if the temperature falls to 0 °C? The linear expansivity of steel is 12×10^{-6} K^{-1}.
3. An aluminium component is to be formed to shape by sand casting and must have a length of 300 mm when it has cooled down to a temperature of 20 °C. If the temperature at which aluminium solidifies is 660 °C, what must be the length of the moulding pattern to allow for contraction when cooling? The linear expansivity of aluminium is 24×10^{-6} K^{-1}.

3.6 Electrical systems

Power stations convert the energy available in fossil fuels, nuclear materials, the wind and water into electrical energy. In this form it can be moved over large distances via the National Grid for use in homes and industry. Smaller amounts of electrical energy can be obtained from the chemicals contained in batteries which are used to power calculators, watches and portable stereos.

Electrical systems may be investigated while considering the descriptions and definitions of:

- conductors and insulators
- electrical circuit symbols
- Ohm's law
- electrical power
- the effects of temperature on conductors
- resistor combinations
- alternating current

Conductors and insulators

Metals are good conductors of electricity, particularly silver, copper and aluminium. Non-metallic substances, such as plastics and ceramics, are bad conductors of electricity but they are of equal importance in the design of electrical systems where they are used as insulators.

Metals have atoms from which some of the orbiting electrons can easily become detached. They are then known as **free electrons** and like all electrons, they carry a small negative charge. When an electrical potential difference is applied between the ends of a conductor, the negatively charged free electrons are attracted to the positive end. It is this slow drift of free electrons which constitutes a direct electric current.

It was originally thought that an electric current flowed in a direction from the positive to the negative potential. Although it is now known that the opposite is the case, and electrons flow towards the positive potential, the old convention is still used to indicate the direction of current flow on circuit diagrams.

The charge on a single electron is very small. The SI unit of electrical charge is the **coulomb** where:

$$\textbf{1 coulomb} = \mathbf{6.24 \times 10^{18}} \textbf{ electron charges}$$

When 1 coulomb of charge passes a point in 1 second, the current flowing is 1 ampere.

Electrical circuit symbols

Figure 3.36 shows some of the symbols commonly used in circuit diagrams as specified in BS 3939.

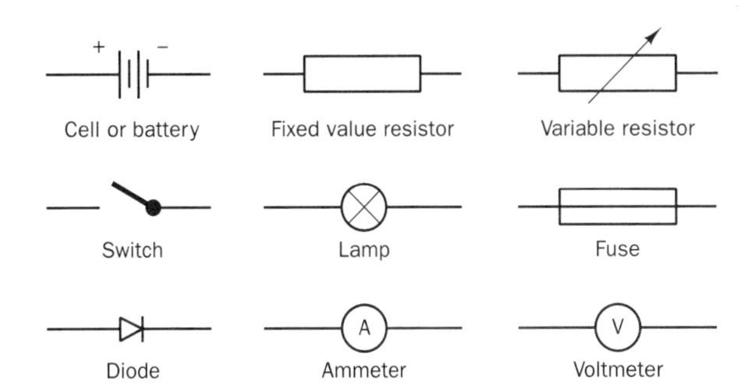

Figure 3.36 British Standard 3939 circuit symbols

Typical circuits in which these symbols are used are shown in Fig. 3.37.

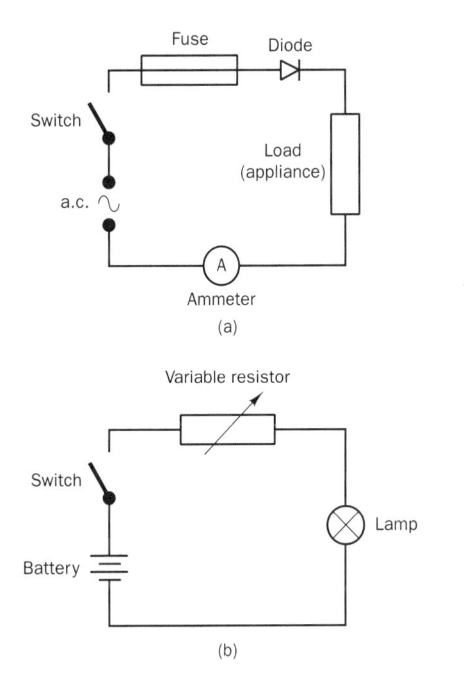

Figure 3.37 Two typical circuits: (a) a.c. to d.c. conversion; (b) lamp dimmer

Ohm's law

Electrical potential difference is also called electro-motive force because it is this which forces current around a circuit. The SI unit of potential difference is the **volt** and the SI unit of current is the **ampere**.

The relationship between potential difference and current is given by Ohm's law which states that **'the current flowing in an electrical conductor is proportional to the potential difference between its ends'**.

It assumes that physical conditions, such as temperature, do not change, or that if they do, they have no effect on the current. A graph of potential difference V against the current I

which is flowing in a particular circuit will be a straight line through the origin (Fig. 3.38).

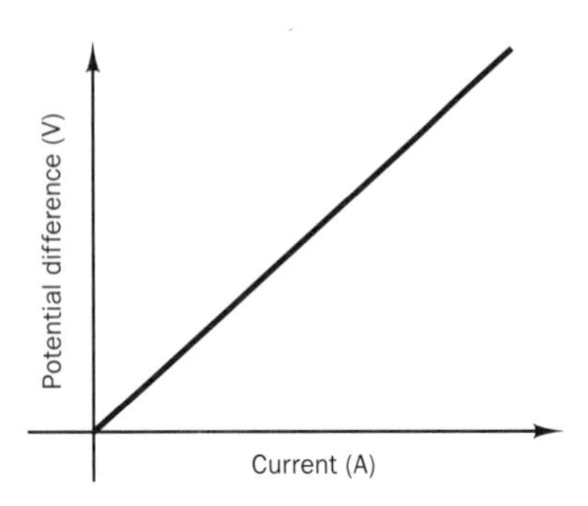

Figure 3.38 Graph of potential difference against current

The graph indicates that the current is proportional to potential difference, or that:

Potential difference = Constant × Current

The constant is the resistance R of the conductor. It is measured in Ohms whose symbol is Ω (the Greek capital letter omega). The potential difference V and the current I are the variables and the formula connecting them is written as:

$$V = IR$$

It follows from Ohm's law that when a potential difference of 1 volt causes a current of 1 ampere to flow in a conductor, its resistance must be 1 ohm.

Examples

1. A 6 volt battery supplies current to a lamp of resistance 15 ohms. What is the current flowing in the circuit?
   Answer:
   Finding the current:

   Since $V = IR$

   $$I = \frac{V}{R}$$

   $$I = \frac{6}{15}$$

   $$\mathbf{I = 0.4\,A}$$

2. When the buzzer sounds in an alarm circuit, the current flowing through it is 200 mA and the potential difference across it is 12 volts. What is the resistance of the buzzer?
   Answer:
   Finding the resistance:

   Since $V = IR$

   $$R = \frac{V}{I}$$

   $$R = \frac{12}{200 \times 10^{-3}}$$

   $$R = 60\,\Omega$$

Electrical power

When current flows in an electrical system, energy conversion takes place. The electrical energy may be converted into heat, light, sound or mechanical energy.

The rate at which energy is converted is the **power rating** of the system. This is given by the formula:

Power = Potential difference × Current
Power = VI (watts)

As with mechanical and thermal systems, the SI unit of power in electrical systems is the **watt**. A circuit has a power rating of 1 watt when energy is being converted at the rate of 1 joule per second.

From Ohm's law, $V = IR$, substituting for V in the equation above gives another formula for calculating power:

$$\text{Power} = IR \times I$$
$$\textbf{Power} = I^2R$$

Also from Ohm's law, $I = V/R$, substituting for I in the equation gives yet another formula:

$$\text{Power} = V \times \frac{V}{R}$$

$$\textbf{Power} = \frac{V^2}{R}$$

Examples

1. A current of 0.5 ampere flows through a lamp when a potential difference of 120 volts is applied to it. Calculate its resistance and the rate at which it gives off energy in the form of heat and light.
   Answer:
   Finding resistance of lamp:

   $$R = \frac{V}{I}$$

   $$R = \frac{120}{0.5}$$

   $$\mathbf{R = 240\,W}$$

   Finding rate of energy conversion, i.e. power:

   $$\text{Power} = VI$$
   $$\text{Power} = 120 \times 0.5$$
   $$\textbf{Power} = \mathbf{60\,\Omega}$$

2. An electric heater has an element of resistance 650 ohms through which a current of 1.5 amperes is flowing. How much heat energy is the heater giving off per second?
   Answer:
   Finding heat energy given off per second, i.e. power:

   $$\text{Power} = I^2R$$
   $$\text{Power} = 1.52^2 \times 650$$
   $$\textbf{Power} = \mathbf{1.46 \times 10^3\,W} \quad \text{or} \quad \mathbf{1.46\,kW}$$

Self-assessment tasks 3.18

1. What will be the potential difference across a 220 W resistor whose purpose is to limit the current in a circuit to 3.5 A?
2. What will be the current through a lamp of resistance 150 W when the potential difference across it is 24 V?
3. An electric heater carries a current of 4.5 A when a potential difference of 120 V is applied. Calculate (a) its resistance; (b) the power output.
4. A security light operates from a 220 V supply and is rated at 750 W. Calculate (a) the current which flows through it; (b) its resistance.
5. An audio warning siren has a resistance of 470 Ω and carries a current of 500 mA. Calculate (a) the power it absorbs; (b) the potential difference across it.

The effects of temperature on conductors

It is found for many materials, and particularly metals, that an increase in temperature causes an increase in electrical resistance. Such conductors are said to be non-ohmic. Graphs of resistance against temperature for non-ohmic and ohmic conductors are given in Fig. 3.39.

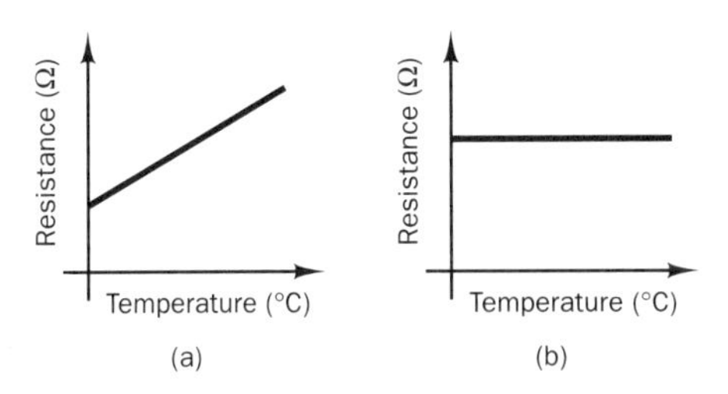

Figure 3.39 Graphs of resistance against temperature: (a) metal conductor (non-ohmic); (b) carbon conductor (ohmic)

Carbon is an ohmic conductor whose resistance changes very little with temperature. This is one reason why carbon compound resistors are widely used in electronic circuits.

When a current flows in a metal conductor, heat is generated. If the current is increased, the temperature rises, and so does the resistance. Small increases in current may have little effect on resistance, but if a graph is plotted of potential difference against current over a wide range of values the result may not be a straight line.

The effect may be demonstrated by comparing the current–voltage graphs for two lamps, one with a metallic filament and one with a carbon filament (Fig. 3.40).

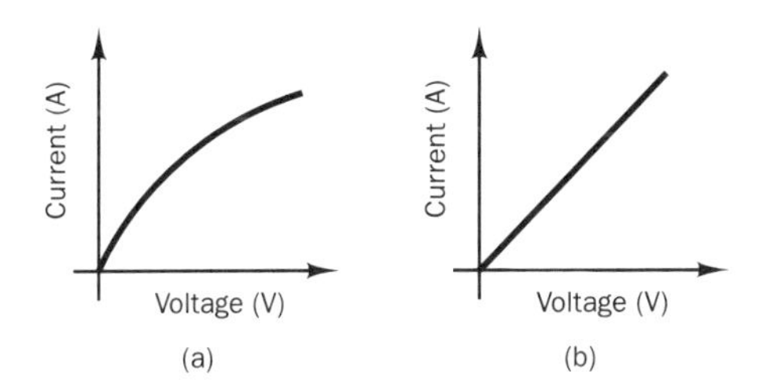

Figure 3.40 Graphs of current against voltage: (a) metak filament lamp; (b) carbon filament lamp

As can be seen, the graph for the metal filament lamp is a curve, showing that current is not proportional to voltage. The graph for the carbon filament lamp is a straight line, showing that current is proportional to voltage and that Ohm's law is obeyed irrespective of the temperature of the filament.

Resistor combinations

The total resistance of an electric or electronic circuit depends on the resistance values of the separate components. In addition to the carbon and wire-wound resistors, lamps, heaters, buzzers, motors, etc. all have resistance.

The total resistance also depends on the way in which the components are connected together. Depending on the particular application, the components may be connected:

- in series
- in parallel

Resistors in series

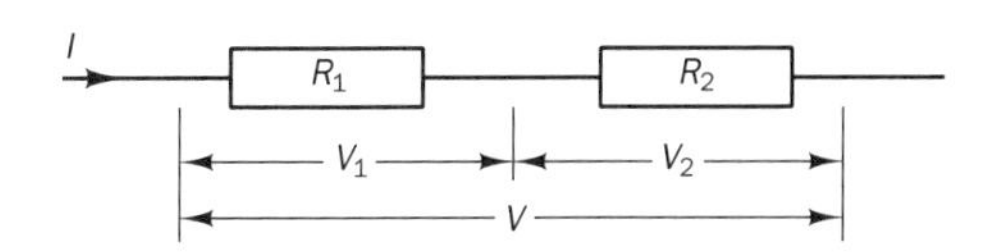

Figure 3.41 Resistors in series

The resultant resistance R of a number of resistors connected in series (Fig. 3.41) is given by:

$$\boldsymbol{R = R_1 + R_2}$$

The same current flows through each of the resistors in series and the sum of the potential differences across each one is equal to the total potential difference applied.

$$V = V_1 + V_2$$

The two voltages can be found by applying the formula $V = IR$ to each resistor in turn:

$$V_1 = \boldsymbol{IR_1} \quad \text{and} \quad V_2 = \boldsymbol{IR_2}$$

If a required value of resistor is not available for a particular circuit, the value can often be obtained by connecting a number of smaller resistors in series.

Example

Two small lamps with resistances of 3 Ω and 5 Ω are connected in series to a 12 V supply, as shown in Fig. 3.42. Determine the current flowing in the following circuit and the potential difference across each lamp?

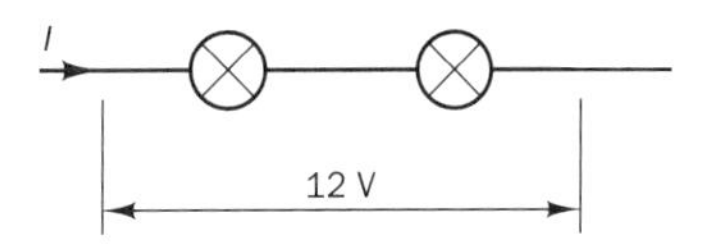

Figure 3.42 Two lamps in series

Finding total resistance of circuit:

$$R = R_1 + R_2$$
$$R = 3 + 5$$
$$\boldsymbol{R = 8\,\Omega}$$

Finding current flowing in circuit.

$$I = \frac{V}{R}$$
$$I = \frac{12}{8}$$
$$\boldsymbol{I = 1.5\,A}$$

Finding potential difference across 3 W lamp:

$$V_1 = IR_1$$
$$V_1 = 1.5 \times 3$$
$$\boldsymbol{V_1 = 4.5\,V}$$

Finding potential difference across 5 W lamp:

$$V_2 = IR^2$$
$$V_2 = 1.5 \times 5$$
$$V_2 = 7.5\,V$$

Resistors in parallel

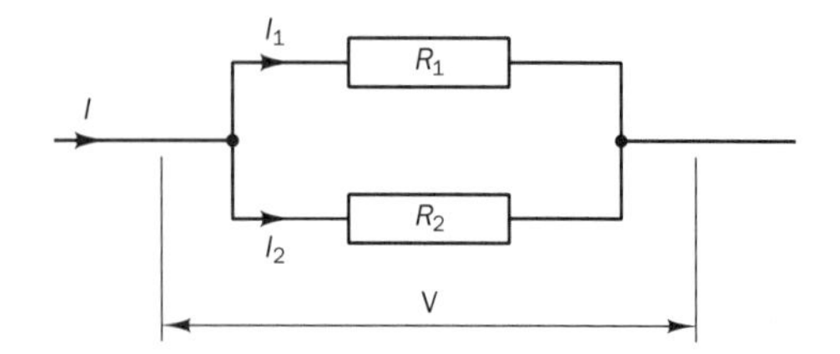

Figure 3.43 Resistors in parallel

The resultant resistance R of a number of resistors connected in parallel, as shown in Fig. 3.43, is obtained from the formula:

$$\frac{1}{R} = \frac{1}{R_1} + \frac{1}{R_2}$$

The potential difference across each of the resistors in parallel is the same, and the sum of the currents flowing in each is equal to the total current flowing in the circuit:

$$I = I_1 + I_2$$

The two currents can be found by applying the formula $I = V/R$ to each resistor in turn:

$$I_1 = \frac{V}{R_1} \quad \text{and} \quad I_2 = \frac{V}{R_2}$$

When connected in parallel, the resulting resistance is always less than the smallest value in the system. If a required value of resistor is not available for a particular circuit, the value can often be obtained by connecting a number of larger values in parallel.

Example

Because a particular value of resistor is not available a 40 Ω and a 60 Ω resistor are connected together in parallel to make the value. The combination is connected to a 20 V supply as shown in Fig. 3.44. What is their combined resistance and what is the current flowing through each one?

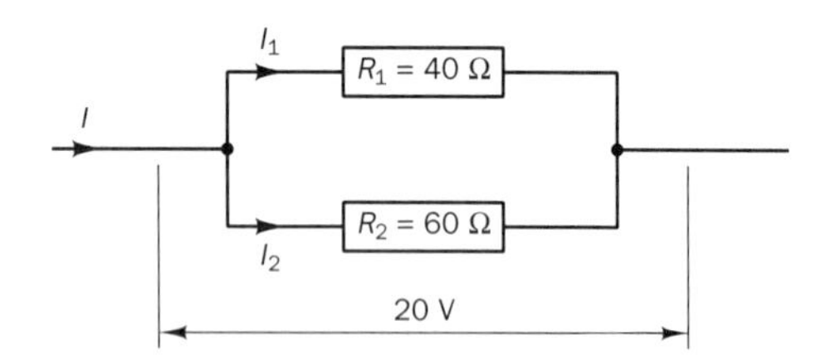

Figure 3.44 Combined resistance in parallel

Finding combined resistance:

$$\frac{1}{R} = \frac{1}{R_1} + \frac{1}{R_2}$$

$$\frac{1}{R} = \frac{1}{40} + \frac{1}{60}$$

Note: The value of R can be found most easily by using the '1/x' key on your electronic calculator. The procedure is as follows:

[40] [1/x] [+] [60] [1/x] [=] [1/x]

The value shown on your calculator display should be

$$R = 24\,\Omega$$

Finding current in 40 W resistor:

$$I_1 = \frac{V}{R_1}$$

$$I_1 = \frac{20}{40}$$

$$I_1 = 0.5\text{ A}$$

Finding current in 60 W resistor:

$$I_2 = \frac{V}{R_2}$$

$$I_2 = \frac{20}{60}$$

$$I_2 = 0.33\text{ A}$$

Self-assessment tasks 3.19

1. A lighting circuit contains two lamps of resistance 10 Ω and 15 Ω connected in series. If a current of 3 A flows through them, calculate (a) the total resistance of the circuit; (b) the voltage supplied to the circuit.
2. An alarm circuit contains a warning lamp of resistance 15 Ω and a buzzer of resistance 20 Ω which are connected in series to a 9 V supply. Calculate (a) the total resistance of the circuit; (b) the current which flows.
3. Two heaters of resistance 80 Ω and 120 Ω are connected in parallel to a 230 V supply. Calculate (a) the resistance of the circuit; (b) the current flowing through each heater.
4. Part of an electronic circuit contains a 220 Ω resistor and a 470 Ω resistor connected in parallel. If the potential difference across the combination is 9 V, calculate (a) the resistance of the parallel branch; (b) the power which is absorbed in this part of the circuit.

Alternating current

In a conductor carrying a direct current, the flow of electrons is maintained in one direction by a steady potential difference. In a conductor carrying an alternating current, the potential difference is constantly changing so that each end becomes alternately positive and negative. The electrons move first in one direction and then the other, vibrating from side to side (Fig. 3.45).

The current generated in power stations is alternating current which changes direction fifty times per second. It is said to have a frequency of 50 hertz. Alternating currents

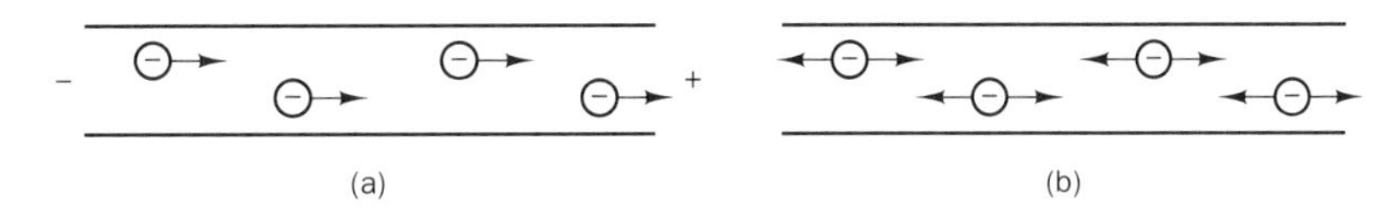

Figure 3.45 Electrons in a conductor: (a) d.c. – free electrons move to positive potential; (b) a.c. – free electrons vibrate from side to side

can be carried over long distances with much lower energy losses than direct currents. Furthermore, the high transmission voltages which are required can easily be stepped down to the level required in homes and industry by means of transformers. This could not be done with direct current.

A graph plotted to show how an alternating current or voltage varies with time is as shown in Fig. 3.46.

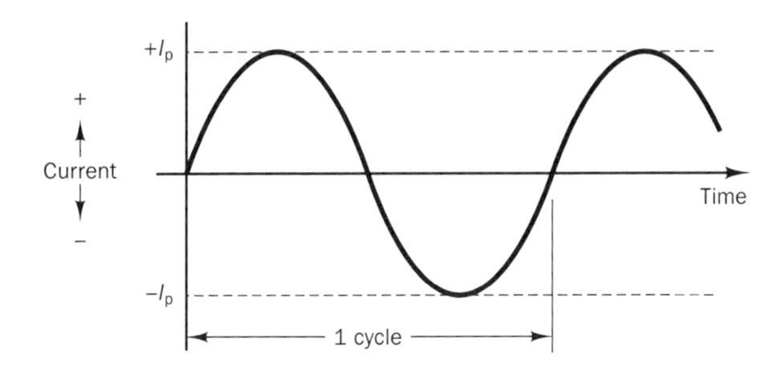

Figure 3.46 Variation of an alternating current

The time for one complete cycle is called the **periodic time**. In the case of the a.c. mains supply, this is 150th of a second. The peak values to which the current rises are $+I_p$ and $-I_p$. Just like a direct current, an alternating current can produce heat and light and drive electric motors. In order to calculate the power produced by an alternating current it is necessary to know the value of the direct current to which it is equivalent, i.e. the direct current which would produce the same power. This is called the 'root mean square' value, I_{rms}, of an alternating current. It can be shown that:

$$I_{rms} = \frac{I_p}{\sqrt{2}}$$

or $I_{rms} = 0.707\, I_p$

The same formula can be applied to find the root mean square value of the alternating voltage:

$$V_{rms} = 0.707\, V_p$$

In the case of the domestic a.c. mains, the root mean square supply voltage is 230 volts. The formulae derived from Ohm's law, and for the power absorbed in d.c. circuits, can be applied in just the same way to a.c. circuits which have pure resistance:

$$V = IR$$

and $\text{Power} = VI$ or $\text{Power} = \dfrac{V^2}{R}$ or $\text{Power} = I^2R$

Where V and I are the root mean square values of alternating voltage and current.

Unless otherwise stated, it can be assumed that any voltage and current values given for a.c. circuits are the root mean square values and there is no need to write the 'rms' subscript after V and I.

Examples

1. Calculate the peak value of the domestic a.c. supply voltage which has a root mean square value of 230 volts.
   Answer:
   Finding V_p, the peak value of the supply voltage:

   $V = 0.707 \times V_p$ where V is the root mean square value

   $$V_p = \frac{V}{0.707}$$

   $$V_p = \frac{230}{0.707}$$

   $$\mathbf{V_p = 325\ V}$$

2. An a.c. circuit has a resistance of 300 ohms and a power supply whose e.m.f. has a peak value of 150 volts. Calculate the root mean square values of the e.m.f., and the current for the circuit and the power absorbed.
   Answer:
   Finding V, the root mean square value of the e.m.f.:

   $$V = 0.707 \times V_p$$
   $$V = 0.707 \times 150$$
   $$\mathbf{V = 106\ V}$$

   Finding I, the root mean square value of the current:

   $$I = \frac{V}{R}$$

   $$I = \frac{106}{300}$$

   $$\mathbf{I = 0.354\ A}$$

   Finding power absorbed.

   $$\text{Power} = VI$$
   $$\text{Power} = 106 \times 0.354$$
   $$\textbf{Power} = \mathbf{37.5\ W}$$

3. An electric kettle contains 1 kg of water at a temperature of 200 °C. The supply voltage and current are measured to have root mean square values of 230 volts and 8 amperes respectively. The time taken for the kettle to reach a temperature of 900 °C is 3 minutes. Calculate (a) the electrical power input; (b) the output power to the water; (c) the efficiency of the kettle. The specific heat capacity of water is 4.178 $\text{kJkg}^{-1}\text{K}^{-1}$.
   Answer:
   (a) Finding heat energy received by water:

   $$Q = mc(T_2 - T_1)$$
   $$Q = 1 \times 4.178 \times 10^3 \times (90 - 20)$$
   $$\mathbf{Q = 292 \times 10^3\ J} \quad \text{or} \quad \mathbf{292\ kJ}$$

   Finding power output to water:

   $$\text{Output power} = \frac{Q}{t}$$

   $$\text{Output power} = \frac{292 \times 10^3}{3 \times 60}$$

   $$\textbf{Output power} = \mathbf{1624\ W} \quad \text{or} \quad \mathbf{1.62\ kW}$$

(b) Finding input power to kettle:

Input power $= VI$

Input power $= 230 \times 8$

Input power = 1840 W or 1.84 kW

(c) Finding efficiency of the kettle:

$$\text{Efficiency} = \frac{\text{Output power}}{\text{Input power}}$$

$$\text{Efficiency} = \frac{1624}{1840}$$

Efficiency = 0.883 or 88.3%

Self-assessment tasks 3.20

1. An alternating current supply unit gives a root mean square voltage of 120 V. What is the peak voltage value?
2. An alternating current has a peak value of 300 mA. What is its root mean square value?
3. A heater is supplied with alternating current. If the peak values of voltage and current are 300 V and 5 A, calculate (a) the root mean square value of the voltage; (b) the root mean square value of the current; (c) the power output.
4. A lighting circuit contains two lamps, each of resistance 20 W which are connected in parallel. If the circuit is supplied with alternating current of peak voltage is 120 V, calculate (a) the resistance of the circuit; (b) the root mean square value of the voltage; (c) the power output.

3.7 Measurement of physical quantities

Topics covered in this section are:

- identification of appropriate devices for measuring physical quantities
- description of the functions of each device
- measurement of physical quantities using appropriate devices
- correct recording of the measurements of physical quantities

Measuring devices and their functions

The most appropriate instrument for measuring a variable depends largely on the degree of accuracy which is required. This, together with the working environment and the skill of the user, are factors which must be taken into account before selecting a measuring device for a particular variable.

The devices identified and described below are in common use for measuring:

- mass
- length
- temperature
- time
- electrical quantities

Mass measuring devices

The most common mass measuring devices are based on these principles:

- spring balance
- dead-weight balance

Both types are made with different degrees of complexity to suit different engineering applications.

Spring balance devices

The simplest form of spring balance consists of a tension spring on which the mass is hung and a pointer which moves on a linear scale calibrated in kilograms (Fig. 3.47(a)).

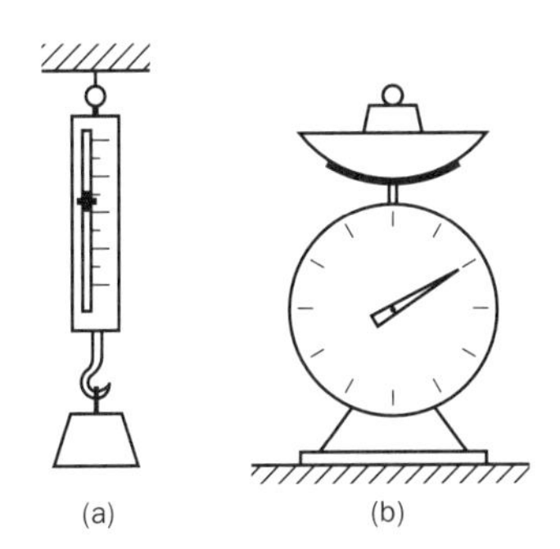

Figure 3.47 Balances: (a) tension spring; (b) top pan spring

The top-pan spring balance is slightly more complex (Fig. 3.47(b)). The mass is placed in a pan mounted on top of a compression spring. As the spring is compressed it operates a rack and pinion. A pointer attached to the pinion shaft rotates on a circular scale which is calibrated in kilograms.

Both types of spring balance may also be calibrated in newtons to measure the weight of an object or substance. They are quick and easy to use and well suited for applications where a high degree of accuracy is not required.

For more accurate measurement, electronic balances have been developed which incorporate electrical resistance strain gauges. These consist of a fine grid of wire or metal foil no more than a few millimetres square, whose electrical resistance changes if they are distorted in any way.

They are connected to the spring element of the balance and when a mass is placed on it, the spring deflects and their resistance changes. This causes the current passing through them to change slightly.

The change is amplified and the signal is used to indicate the mass in grams or kilograms on a digital display. They may also be interfaced with a computer which enables a record to be kept of the measuring operations.

Dead weight balance devices

The simplest form of dead weight balance consists of a balance beam, pivoted at its centre with a pan at one end to hold the mass to be measured and a platform at the other to hold the known dead weights (Fig. 3.48).

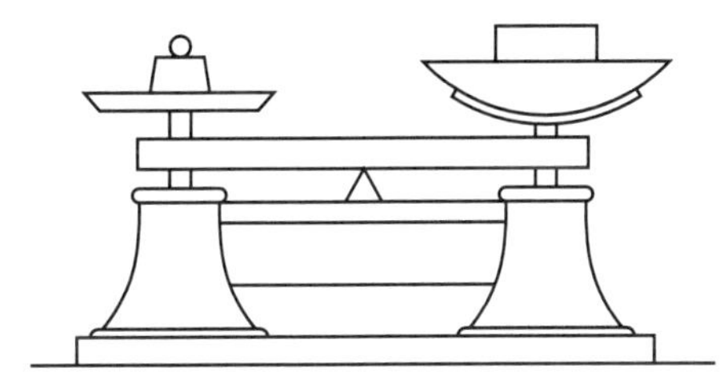

Figure 3.48 Dead weight balance

A slightly more complex design employs a fixed-value dead weight with a pointer and scale. The balance beam which carries the dead weight is attached to the load platform through a four-bar linkage (Fig. 3.49).

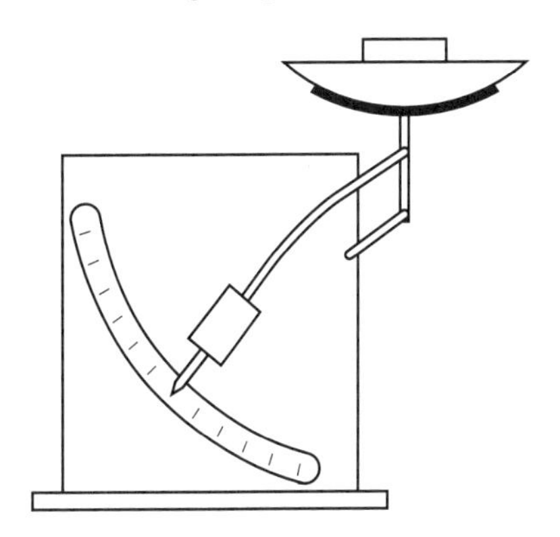

Figure 3.49 Lever balance with fixed-value dead weight

Although electronic balances are now widely used for the accurate measurement of mass, the dead weight chemical balance is still to be found in some laboratories and factories. This has a balance beam, pivoted at its centre on a knife edge, with pans hung from each end. The beam can be lowered so that the pans rest on supports when loading and unloading (Fig. 3.50).

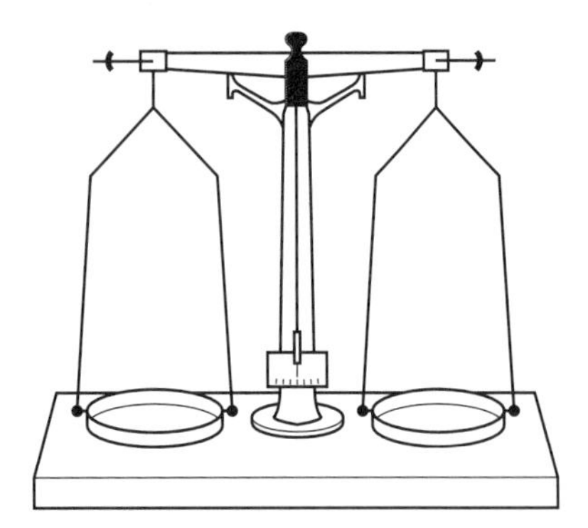

Figure 3.50 Chemical balance

The base contains a spirit level and level adjusting screws. The beam contains balance adjusting screws by which the central pointer is set to zero before measuring begins. Chemical balances are supplied with sets of calibrated precision dead weights which should be handled only with tweezers when loading and unloading them.

Length measuring devices

Three of the most commonly used length measuring devices in engineering workshops and laboratories are the:

- engineer's rule
- micrometer
- Vernier calliper

Engineer's rule

The engineer's rule is widely used for the measurement of length to an accuracy of within 0.25 mm. It can be used directly on components or in conjunction with outside and inside callipers for measuring shaft and hole sizes (Fig. 3.51).

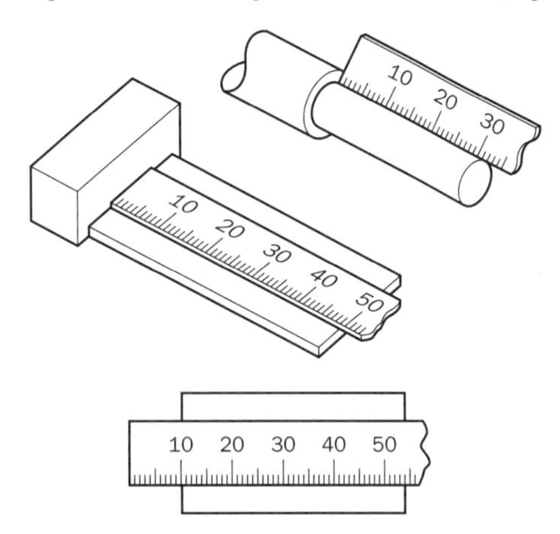

Figure 3.51 Use of engineer's rule

Engineer's rules are made from hardened and tempered spring steel in lengths from 150 mm to 1 metre. They are usually graduated in millimetres and half-millimetres.

When taking a measurement, the end of the rule should, where possible, be located against a shoulder or perpendicular face. Otherwise, the 10 mm graduation line should be placed on the edge of the work and after taking a reading, 10 mm must be subtracted from it. Parallax error can be avoided by looking down directly above the point on a rule where a reading is to be taken.

Micrometer

Micrometers are precision instruments for measuring lengths to accuracy of within 0.01 mm. They are made in many shapes and sizes for different measuring applications. The most common type is the outside micrometer which is available in increasing sizes from 0 mm to 25 mm, in steps of 25 mm (Fig. 3.52).

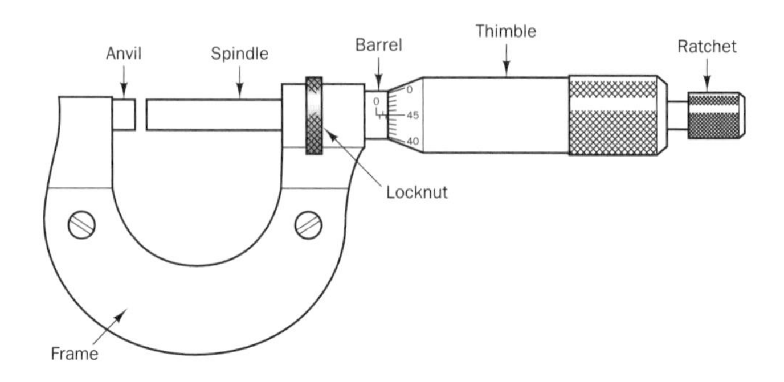

Figure 3.52 The micrometer

Micrometers contain a precision screw thread whose pitch is 0.5 mm. One turn of the thimble will thus move the spindle a distance of 0.5 mm. The barrel scale is graduated in millimetres and half-millimetres and the rotating thimble scale has 50 divisions.

Rotating the thimble through one scale division, or one 50th of a turn, moves the spindle through a distance of $0.5 \text{ mm} \div 50 = 0.01 \text{ mm}$.

To take a measurement with an outside micrometer, bring the anvil and spindle in contact with the component using the ratchet. This avoids over-tightening. The reading is then made as follows (Fig. 3.53):

1. Note the number of whole millimetres showing on the barrel scale.
2. If there is a half-millimetre division showing after the last whole millimetre, add this on.
3. Note the number of the thimble scale division which is nearest to the datum line of the barrel scale. This indicates the additional hundredths of a millimetre which must be added to the barrel scale reading, i.e. if the thimble scale reads 24 divisions add on $24 \times 0.01 = 0.24$ mm.

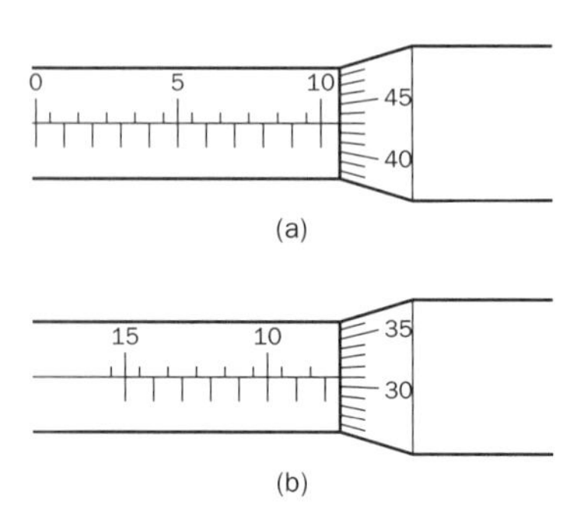

Figure 3.53 Micrometer readings: (a) 10.43 mm; (b) 22.81 mm

In recent years electronic digital micrometers have been introduced which are more expensive but easier to use. They have constant force plunger systems which always ensures that the

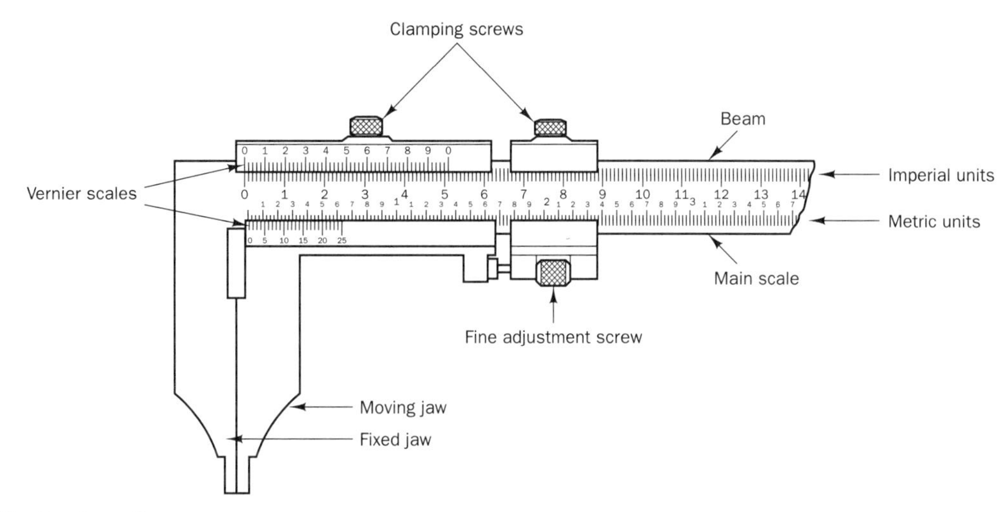

Figure 3.54 The Vernier callipers

same pressure is applied to a component when taking a measurement. The reading is displayed directly on a digital readout.

Vernier calliper

Vernier callipers are precision instruments for measuring lengths to an accuracy of within 0.02 mm. They are not as easy to use as micrometers and require good eyesight and a sensitive touch to achieve this degree of accuracy (Fig. 3.54).

The Vernier calliper has two scales. The beam contains the main scale which has millimetre and half-millimetre graduations. The moving jaw contains the Vernier scale which has 25 divisions each one of which is 0.02 mm smaller than the 0.5 mm main scale graduations.

When taking a reading the moving jaw is brought close to the component face and the fine adjustment clamp is tightened. The fine adjustment nut is then turned to bring the moving jaw into light contact with the component. The clamping screw is then tightened to hold the moving jaw in position and the reading is taken as follows (Fig. 3.55):

1. Note the number of whole millimetres showing on the beam scale up to the start of the Vernier scale.
2. If there is a half millimetre division showing after the last whole millimetre add this on.
3. Note the division number of the moving scale which best lines up with a division on the main scale.
4. Multiply the number by 0.02 mm and add it to the main scale reading, i.e. if the 14th division is in line add on $14 \times 0.02 = 0.28$ mm.

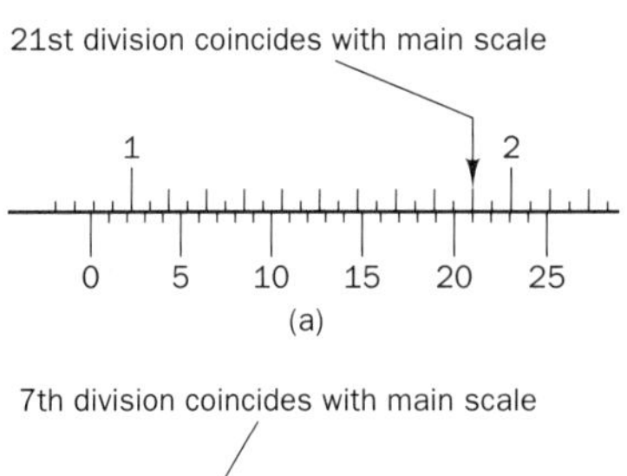

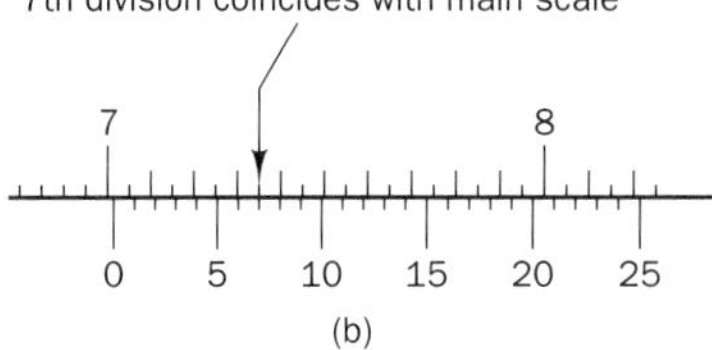

Figure 3.55 Vernier calliper readings: (a) 7.92 mm; (b) 71.64 mm

Vernier callipers have the advantage that they can measure a wider range of length on one instrument than the micrometer. They can also measure inside dimensions on the same instrument by using the outside of the jaws. These are specially rounded for locating on the inside of holes. When used in this way the width of the jaw tips, which is usually marked on the instrument, must be added to the reading.

Temperature measuring devices

Many devices have been developed for measuring temperature, two of the most common are:

- liquid-in-glass thermometers
- thermocouples

Liquid-in-glass thermometers

Within limits, liquids expand uniformly with temperature rise. Within liquid-in-glass thermometers the expansion takes place along a thin capillary tube which magnifies the effect. The tube is graduated in kelvins or degrees Celsius (Fig. 3.56).

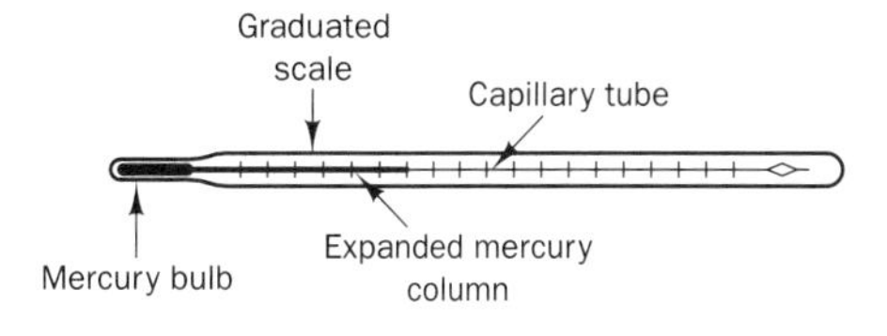

Figure 3.56 A mercury glass thermometer

The most common liquid used is mercury which operates at temperatures from −39 °C, below which it starts to solidify, to 350 °C, above which it starts to boil. Coloured alcohol is often used in some applications and is particularly good for low temperature measurement having a range from −80 °C to 70 °C.

Liquid-in-glass thermometers are made in a variety of sizes and temperature ranges, of which the 0–100 °C is perhaps the most common. They are also made with different standards of accuracy for different applications, i.e. laboratory work, process monitoring, etc.

Thermometers are well suited for measuring the temperature in fluids but not so good for measuring the surface or internal temperatures of solid materials.

Liquid-in-glass thermometers tend to be slow in their response to temperature change and they should always be allowed time to settle before taking a reading. Some thermometers are designed for bulb immersion in fluids and others for total immersion. The correct type should be used where accurate measurements are required.

As when reading other instrument scales, care should be taken to avoid parallax error by viewing a thermometer scale level with the top of the mercury or alcohol.

Thermocouples

Thermocouples operate on the principle that when two lengths of wire, made from different metals, are joined at their ends and one end is heated, an electro-motive force is set up which causes a current to flow around the circuit (Fig. 3.57).

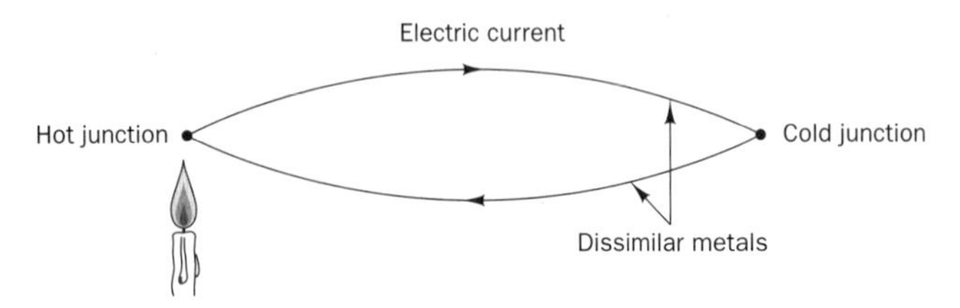

Figure 3.57 The thermocouple principle

This conversion of heat energy into electrical energy is called the **Seebeck effect.**

The electro-motive force generated is directly proportional to the temperature difference between the hot and cold junctions and if a millivoltmeter is positioned at the cold junction, its scale can be calibrated in kelvins or degrees Celsius. The hot junction is usually encased in a metal or ceramic sheath to protect it from oxidation when measuring high temperatures (Fig. 3.58).

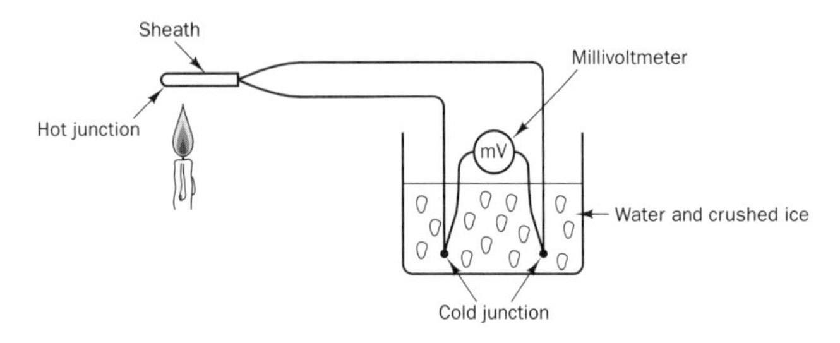

Figure 3.58 The thermocouple

The cold junction should be held at 0 °C. This can be achieved in the laboratory by placing it in a mixture of crushed ice and water. With portable instruments and in industrial applications this is impractical and electronic compensation circuits have been developed to give the same effect. They measure the cold junction temperature and add some extra e.m.f. to that which is generated at the hot junction. This simulates a cold junction temperature of 0 °C.

Although they are more expensive, thermocouple devices have many advantages over liquid-in-glass thermometers. They can be used to measure temperature in excess of 1000 °C where a glass thermometer would have melted. They have a much faster response to temperature change. The electrical signal which they generate enables the temperature to be read or recorded at long distances from where it is measured and it can also be used for temperature control purposes.

Time measuring devices

Time is measured using a variety of clocks, watches and timers which may be mechanically or electrically powered. Two of the most commonly used devices for accurate time measurement are:

- mechanical stopwatches
- electronic stopwatches

Mechanical stopwatches

These are powered by a coiled spring which needs careful rewinding from time to time. They are usually fitted with a sweep second finger and a circular scale graduated in 0–60 or 0–30 seconds per complete revolution.

The winding knob is also used to start the finger by depressing it and to stop the finger by depressing it a second time. After taking the reading, the finger can be returned to the zero position by depressing the knob a third time. A smaller scale and finger is used to record the minutes when longer time periods are being measured.

Electrical stopwatches

These are powered by a small battery and usually have a digital liquid crystal display. The basic types have three control buttons, one of which selects the different modes of operation. By depressing this the instrument can operate as a clock, recording hours and minutes and incorporating an alarm, as a calendar, recording the day, date and month, or as a stopwatch. When in stopwatch mode, the other two buttons are depressed to start, stop and re-zero the instrument.

Devices for measuring electrical quantities

Some of the most commonly used devices for measuring the variables in electric and electronic circuits are the:

- ammeter
- voltmeter
- multimeter
- cathode ray oscilloscope

Ammeter

Ammeters are used to measure the amount of electric current flowing in a circuit. There are two main types, the moving-coil analogue type with a scale and pointer, and the electronic type with a digital display.

Some ammeters are designed to measure both alternating and direct current but others are designed to measure one kind of current only. Care should be taken to select the correct type of ammeter for a particular application.

Ammeters are connected in series with the circuit components through which it is required to measure the flow of current (Fig. 3.59).

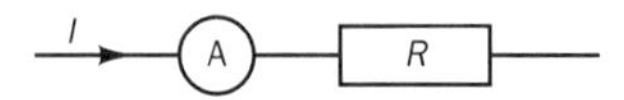

Figure 3.59 Ammeter measuring current through a resistor

Ammeters have low resistances. As a result, the potential difference across them is small and they will consume only a small amount of power from the circuit.

Voltmeter

Voltmeters are used to measure the potential difference across a circuit, or across a particular circuit component. Like ammeters they may be of the moving-coil analogue type or the electronic digital type. Also like ammeters, they may be designed for use with one kind of current only and care should be taken to select the correct type.

Voltmeters are connected between the two points in a circuit where it is required to measure the potential difference. If it is required to measure the potential difference across a particular component, the voltmeter is connected in parallel with it (Fig. 3.60).

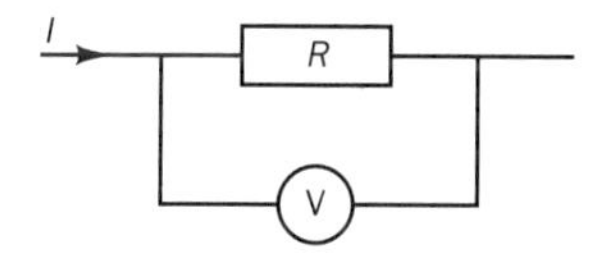

Figure 3.60 Voltmeter measuring potential difference across a resistor

Voltmeters have high resistances. This is so that the current flowing through them will be very small and like ammeters, they will only consume a small amount of power from the circuit.

Multimeter

As with ammeters and voltmeters, there are analogue and electronic multimeters. Both types are designed to measure potential difference, current and resistance over a number of ranges in both d.c. and a.c. circuits.

The insulated probes which are used with a multimeter should be of an approved type and connected the correct way round, i.e. red probe to the positive and black probe to the negative terminal or socket.

The required range, the variable to be measured and the current form are selected using a combination of switches or push buttons. This can seem complicated and if there is any doubt, the user's manual or a more experienced operator should be consulted before attempting to take a measurement.

Multimeters incorporate a battery, but in the moving-coil type this is only used when measuring resistance. It will be seen that the scale for resistance (ohms) has its zero on the right-hand side and increasing values are marked from right to left. The following procedure is used when measuring resistance.

1. Make sure there is no current flowing in the circuit or component whose resistance is to be measured.
2. Set the switches or push buttons on the multimeter to the required resistance range.
3. For moving coil instruments, touch the instrument probes together and set the pointer to zero on the resistance scale using the adjustment knob which is marked 'ohms' or 'resistance'.
4. Touch the probes across the circuit or component whose resistance is required.
5. Read the resistance on the meter scale.

Operations 3, 4 and 5 should be completed as quickly as possible to prolong the life of the battery.

Cathode ray oscilloscope (CRO)

The cathode ray oscilloscope has many uses. Two common ones are the measurement of direct or alternating voltages and the measurement of the frequency of alternating voltages.

The instrument incorporates a cathode ray tube in which a beam of electrons is focused to produce a bright spot where it strikes a fluorescent screen.

The beam may be deflected in the horizontal, or x-direction and in the vertical, or y-direction, when a potential difference is placed across the deflector plates through which the beam passes.

A graticule, or grid, of centimetre squares covers the screen and the central x and y axes may also be graduated in millimetres to enable the deflection of the spot to be measured (Fig. 3.61).

Before it is applied to the x or y deflector plates a potential difference may be amplified. The control knobs marked 'x-gain' and 'y-gain' (volts per centimetre) have a number of different positions for this purpose. To measure a steady direct voltage the following procedure is followed:

1. Focus the beam to a sharp spot using the focus and brightness controls.
2. Centralise the spot on the screen using the x-shift and y-shift controls.
3. Select the appropriate value of y-gain in volts per centimetre.
4. Connect the leads from the voltage source to the input terminals for the y-deflector plates.
5. Measure the vertical deflection of the spot in centimetres.
6. Calculate the voltage using the formula.

Voltage = **Spot deflection** (cm) × **y-gain** (volts/cm)

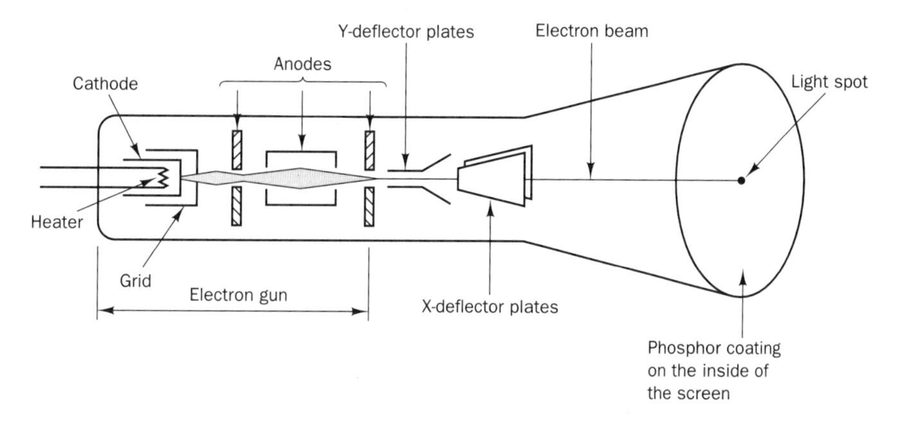

Figure 3.61 The cathode ray tube

The same procedure is followed to measure the peak-to-peak value of an alternating voltage. Here the spot is deflected upwards and downwards above and below the centre of the screen each time the voltage changes direction. This produces a vertical line trace on the screen whose length is measured. The peak-to-peak voltage is then calculated using the formula;

Peak-to-peak voltage = **Length of line** (cm) × ***y*-gain** (volts/cm)

The actual peak voltage will of course be half of this.

If it is required to measure the frequency of an alternating voltage the x-deflector plates are connected to the timebase circuit of the oscilloscope. This produces a voltage which rises slowly and falls back quickly. When plotted on a graph it has what is known as a 'saw-tooth' waveform (Fig. 3.62).

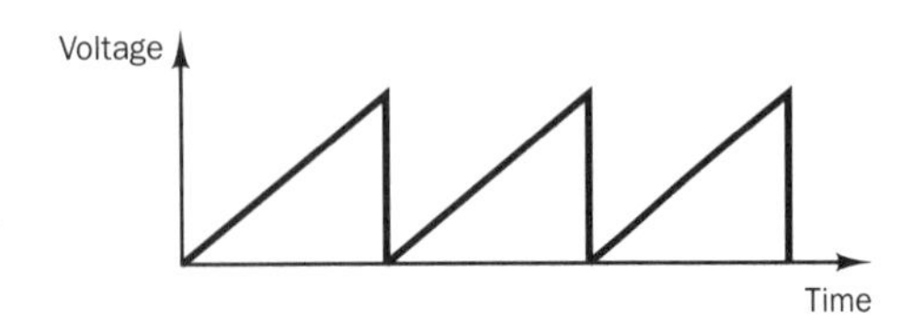

Figure 3.62 Graph of timebase voltage against time

As the voltage rises, the spot is deflected slowly from left to right across the screen and as the voltage falls it is quickly returned back to the left hand side of the screen. The speed of the spot from left to right can be varied using the timebase control which is calibrated in seconds.

To obtain the frequency of an alternating voltage the procedure is as follows:

1. Switch on the timebase circuit and increase the speed of the spot, using the timebase control until the trace appears as a straight horizontal line.
2. Focus the line for sharpness and brightness using the focus and brightness controls.
3. Connect the alternating voltage to the y-deflector plates so that as the spot moves from left to right it is also pulled up and down. As a result the spot traces out a waveform on the screen.
4. Adjust the y-gain control so that the peak-to-peak height of the waveform is approximately half the height of the screen.
5. Adjust the timebase control until the trace is stationary and a complete cycle of the waveform is displayed.
6. Note the length of the complete cycle measured in centimetres and the timebase control setting in milliseconds per centimetre division.
7. Calculate the frequency of the alternating voltage using the formula:

$$\textbf{Frequency } f(\text{Hz}) = \frac{1}{\textbf{Divisions per cycle} \times \textbf{Timebase setting (s/cm)}}$$

Example

Determine the frequency of the voltage whose waveform is shown on the cathode ray oscilloscope in Fig. 3.64 if the timebase setting is 2 ms/cm.

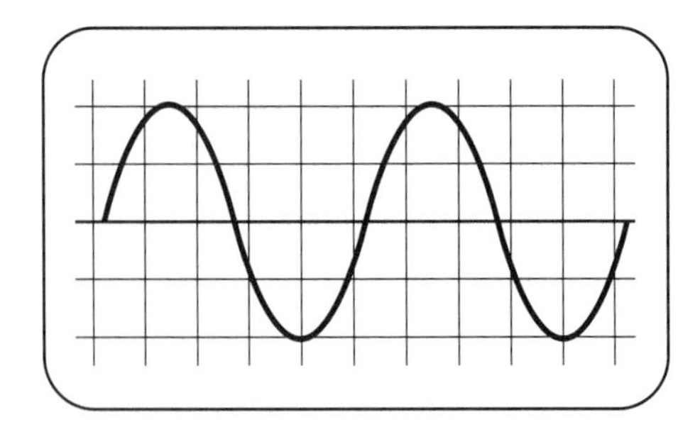

Figure 3.63 Alternating current waveform

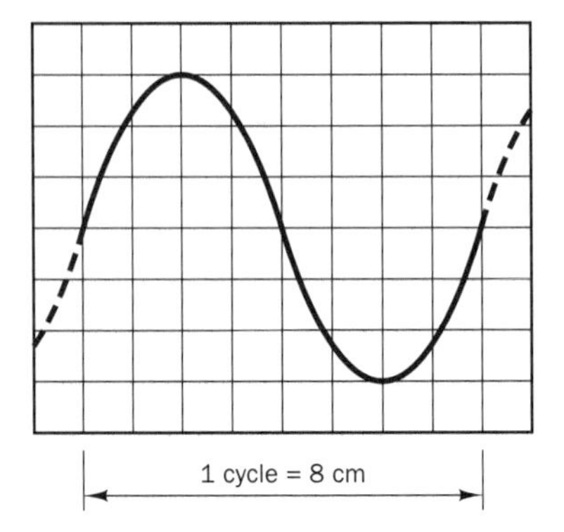

Figure 3.64 Example oscilloscope waveform

$$\text{Frequency} = \frac{1}{\text{Divisions per cycle} \times \text{Timebase setting}}$$

$$f = \frac{1}{8 \times 2 \times 10^{-3}}$$

$$f = 62.5 \text{ Hz}$$

Data measurement, recording and manipulation

Having identified the fixed and variable physical quantities in an engineering system, and the laws and principles which relate them, the relationships may be tested and verified by experimental investigation.

This should be done carefully and methodically using standard procedures for data measurement, data recording, data manipulation and for reporting on the conduct and outcome of the investigation.

The following are investigation briefs which seek to test and verify some of the laws and principles described in the Unit. Each one requires the identification and use of appropriate devices for measuring physical quantities from among those which have also been described earlier.

Assignment brief 1

Investigate the behaviour of a close-coiled helical spring when subjected to increasing loads.

Objectives

- To identify an appropriate method of applying load to a close-coiled spring.
- To identify an appropriate method of measuring the extension of the spring when loaded.
- To obtain a relationship between the load applied and the extension of the spring.

Equipment

Close-coiled helical spring supplied. Other items of equipment to be selected as appropriate.

Assignment tasks

1. Identify an appropriate method of supporting and applying loads to a close-coiled helical spring.
2. Identify an appropriate method of measuring the change in length of the spring when loaded.
3. Apply increasing values of load to the spring and measure the change in length. Obtain at least six sets of readings and record them in a suitable tabular form.
4. Plot a graph of applied load against change in length and from it obtain a mathematical expression which relates them.
5. Submit a report of your activities which contains:
   - A description of the equipment which you selected together with a diagram showing the way in which it was arranged.
   - A description of the procedure which you followed when applying loads to the spring and measuring its change in length.
   - Your tabulated results, graph and derivation of the mathematical expression relating the physical quantities of the system.
   - An evaluation of your activities and findings.

Assignment brief 2

Investigation of a trolley when moving under the action of a constant force.

Objectives

- To investigate how the distance travelled by the trolley increases with time.
- To investigate how the velocity of the trolley increases with time.
- To show that Newton's Second Law of Motion is obeyed by the trolley.

Equipment

Dynamics trolley which runs on a track fitted with a pulley at one end. Ticker timer, ticker tape and a low-voltage a.c. power supply unit. Hanger, slotted masses and a length of cord of approximately the same length as the track. Other items of equipment to be selected as appropriate.

Assignment tasks

1. Place the track on a laboratory bench or high table such that there is at least a metre above the floor. Attach one end of the cord to the trolley and attach the weight hanger to the other end. Position the trolley at the start of the track and run the cord and hanger over the pulley at the opposite end.
2. Note the frequency at which the ticker timer makes dots on the tape and connect it to the power supply unit. Attach a length of ticker tape, of approximately the same length as the track, to the trolley and pass it through the ticker timer. Make sure that the tape passes under the carbon paper disc.
3. Switch on the ticker timer and slowly incline the track until friction is just overcome, and the trolley moves down it with uniform velocity. This will be indicated by a row of equally spaced dots on the ticker tape. Keep the track in this inclined position by means of packing or a wedge.
4. Switch off the timer, replace the ticker tape and reposition the trolley at the start of the track. While holding the trolley stationary, place a known slotted mass on the weight hanger and switch on the timer. Release the trolley, allowing it to accelerate freely down the track.
5. Remove the ticker tape and cut off the first few dots which were made before the trolley started to accelerate. Cut the remaining tape into ten-dot lengths, which correspond to equal intervals of time, and measure the length of each one. Calculate and record the average velocity of the trolley over each ten-dot length.
6. Plot graphs of distance travelled against time and velocity against time and observe their shapes. Use the graph of velocity against time to determine the acceleration of the trolley.
7. Measure the mass of the trolley and use this, together with the mass which was placed on the hanger, to calculate the theoretical value of the acceleration as predicted by Newton's Second Law of Motion.
8. Submit a report of your activities which includes:
   - A description of the equipment which you used together with a sketch or diagram showing the way in which it was arranged.
   - A description of the procedure which you followed, and the precautions which you took, when setting up the equipment and taking readings.
   - Your lengths of ticker tape, displayed in a suitable way, and your graphs of distance travelled against time and velocity against time.
   - Your calculations of the acceleration of the trolley.
   - An evaluation of your activities and findings.

Assignment brief 3

Investigate the vaporisation of water in an electric kettle.

Objectives

- To identify appropriate methods of measuring the mass of water evaporated by the kettle, its temperature and the time taken.
- To determine the specific latent heat of vaporisation of water.

Equipment

Supplied are a jug-type mains electric kettle incorporating a level indicator and a 230 V power supply unit incorporating an ammeter and voltmeter or a wattmeter. Other items of equipment to be selected as appropriate.
Note: Mains electricity and boiling water can be dangerous. The equipment should only be operated by a suitably qualified person. When handling the equipment it is advisable to wear heat resistant rubberised gloves.

Assignment tasks

1. Select appropriate devices for measuring the mass of the kettle and its water content, the electrical power supplied, the temperature of the water and the time for which the kettle will be boiling.
2. Devise a means of reducing the amount of heat energy lost by the kettle to its surroundings.
3. Fill the kettle to approximately $\frac{3}{4}$ full. Measure and record the mass of the kettle and its contents.
4. Connect the kettle to the power supply and switch on. When the kettle starts to boil, start the timing device and record the water temperature.
   Note: The kettle should be positioned near to an open window to prevent the accumulation of water vapour in the room. All electrical terminations should be suitably guarded and connection to the mains supply should be made by a suitably qualified person.
5. Record the time taken for the water level to fall to approximately half full. Remove the kettle from the power supply unit and again measure and record the mass of the kettle and its contents.
6. Calculate the mass of water evaporated by the kettle, the energy supplied during the recorded time period and the specific latent heat of evaporation of the water.
7. Submit a report of your activities which contains:
8. A description of the equipment which you selected together with a sketch or diagram showing the way in which it was arranged.
9. A description of the procedure which was followed, and safety precautions which were taken, when setting up the equipment and taking readings.
10. Your readings and calculations of the energy supplied to the kettle in the recorded time period, and the specific latent heat of evaporation of the water.
11. An evaluation of your activities and findings.

Assignment brief 4

Investigate the current flowing in a d.c. circuit.

Objectives

- To identify appropriate methods of measuring potential difference, current and resistance.
- To obtain a relationship between the current flowing in a circuit and the applied potential difference.
- To verify the quoted value of a fixed resistor.

Equipment

Variable low-tension d.c. power supply unit, fixed resistor and connecting wire. Other items to be identified and selected as appropriate.

Assignment tasks

1. Select an appropriate device for measuring resistance. Measure the value of the fixed resistor and compare it with its nominal value.
2. Select appropriate devices for measuring potential difference and current in a d.c. circuit.
3. Connect the measuring devices and the fixed resistor to the power supply unit in such a way that the potential difference across the resistor and the current flowing through it can be measured.
   Note: Have your circuit checked and the power supply connected to the mains by a suitably qualified person.
4. With the voltage control set to zero, connect the power supply unit to the mains and switch it on. Slowly increase the voltage until a small potential difference is shown to be present across the resistor. Record its value together with that of the current flowing through the resistor.
5. Increase the potential difference across the resistor in small increments and for each one, record the values of potential difference and current in a suitable tabular form.
6. Plot a graph of potential difference against current and observe its shape. From the graph derive an expression which relates the two variables. Determine also the value of the resistor and compare it to its nominal value and the value measured in task 1.
7. Submit a report of your activities which includes:
   - A description of the equipment which you selected together with a circuit diagram showing the way in which it was arranged.
   - A description of the procedure which you followed, and the safety precautions which you took, when setting up the equipment and taking readings.
   - Your tabulated results, graph, measured and calculated values of resistance and the mathematical expression which relates the physical quantities in the circuit.
   - An evaluation of your activities and findings.

Assignment brief 5

Measurement of voltage and frequency using the cathode ray oscilloscope.

Objectives

- To measure the value of a direct voltage.
- To measure the value of an alternating voltage.
- To determine the frequency of an alternating voltage.

Equipment

Cathode ray oscilloscope and an a.c./d.c. low-tension power supply unit. Other items of equipment to be selected as appropriate.
Note: Mains connectors should be made by a suitably qualified person.

Assignment tasks

1. Note the positions of the oscilloscope controls and input terminals. Switch on the oscilloscope and turn the timebase control switch to the 'off' position. Adjust the x-shift, y-shift, focus and brightness controls to obtain a clearly defined spot which is approximately in the centre of the screen grid.
2. Switch the oscilloscope input to d.c., short out the y-input terminals and, if necessary, reposition the spot in the centre of the screen. Remove the shorting link and apply a direct voltage from the power supply unit to the y-input terminals.
3. Note the number of centimetres which the spot has moved up or down from the centre of the screen together

with the y-gain control setting. Calculate the value of the applied direct voltage.

4. Disconnect the power supply unit and switch the oscilloscope input to a.c. Short out the y-input terminals and, if necessary, reposition the spot in the centre of the screen. Remove the shorting link and apply an alternating voltage from the power supply unit to the y-input terminals.
5. Note the length in centimetres of the vertical trace which appears on the screen together with the y-gain control setting. Calculate the peak value and the root mean square value of the applied alternating voltage.
6. With the alternating voltage still connected to the y-input terminals, switch on the timebase and adjust its setting until a little more than one complete cycle of its waveform is displayed on the screen.
7. Note the length in centimetres of one complete cycle together with the timebase control setting. Calculate the frequency of the alternating voltage.
8. Submit a report of your activities which contains:
   - An illustrated description of the cathode ray oscilloscope and its mode of operation.
   - A description of the procedure which you followed, and the safety precautions which you took, when setting up the equipment and taking readings.
   - Your readings and calculations.
   - An evaluation of your activities and findings.

3.8 Unit test

Test yourself on this unit with these sample multiple-choice questions.

1. Which physical quantity is monitored on the display of a car radio?

(a) voltage
(b) current
(c) frequency
(d) resistance

2. Which physical quantities are monitored in a baking oven?

(a) temperature and time
(b) voltage and resistance
(c) current and mass
(d) frequency and length

3. In an electric circuit which obeys Ohm's law the fixed physical quantity is

(a) voltage
(b) current
(c) time
(d) resistance

4. A metal bar is heated in a furnace. Which physical quantity is fixed?

(a) length
(b) mass
(c) temperature
(d) time

5. Which physical quantity is a variable for a spring?

(a) mass
(b) stiffness
(c) length
(d) resistance

6. The relationship between potential difference, current and resistance in electric circuits is given by

(a) Hooke's law
(b) the principle of moments
(c) Newton's laws of motion
(d) Ohm's law

7. Newton's second law of motion describes the relationship between

(a) length, temperature rise and expansion
(b) force, mass and acceleration
(c) spring stiffness, load and extension
(d) voltage, current and resistance

8. The force per square metre which a gas exerts on the walls of its container is its

(a) pressure
(b) power
(c) work done
(d) efficiency

9. The temperature in ovens and furnaces is measured by means of a

(a) multimeter
(b) cathode ray oscilloscope
(c) thermocouple
(d) micrometer

10. The diameter of a machined component can be accurately measured using

(a) an ammeter
(b) an electronic balance
(c) a thermometer
(d) a Vernier calliper

11. A function of a cathode ray oscilloscope is to measure

(a) time
(b) frequency
(c) current
(d) temperature

12. A multimeter may be used to measure

(a) frequency
(b) pressure
(c) resistance
(d) force

13. The mass of an object can be measured using

(a) a pressure gauge
(b) a mercury-in-glass thermometer
(c) a dead weight balance
(d) an engineer's rule

14. The current flowing in an electric circuit can be measured by means of a

(a) micrometer
(b) electronic balance
(c) multimeter
(d) cathode ray oscilloscope

15. The weight of an object can be measured using a

(a) spring balance
(b) Vernier calliper
(c) thermocouple
(d) stopwatch

16. If the current is proportional to the potential difference, what are the two missing readings below?

Potential difference (volts)	1.5	3.0	4.5	6.0	7.5	—
Current (amperes)	0.5	1.0	1.5	2.0	—	3.0

(a) 8.0 V and 3.5 A
(b) 8.5 V and 2.5 A
(c) 9.0 V and 2.5 A
(d) 9.5 V and 3.5 A

17. The force F (N) required to stretch an elastic spring of stiffness k (Nm^{-1}) through a distance of x (m) is given by the formula

(a) $F = k \times x$
(b) $F = kx$
(c) $F = k + x$
(d) $F = \frac{x}{k}$

18. The acceleration a (ms^{-2}) of a vehicle which speeds up uniformly from u (ms^{-1}) to v (ms^{-1}) in time t (seconds) is given by the formula

(a) $a = (v - u)t$
(b) $a = \frac{vu}{t}$
(c) $a = v + ut$
(d) $a = \frac{(v - u)}{t}$

19. When the formula $V = IR$ is transposed to make I the subject, it becomes

(a) $I = VR$
(b) $I = \frac{R}{V}$
(c) $I = \frac{V}{R}$
(d) $I = \frac{1}{VR}$

20. Transposing the formula $p = F/A$ to make A the subject gives

(a) $A = \frac{1}{pF}$
(b) $A = \frac{p}{F}$
(c) $A = p + F$
(d) $A = \frac{F}{p}$

21. When the formula $v = u + at$ is transposed to make u the subject the result is

(a) $u = at - v$
(b) $u = \frac{v}{at}$
(c) $u = v - at$
(d) $u = \frac{at}{v}$

22. The heat received by a substance when its temperature rises is given by the formula $Q = mc(T_2 - T_1)$. Transposing to make c the subject gives

(a) $c = \frac{Q}{m(T_2 - T_1)}$
(b) $c = \frac{Q(T_2 - T_1)}{m}$
(c) $c = \frac{m(T_2 - T_1)}{Q}$
(d) $c = Qm(T_2 - T_1)$

23. The change in length of a material whose temperature changes is given by the formula $x = l\alpha(T_2 - T_1)$. Transposing this to make T_2 the subject gives

(a) $T_2 = \frac{x}{l\alpha T_1}$
(b) $T_2 = \frac{l\alpha}{x} + T_1$
(c) $T_2 = T_1 + l\alpha x$
(d) $T_2 = \frac{x}{l\alpha} + T_1$

24. The final velocity of an accelerating machine slide is given by the formula $v = u + at$. If $u = 2.0\,ms^{-1}$, $a = 1.5\,ms^{-2}$ and $t = 3.25\,s$, the final velocity correct to three significant figures will be

(a) $6.88\,ms^{-1}$
(b) $4.74\,ms^{-1}$
(c) $2.88\,ms^{-1}$
(d) $6.25\,ms^{-1}$

25. The heat energy received by the water in a boiler is given by the formula $Q = mc(T_2 - T_1)$. If $m = 50$ kg, $c = 4187\,Jkg^{-1}K^{-1}$, $T_2 = 95\,°C$, and $T_1 = 20\,°C$, the heat energy received, correct to three significant figures, will be

(a) 19.9 MJ
(b) 15.7 MJ
(c) 4.19 MJ
(d) 2.39 MJ

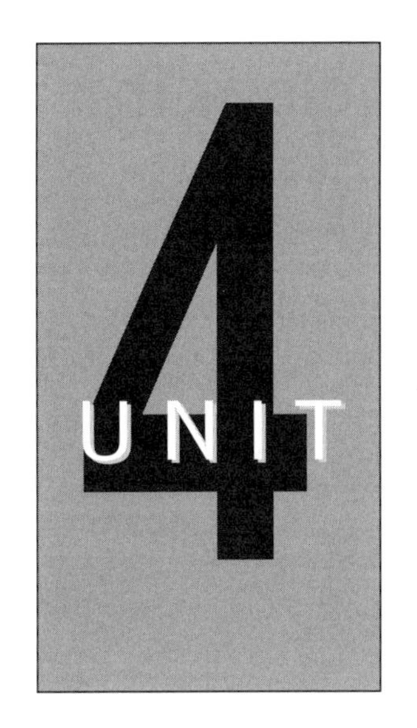

Engineering in Society and the Environment

Roger Timings

This Unit consists of two main sections. The first section considers the application of engineering not only to industry but also to a range of less obvious sectors including health, leisure and the home. It examines the environmental impact of engineering activities in these sectors, and the requirements for the management of this environmental impact.

The second section considers the career options and pathways in the engineering industry, including topics such as personal ambitions, skills and qualifications required, the importance of writing a well-presented CV and, finally, interview techniques.

4.1 The application of engineering technology in society

After reading this section you will be able to:

- Describe the application of engineering technology in home and leisure activities.
- Describe the application of engineering technology in industry and commerce.
- Describe the application of engineering technology in health and medicine.
- Identify the environmental impact of engineering technology.
- Describe the environmental impact of engineering activities.
- Describe the requirements for managing the impact of engineering activities on the environment.

Almost everything you want or need depends to some extent upon engineering and technology. From the plants that produce the baked beans and crisps that you eat, and the containers in which they are packaged, to the bike, car, bus or train in which you travel, and the roads and bridges you cross, engineering and technology has been involved in their design, manufacture and construction.

Engineering technology in the home

As soon as metals were discovered, objects made from metal found their way into the home. Initially these were simple but useful objects such as cooking pots. The more wealthy and influential members of early societies also had metal ornaments and trinkets.

Nowadays, we still use simple metal objects such as cooking utensils, scissors, knives, forks and spoons in the home. However, most households also own and use highly sophisticated engineering devices. Such devices are referred to generally as **consumer durables.** They are used by the people like you and me, people who are 'consumers'. Also such devices last quite a long time so they are 'durable', hence the name given to them. An example is a motor car. Consumer durables used inside the home are more likely to be called **domestic appliances.**

Examples of such appliances are:

- washing machines
- vacuum cleaners
- microwave ovens

Self-assessment task

A 'tree' diagram for consumer durables and domestic appliances is shown below. I have started you off by adding the devices we have discussed so far. Complete the diagram by adding *two* more consumer durables and *three* more domestic appliances.

Consumer durables
Motor car
Domestic appliances
Washing machine
Microwave oven
Vacuum cleaner

Sophisticated engineering and information technology is becoming increasingly used in domestic appliances. This is necessary in order to make them more:

- versatile, yet easier to operate
- energy efficient
- environmentally acceptable
- cost effective

Washing machines

Let's look at something familiar, such as a washing machine, to see how the improvements listed above have been achieved. Until a few years ago all the control systems were operated by a mechanical time clock. This was quite a complex device full of moving parts such as gear wheels, escapements and cams which opened and closed switches.

- Making such clocks was labour intensive.
- Because of all the moving parts they wore out and had to be replaced.
- Only the duration of the washing cycle could be controlled.
- The temperature control was limited to a switch with high, medium and low settings.
- The volume of water used was fixed irrespective of the washing load.

Then came micro-electronics based on solid state silicon devices such as transistors, integrated circuit chips, microprocessor chips and similar devices. These devices made possible the small but powerful computers and control systems we have today. There are two sorts of computers:

1. **Personal, desktop** or **laptop computers** – Like the one I am using to write this chapter and like one you may have at home. You can load all sorts of software into this type of computer and use it for word processing, for playing games and many other things.
2. **Dedicated computers.** These only do one job. You can't load your favourite software into them. Many modern cars have an engine management computer. Many modern domestic appliances such as washing machines and microwave ovens also have dedicated computers in them. So as not to frighten the purchaser, while still sounding up-market, they are said to be **microprocessor controlled.** The computer memory chip can store many more wash programmes than the mechanically operated controller. It can also compare the inputs from the various sensors with the programme selected and adjust the water volume, time and temperature accordingly. Also, microchips have no moving parts to wear out.

A schematic diagram for a typical microprocessor-controlled washing machine is shown in Fig. 4.1. So let's look at the advantages of this system compared with an earlier machine fitted only with a simple mechanical programming system.

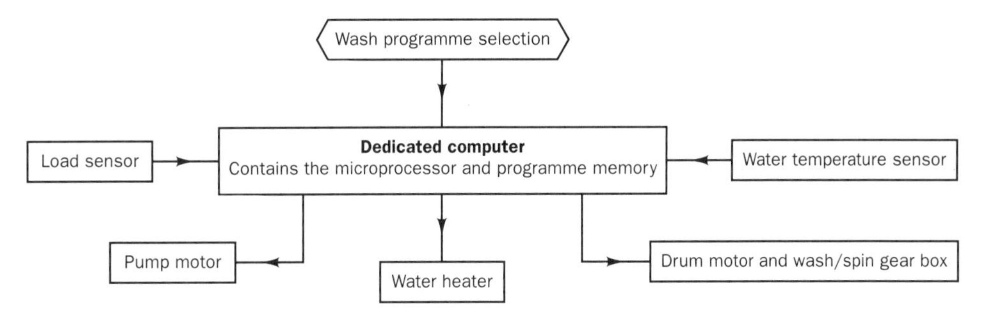

Figure 4.1 Schematic diagram of a washing machine

The user merely has to enter the required or recommended wash programme information and switch on. In some cases the fabric code printed on the garment label is all the information that is necessary.

- The load sensor automatically 'weighs' the load and inputs this information into the computer.
- The computer stores this information so that, at each stage in the washing cycle, only the minimum necessary volume of water is used.
- The temperature sensor inputs the water temperature into the computer. The computer compares this with the programme data and heats the water accordingly.
- The computer controls the washing cycle and keeps the amount of energy and the volume of water used to the minimum necessary for the load and type of fabric.
- Some washing machines use an atomised water spray to reduce the volume of water used still further. Low water pressure compensation and foam sensing is also often incorporated.

The advantages of keeping the volume of water to a minimum are:

- Reduced water charges if the water supply is metered.
- Reduced water usage (better for the environment).
- Less waste water to be disposed (better for the environment).
- Less energy required to heat the water.
- Less energy required to rotate the drum since the mass of the load plus water is lower.
- Less energy required to pump the water out.

This overall reduction in energy usage results in lower operating costs and less pollution at the power station (better for the environment). So our microprocessor control results in a more efficient, flexible and environmentally friendly appliance. At the same time, the load is washed more effectively and with less wear and damage to the fabrics. Further, the microprocessor control unit has no moving parts so it does not wear out like a mechanical clock.

Mechanically, the 'insides' of the appliance are better engineered, more efficient and have a longer life because of the increasing use of computer controlled design and manufacture. This will be considered in a future section.

There is one further device I want to consider: the temperature sensor. The simplest form of temperature sensor is a bi-metal thermostat. These are not very sensitive and there is often a considerable gap between the temperature at which the contacts open and the temperature at which they close again. It does have the advantage, however, that it can switch devices, such as motors or heaters, on and off directly.

Self-assessment task

Examine a bi-metal strip thermostat and explain, with the aid of sketches, how it works.

Nowadays a **thermistor** is more likely to be used. This is a solid state device made from silicon that has been prepared to make it temperature sensitive. The resistance of the thermistor falls considerably when the temperature increases and it can therefore pass more current. Similarly, it passes less current when its temperature decreases. This change in current operates a control circuit which, in turn, can turn motors, heaters or other devices on or off as required. Compared with a bi-metal strip thermostat it has the advantage of being more sensitive and precise in its operation so that it gives closer control. Further, there are no moving parts to wear out and it is more compact.

I have described the engineering of a domestic washing machine in some detail to show how complex engineering devices are taken into our homes today, how they are being made more environmentally friendly, and how such devices are largely taken for granted because of the ease with which they can be operated.

Toaster

Traditionally, 'pop-up' toasters used a bi-metal strip thermostat to operate the various spring loaded latches and levers that controlled the toasting process. Many are still made this way. However, the latest toasters are more likely to use microchip control in conjunction with solid state sensors. This not only gives more consistent browning but reduces the number of mechanical moving parts that are liable to get out of adjustment or simply wear out.

Vacuum cleaners

Originally vacuum cleaners used a cloth bag to act as a filter and also to retain the dust. They were not very efficient and much of the dust escaped. Further, emptying the bag was a dirty and unpleasant job and, again, much dust escaped. The first improvement was the disposable porous paper bag. This could be thrown away without having to empty it. Next came the microporous filter that could not only prevent dust from re-entering the atmosphere, but could also filter out pollen and house mites. This is of great advantage to people who suffer from hay fever and asthma.

Refrigerators and freezers

These now incorporate motor/compressor units that are sealed for life and require no maintenance. Modern refrigerators and freezers benefit from having an automatic 'de-frost' cycle that removes the need for the time consuming and messy chore of regularly defrosting the appliance manually.

Food processors

Anyone who has used the traditional wooden spoon and mixing bowl will realise the hard work involved and what a welcome labour-saving device the power-driven food mixer was. The **food processor** takes this labour saving technology several steps forward. As well as mixing, it can cut, chop, blend and mince, with sufficient power to mix dough and cake mixtures. It is microchip controlled to provide variable speeds to suit the various processes. Once set, it maintains a constant speed as the texture of the food changes during processing.

Self-assessment tasks

1. Compare a modern microwave oven with a traditional cooking oven to show how:
   (a) The microwave cooker is a technically sophisticated device.
   (b) It is more environmentally friendly.
2. (a) Name *three* domestic appliances that are electronically controlled.
   (b) Name *three* domestic appliances that are not electronically controlled.

Modern technology has not only been incorporated into the control of domestic appliances, it is also used in the development of the materials from which the appliances are made.

Refrigerants

Refrigerators and freezers have undergone changes to make them more environmentally friendly. The chloro-fluoro-carbon (CFC) refrigerants, used until recently, damaged the ozone layer if they escaped. Without the ozone layer, life on planet earth cannot exist because of the harmful radiations from the sun which the ozone blocks. New refrigerants are now used which do not affect the ozone layer.

Stainless steel

This material is widely used for making cooking utensils, cutlery and sink units. It does not corrode and its shiny surface can be easily cleaned to keep it hygienic. It is unaffected by detergents and boiling water.

Plastics

These are used for the handles of cooking utensils as they are easily cleaned and, being poor conductors of heat, stay cool. Plastic worktops are used in most kitchens as they can be easily wiped down and disinfected. Polytetrafluoroethylene (PTFE), also known by the trade name 'Teflon', is widely used for lining cooking pans. It is not only heat resistant but is also the most slippery substance known and food being cooked does not stick to it and burn. This makes it easy to clean.

Ultra-high density polyvinyl chloride (uPVC) is now used widely for window and door frames and even for the doors themselves. This material is long-lasting without going shabby and never requires painting to keep it from rotting. This reduces maintenance costs for homeowners.

Where wood has to be painted, varnishes and gloss paints are now based on a synthetic rubber called polyeurethane. This seals the wood against moisture and rotting and, being a rubber, it is impact resistant and doesn't chip off.

Self-assessment tasks

1. Find out what materials were used for cooking utensils, sinks and work surfaces early in this century.
2. (a) How were they kept clean?
   (b) Was this easier or more difficult than keeping modern materials clean?
3. Find out how solar panels can be used to make a house more energy efficient.

Engineering technology in leisure activities

State of the art technology and engineering is widely used in the hardware and software associated with many of our leisure activities. Some examples are given below.

Communications

Telecommunication (communication at a distance) is widely used for leisure as well as for work and has been greatly improved by modern technology. Let's look at some examples.

- Telephones are now linked by digital exchanges that offer quicker interconnection with less background noise and greater clarity. Fibre-optic trunk connections ensure improved sound quality and greater reliability. Satellites enable calls to be made overseas that are as clear and reliable as a local call. In addition, mobile phones are essential to people who travel about a lot and cannot always have access to a public telephone.
- Radio sets have been miniaturised by the use of transistors and improved loud speaker design. At the same time, sound quality has been greatly improved. This is especially so when receiving frequency modulated (FM) stereo transmission. The power consumption of modern sets is so low that two small flashlight batteries is all that is required. This has made possible the personal stereo receiver. These often contain a tape deck.
- Television sets have become much more reliable and compact by replacing valves with transistors and integrated circuits (ICs). Stereo sound is available and many sets have teletext facilities. This enables such sets to be used for entertainment and as a source of information. When coupled to a video cassette recorder (VCR) programmes may be recorded for future viewing. Alternatively, pre-recorded cassettes can be hired or bought for entertainment and instructional purposes. Most television sets can now receive teletext, which provides an up-to-the-minute service of news bulletins, sports results, weather forecasts, travel information and much other useful information.

- Camcorders are now widely used in place of the traditional cinecamera for making home movies. The moving images and sound are stored on tape in the same way as in a VCR and can be replayed instantly through the camera's screen for checking the shot. For greater clarity and general viewing the tape can be played through a television set. Editing facilities and special effects are also available.
- Satellite television enables additional programmes to be beamed directly into our homes by geostationary satellites. Many of the programmes are live sport from abroad. Special high-gain aerials are required and an interface unit between the aerial and the receiver. The quality of reception can be affected by the weather and the positioning of the aerial. An additional fee has to be paid to the providers of satellite services.
- Cable television does not require an aerial. The set is plugged into an underground cable carrying a wide range of programmes. Services other than just television are also provided (e.g. telephone and computer links). Again, an additional fee is required for the use of the service. One advantage of this system is that it frees up radio frequencies that can then be used for other purposes.

Music

Music as a means of entertainment has been revolutionised by modern electronics, both for the performer and for the listener.

- High fidelity (Hi-fi) music centres are available that will reproduce music from record, tapes and compact discs. These are engineered to provide stereo sound of very high quality. The highest quality is provided by digital recording which is largely free from background noise and distortion. Compact discs are scanned by a small, low-power laser and the signal is processed by a miniature computer into a very close replica of the original sound. Personal CD players as small as personal radios and are a masterpiece of electronic miniaturisation.
- Electronics have been used to amplify conventional instruments for some time, for example electric guitars. However, small dedicated computers are now used to produce a new generation of instruments in their own right. These are usually keyboard instruments. For example, the synthesiser can imitate a wide variety of conventional instruments. Electronic 'pianos' such as the Yamaha 'Clavinova' can closely imitate the real thing. For the less accomplished performer it can also add a full accompaniment while the player only has to pick out the tune. It also has the advantage that the player can practice while wearing a head set. This avoids annoying the neighbours.

Computers

Desktop and laptop personal computers are now found in most homes. They are used for a wide variety of purposes including computer games, e-mail and the Internet. More recently the advent of 3D virtual reality head sets has greatly enhanced the realism of computer games. Educational packages are also available. In addition, hand-held computer games are available that come in pocket-size packages that can be used to relieve the boredom of long journeys when travelling.

Personal transport

Personal transport such as cars and motor-cycles are now being made more environmentally acceptable and safe to drive by the use of:

- Electronic ignition and the use of fuel injection in place of the traditional carburettor. This gives enhanced reliability and performance and less exhaust pollution by ensuring a correct fuel–air mixture and complete combustion under all load conditions.
- Catalytic converters to further clean up the exhaust emissions and reduce the level of pollution even further. Catalytic converters can only be used in engines fitted with fuel injection. Most new cars are fitted with engine management computers to control the ignition and injection systems.
- Antilock braking systems (ABS) and traction control systems to improve safety when driving on wet and icy roads. Some high performance cars also have active suspension systems to improve comfort and controllability when driving fast. All these devices rely upon electronics and on-board dedicated computers. The use of computers to improve tyre design and the use of improved tyre materials has added greatly to the safety of road vehicles.
- Air bags which are automatically deployed in the event of a head-on collision. These are designed to protect the driver and, in some cars, the front passenger, from injuries caused by hitting hard objects such as the steering wheel, instrument panel and the windscreen.
- Recyclable materials such as plastics.

Bicycles have also benefited from modern technology. High-strength, lightweight materials developed for space exploration are now available commercially and some of these have been incorporated into bicycle manufacture. For example, frames can now be made from carbon-fibre materials and wheel rims from magnesium alloys. Polytetrafluoroethylene (PTFE) low-friction bearing materials can also reduce the effort required to propel the modern bicycle.

Public transport

- The development of the jet engine for civil airliners has enabled them to fly at very high speeds at altitudes above weather systems. This shortens flight times and adds to passenger comfort. They are also much more powerful and reliable than the piston engines they have replaced. Airliners also rely heavily on the computerisation of their on-board control systems (avionics) and navigational systems (navaids).
- Computerised systems and electronics are essential for air traffic control and for 'blind' landing systems at some of the bigger and busier airports. Computers, electronics and radio links are also used for booking flights, baggage handling and passenger control.
- Rail transport also relies on modern technology. For example, electric locomotives with their train of brightly lit, air-conditioned coaches are guided safely through the rail network by computerised signalling and track control.
- Road vehicle owners and drivers can now benefit from electronic navigational systems. Tracking devices for tracing stolen vehicles are also available. Closed circuit television is increasingly used to monitor and control traffic flow in conjunction with electronic roadside information displays.

Self-assessment tasks

1. Briefly describe how engineering technology has been used to make motor cars:
   (a) more environmentally friendly
   (b) more energy efficient
   (c) safer to drive
2. Compare the advantages of a compact disc with a pressed vinyl record as a means of recording music.
3. List the main uses for using a computer in the home.

Engineering technology in industry and commerce

The engineering industry itself has been revolutionised by recent changes in technology. The areas mostly affected are:

- The application of computers and computer technology to the commercial and management activities of companies, also to the design of products and the manufacturing processes involved.
- Improvements in the associated hardware (machines) and cutting tool materials, in order to take full advantage of the opportunities offered by computer-aided manufacturing (CAM).

Two aspects of computer-aided engineering (CAE) are particularly important, these are computer-aided design (CAD) and computer-aided manufacture (CAM). The latter can be broken down further as shown in Fig. 4.2.

Computer-aided design (CAD)

CAD can be used to produce two-dimensional orthographic engineering drawings more quickly and easily than by hand. Changes can be made as frequently as required without damage to the drawing 'surface'. Copies can be stored on disk more conveniently than on bulky and fragile tracings. Disks can be sent anywhere in the world easily, safely and quickly. Alternatively the data can be sent directly 'down loaded' to the customer's computer.

CAD can also be used to produce isometric and other pictorial views automatically from orthographic drawings. These views can then be rotated so that they can be looked at from different positions.

CAD can also be used to produce three-dimensional images during the design process. The digitally stored data from such images can be used for costing purposes, for the mathematical modelling required for stress calculations and performance calculations, and for simulation of the behaviour of the design under typical working conditions. Three-dimensional imaging requires very powerful computers.

Computer numerical control (CNC)

Computer numerical control enables machine tools to manufacture components of high and consistent accuracy automatically. A CNC turning centre (lathe) makes turned parts, and a CNC machining centre makes milled, drilled, bored and tapped parts. A dedicated computer replaces the operator in making the decisions required to control the machine. That is, the computer controls such functions as the relative positions of the work and the tool during cutting, turning the machine on and off, changing the speed and feed rates, turning the coolant on and off and automatic tool changing. To do this the computer must be programmed. The **program** required to control the machine can be produced manually, or on a computer using simulator software. The program can then be transferred to the computer in the machine's control unit in various ways: for example by the use of punched tape, magnetic tape or directly down loaded from the programming computer.

Like all automated machines, CNC machines:

- Never become tired and make mistakes.
- Can maintain consistently high dimensional accuracy. Some machines have built in gauging and can even adjust themselves to allow for tool wear.
- Can produce profiles and contours of a complexity and accuracy that would be impossible on manually operated conventional machines.
- Can perform many operations at one setting on a single machine, where previously the work would have to be passed from machine to machine. This eliminates the need to maintain large stocks of work in progress between the machines.

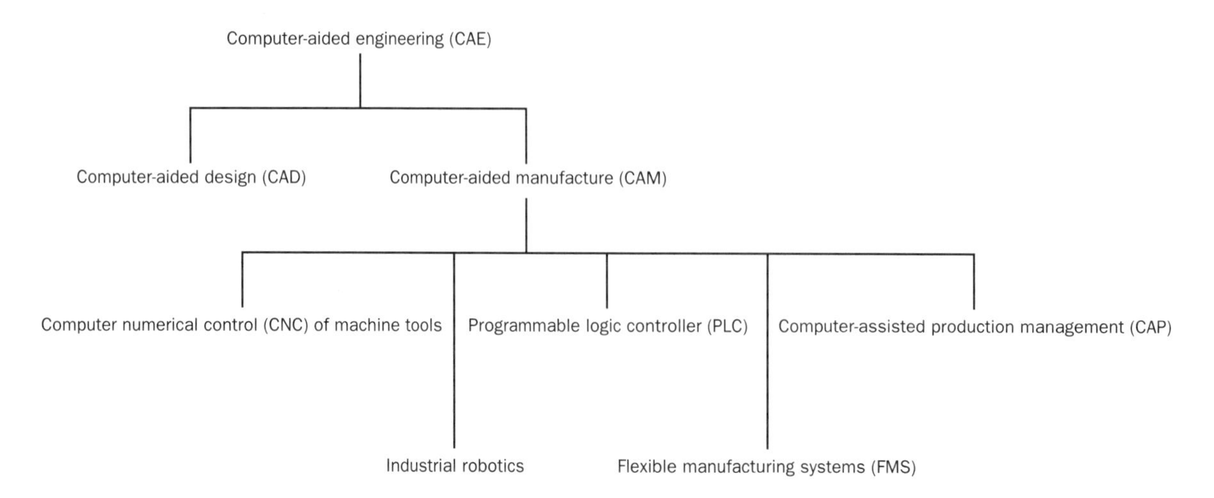

Figure 4.2 Computer-aided engineering, design and manufacture

Industrial robots

Industrial robots have no resemblance to the imaginative creations of science fiction film makers and cartoonists. Industrial robots vary widely in design according to the complexity of the work they have to do. Some examples are shown in Fig. 4.3. They are often shown on television being used in manufacturing situations for:

- automated materials handling
- loading and unloading other machines, assembly fixtures, etc.
- repetitive operations such as paint spraying and welding

There are various ways of programming robots:

- They can be controlled by a program that has been generated manually or on a computer similarly to the programs used on CNC machines.
- They can be 'taught' by a skilled operator physically moving the arm of the robot through the movements it is to make. These movements are automatically recorded in the memory of the computer. This technique is widely used for robots carrying out paint spraying and welding operations.
- The robot can be 'driven' through its movements step by step from a hand-held control keypad coupled to the robot's computer by a trailing cable. Again the movements are retained in the memory of the computer controlling the robot.

Programmable logic controllers (PLCs)

These are used to control automated processes in both the mechanical engineering and the chemical engineering industries. They can be linked with computers and/or other PLCs. Although not an industrial example, you frequently meet PLCs when you come to a road junction controlled by traffic lights. The PLC is usually mounted in the metal cabinet sited at the side of the road. It controls the sequence of light changes and the time the traffic can flow in a given direction. Because the controller is programmable, the traffic sequence can be changed from time to time to suit changes in traffic demand.

This is only a simple example compared with the complexity of controlling a chemical processing plant or an automated car assembly line catering for models with differing specifications offering you a choice of:

- body colours and interior trims
- engine sizes and types
- various combinations of doors
- optional accessories

This is only possible in volume production cars because of the flexibility of the programmable logic controllers used to control the production line and the processes feeding the production line back at the factory.

Self-assessment tasks

1. Name the type of computer controlled equipment you would use for:
   (a) making the component shown in (a) below
   (b) making the component shown in (b) below
   (c) automatically loading and unloading the machines named in (a) and (b)
2. The software and hardware you would require to produce the drawings shown by CAD.

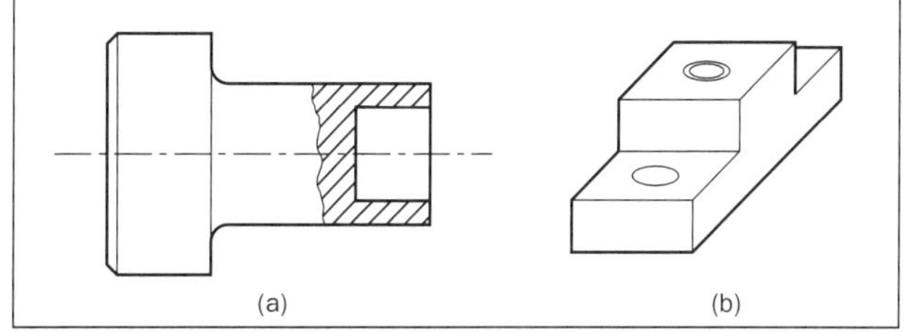

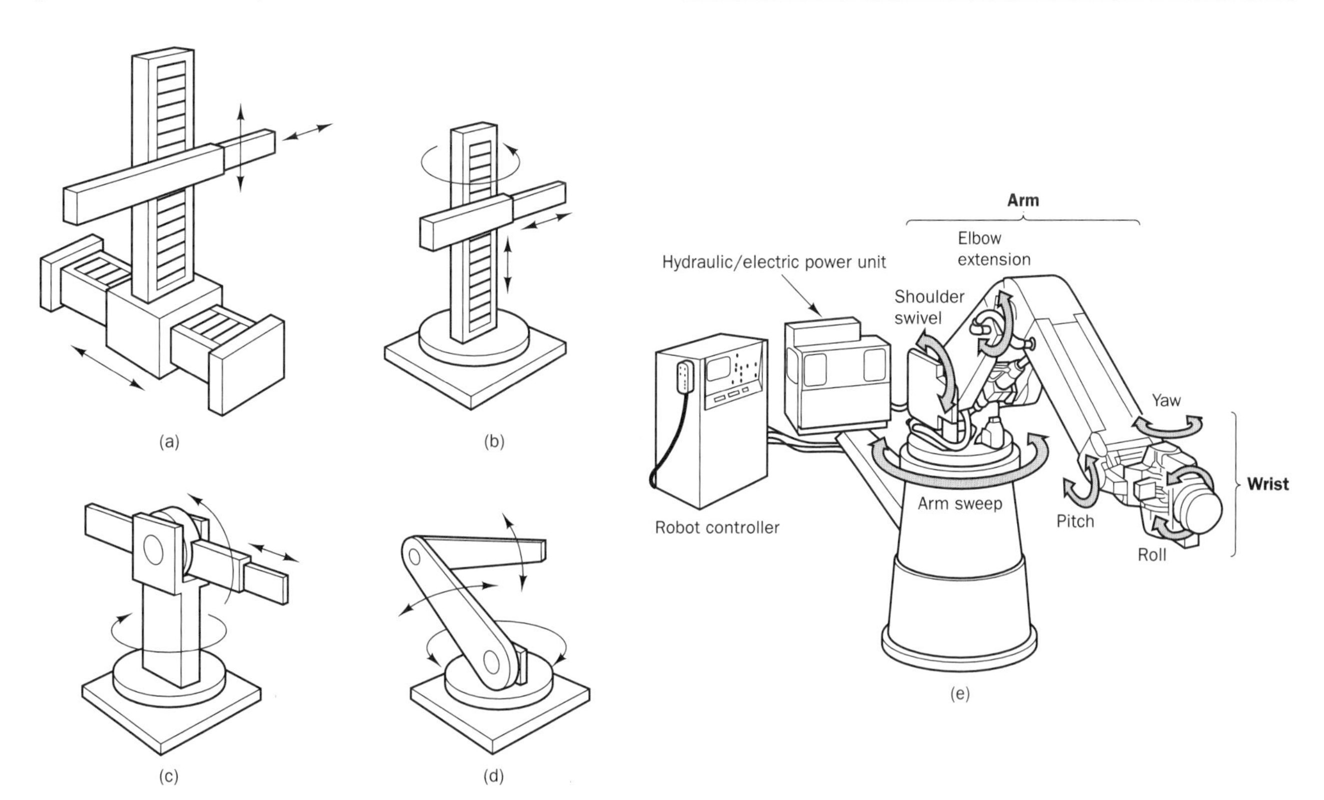

Figure 4.3 Industrial robots: (a) Cartesian coordinate; (b) cylindrical coordinate; (c) spherical (polar) coordinate; (d) revolute (angular coordinate); (e) typical robot and system (*source*: Butterworth Heinemann)

Engineering materials

The materials available to engineers are also constantly becoming more numerous and of higher strength. For manufacturing, most of these are synthetic (plastic) materials with higher strengths and better performance at elevated temperatures. New metal alloys are also constantly being developed for special purposes. For example, materials with higher strength-to-weight ratios for the aerospace industry and alloys that will operate at higher temperatures for use in jet engines.

One of the main areas of development is in cutting tool materials. For example tools made from high-speed steel (HSS) and carbide-tipped tools are now frequently given a thin coating of:

- Titanium carbide to reduce wear.
- Titanium nitride to reduce adhesion (chip welding).
- Aluminium oxide to allow alloy steel tools to operate at higher temperatures without softening.
- Ceramic, which is even harder than carbide-tipped tools, and can remove metal even faster without undue wear. Unfortunately they are brittle and easily chipped. Therefore they are best used on CNC machines where the cutting conditions are very closely controlled.
- Cubic boron nitride, which is the hardest synthetic material, second only to diamond. It is used to coat carbide tool tips and is particularly suitable for machining hardened steels and high temperature alloys.

Diamond, used for high-speed finishing cuts on non-ferrous metals such as aluminium alloys and bronze which are difficult to finish by grinding. Motor car pistons are usually finished by high-speed diamond turning.

Self-assessment tasks

Find out the name of a suitable material for each of the following applications:

1. A tough, non-corrosive plastic suitable for the moulded front and rear bumpers of cars.
2. A strong but lightweight material for tennis racket frame.
3. An anti-wear coating for a highspeed steel twist drill.
4. A cutting tool material suitable for finish turning aluminium lens mounts on a CNC lathe.
5. A suitable coating for a carbide-tipped tool to prevent chip welding when cutting steel.

Information processing is also widely used in commerce. Terminals at each workstation enable office personnel to have up-to-the-minute information. For example sales staff can instantly access stock availability, insurance brokers can give instant quotations over the phone and accounts can be kept accurate and up to date.

- Word processing software allows letters and other documents to be produced accurately with a high-quality appearance if a laser printer is used. Spreadsheet software simplifies sales, purchasing and financial planning and analysis. It also allows charts, graphs and tables to be created quickly. Documents and charts can also be enhanced by the use of colour printers.
- Telephones are interconnected by digital exchanges and satellites are used for international communications. The use of fibre-optics improves the sound quality and reliability of long distance phone calls.
- Mobile phones connect to the exchanges by radio so that they can be used at any location.
- Photocopying machines use laser technology and are electronically operated and controlled by pre-programmed microchips and capable of high quality reproduction.
- Fax machines enable complete documents and illustrations to be copied and sent over the telephone network and reproduced automatically at the receiver.
- E-mail and the Internet gives access to many new services. The user's computer is connected via a device called a **modem** to the international telephone network. This enables the user to be connected to sources of information such as public libraries, universities, banks, stock markets, insurance companies, trade organisations and customers anywhere in the world. They can then communicate with these organisations and complete business and financial transactions using their personal computers.
- The teletext facility of television receivers can also be used for stock market, exchange rates and similar financial and business reports, together with national and international news updates.

Self-assessment tasks

1. Name the electronic office equipment you would use for:
   (a) sending an illustration over the telephone network
   (b) leaving a message on a customer's bulletin board
   (c) making a phone call when off-site
2. Name the software you would require for:
   (a) producing a letter or similar document
   (b) producing a financial analysis chart

Engineering technology in health and medicine

Anyone who has watched the many hospital and medical practice dramas on television will be familiar with the many aspects of technology used in health and medicine. The use of technology can be divided between information technology (IT) and hardware.

Information technology

Computers are widely used for a variety of purposes. For example:

- The business management of general practices. Group practices employing a number of doctors, and medical and secretarial support staff, have control of relatively large budgets. Because of these budgets they are increasingly financially accountable to the local health authority. Computers are essential for collating and analysing all the data required to keep the paperwork up to date.
- The problems of the business management of a general practice is magnified many times over in a large modern hospital. Again, computers are the only means by which the management and accountancy of such an organisation can function in a market-oriented environment.
- For maintaining clinical patient records. Until recently, patients records were hand-written and kept in files in rows and rows of filing cabinets. This took up a lot of room and a lot of time in searching for the files whenever

a patient consulted the doctor. Nowadays the doctor only has to type the patient's name into his or her desktop computer and the patient's record appears instantly on the screen. Further, the doctor can look up data on medicines and drugs and other medical data that may be required. In an emergency the doctor can access a consultant by computer link.

- Appointments can also be booked by computer which will ensure that the patient is automatically routed to the correct doctor.
- For research and the collation and analysis of statistical data. As well as treating patients and performing operations, many consultants also undertake research programmes in their specialist fields. The computer can save much time by searching through data, and grouping it as required, at the touch of a button.
- Monitoring a patient's condition during operations and in intensive care units, for example blood pressure, heart beat, and respiration. As well as presenting a visual readout on screen, audible warning tones can indicate when things start to go wrong.

Hardware

This can range from a nurse's scissors to the complex instruments used by a surgeon. These are essentially mechanical devices. Examples of more complex equipment are:

- Equipment for administering a general anaesthetic so that the correct amount of anaesthetic is safely administered under the control of the anaesthetist, who also uses computerised equipment to monitor the condition of the patient.
- The **endoscope** uses fibre-optic techniques to carry out a visual internal examination without the need for intrusive surgery. A fine, flexible tube carrying glass fibres has a light at the end inserted into the patient's body via the throat or other convenient orifice, and an eyepiece or camera at the viewing end.
- X-ray machines and automated X-ray film processing equipment. Over-exposure to X-rays can cause cancer both in the patient and the radiographer. Considerable development has been made to obtain better pictures with lower radiation levels.
- Scanning equipment. Computerised axial tomography (CT or CAT) creates a three-dimensional image by taking X-rays from a camera that rotates around the body while travelling along it. Very low levels of radiation are used for safety and the image is enhanced by computer. Magnetic resonance imaging (MI or MRI) can distinguish between the different tissues in a body by comparing their magnetic properties. MRI scans can detect tumours as small as a pea. Ultrasound scanning is used in place of X-rays wherever possible as it has no harmful side effects. Sound waves (at too high a frequency for the human ear to hear) are bounced off the various internal organs of the body. A computer analyses the echoes received and builds up a three-dimensional picture. All these scanning techniques combine high technology engineering with advanced computer technology.
- Kidney dialysis equipment. These machines act in place of natural kidneys and cleanse the patient's blood of the toxic substances that are constantly building up in the human body. The patient is coupled up to the machine and his or her blood is pumped into and through the machine and back to the patient's body. This may need to be done at regular intervals or, in extreme cases, continuously. The only permanent cure is a kidney transplant.
- Pacemakers to aid patients with some heart conditions. The pacemaker is inserted into the chest cavity and produces electrical impulses that stimulate the heart and makes it beat regularly and strongly. The device contains its own built-in batteries and micro-electronic circuitry. It has to be replaced every few years when the batteries become weak. Some pacemakers have miniature computers that can sense the load being placed on the heart by the physical work being done and can adjust the strength and frequency of the stimulating pulses accordingly.
- Ambulatory bio-monitors that can transmit body function details to a nurse's workstation while the patient remains free to walk about.
- Ambulances, and helicopter 'air ambulances', used by the paramedics, equipped to give sophisticated on-the-spot treatment to accident victims.
- Physiotherapy equipment that can range from the trainer's cold sponge on the football field to the highly sophisticated equipment used in hospital to rehabilitate seriously injured accident victims.

In addition to the clinical equipment described above, operations to insert artificial joints are now routine for people suffering from arthritis or serious accidents. Artificial limbs offering high degrees of mobility have also been developed. Artificial hearts are also being developed to overcome the shortage of suitable transplant organs.

Self-assessment tasks

1. Name *three* items of equipment used by your dentist that are manufactured by the engineering industry.
2. Name *three* purposes for which your dentist would use a computer.

The environmental impact of engineering technology and activities

All industrial and commercial enterprises affect the environment in various ways. Here are some examples:

- Destruction of the environment by mining and quarrying for fossil fuels, metal ores and building materials.
- Pollution of the environment by the by-products of manufacturing processes.
- Pollution of the environment by transportation.
- Pollution of the environment by accidents, for instance oil spillages at sea and the nuclear disaster at Chernobyl in Russia.
- Disposal of waste, both toxic and non-toxic.

Most companies are becoming increasingly environmentally conscious as a result of Government legislation, shareholder pressure and the perceived advantages of good public relations. Let's now look at some of the measures taken by various sectors of industry, both in the way their businesses are run and in the nature of the products they sell.

Mining and quarrying

Open-cast mining for minerals, and quarrying for sand, gravel and rock must, by its very nature, have a marked impact on the environment. It can only take place where concentrations of useful minerals occur. Extraction is bound to leave the landscape marked and create some degree of nuisance to those living near such workings. However, locating and harnessing the world's resources are essential operations since they form the building blocks for life as we know it today.

There are three stages to every mining or quarrying operation:

- exploration and development
- operation and production
- decommissioning and rehabilitation

The three main problems associated with such operations are water pollution, atmospheric pollution from dust and the disturbance of local flora and fauna to make room for the huge pits and the access roads for the sites. For example the pit at Bingham Canyon in the USA where copper ore is quarried is some 4 kilometres in diameter. In addition, the deep mining of coal and other substances can lead to subsidence and create eyesores such as the winding gear and spoil heaps at the pit head.

The extraction of the metals and other substances from the mineral ore is another major cause of pollution. Sometimes extraction plants are built adjacent to the mine, sometimes such a plant is built on a convenient site where it can be fed from several mines and where there is an adequate supply of energy for the process.

Usually process and drainage water is stored on site in lagoons and constantly recycled so that it does not pollute any adjacent natural water sources. The lagoons are lined so that the water stored in them cannot seep into the ground and cause contamination. Any water released from the site has to undergo treatment and analysis before release.

For several years companies involved in this type of work have invested much money, time and management energy in reducing environmental impact and pollution to a minimum. They have invested in new plant and equipment which is more energy efficient, creates less pollution and less nuisance from dust and noise. Care is also taken to landscape sites so that they have less visual impact.

Although not a mining or quarrying operation in the sense of the examples described above, Richards Bay Minerals in Natal, South Africa mines heavy mineral sands by a sand dredging process. This process is environmentally benign as no chemicals or reagents are used, although it does cause some temporary disturbance. The company's policy is to reform and replant the dunes with pre-mining vegetation as mining proceeds. Figure 4.4 shows a site only 16 years after rehabilitation.

In the past, mines were opened, operated and closed with little thought for the future. Today, planning authorities insist that, as part of the terms of a mining lease, the land is returned to its former condition or to a condition suitable for a planned alternative use. There are two broad aims:

- In the short term, the aim is to create a stable and self-sustaining land surface which can resist erosion by wind and water.
- In the long term, the aim must be to return the land to a condition suitable for other forms of land use, such as the growth of native vegetation into a wildlife habitat, agricultural and forest crops, or alternative industrial use. At the Argyle diamond mine in Australia, soil, overburden and rocky waste is replaced quickly. It is put down behind the operation, in an area that has already been mined, and in the order in which it was removed. It is sown with native seeds that have been collected locally (Fig. 4.5).

Figure 4.4 Rehabilitation of a former quarry

Figure 4.5 Collecting seeds on a site to be mined, and later restored

> **Self-assessment tasks**
>
> 1. State *five* ways in which a traditional coal mine would, in the past, have adversely affected the local community.
> 2. Find out how modern coal extraction processes have largely overcome these problems.

Electrical power generation and distribution

Most of the electricity generated in the UK involves the burning of fossil fuels: either coal, gas or oil. Coal is still the most widely used fuel. When it is burnt, waste products are produced. The following examples show how PowerGen plc has reduced and controlled the waste products released into the environment:

- **Carbon dioxide** – The total emission of this so called 'greenhouse' gas has been reduced from 71 million tonnes per year in 1988 to only 54 million tonnes per year in 1995, by increasing the efficiency of power stations.
- **Particulates** (dusts) – Dusts are removed by electrostatic precipitators (ESPs). These are used at all major coal-fired power stations. A diagram of such a device is shown in Fig. 4.6. The flue gases from the furnaces pass through metal screens electrically charged to a very high voltage. This gives the dust particles a positive electric charge. The collecting plates are given a negative electrical charge. Since unlike charges attract each other, the dust particles are attracted to the collecting plates and are removed from the flue gases passing up the chimneys. The graph in Fig. 4.6 shows how efficient such devices are in cleaning up plant emissions.
- **NO_x emissions** – Compounds of nitrogen (nitrogen oxide and dioxide) tend to create atmospheric ozone and photochemical smog and must be reduced as far as possible. In conventional power stations, the fuel burns with a hot, oxygen-rich flame. This maximises the efficiency of the boiler but also promotes the formation of the harmful NO_x emissions. New means of burning fuel are being devised to reduce this problem.

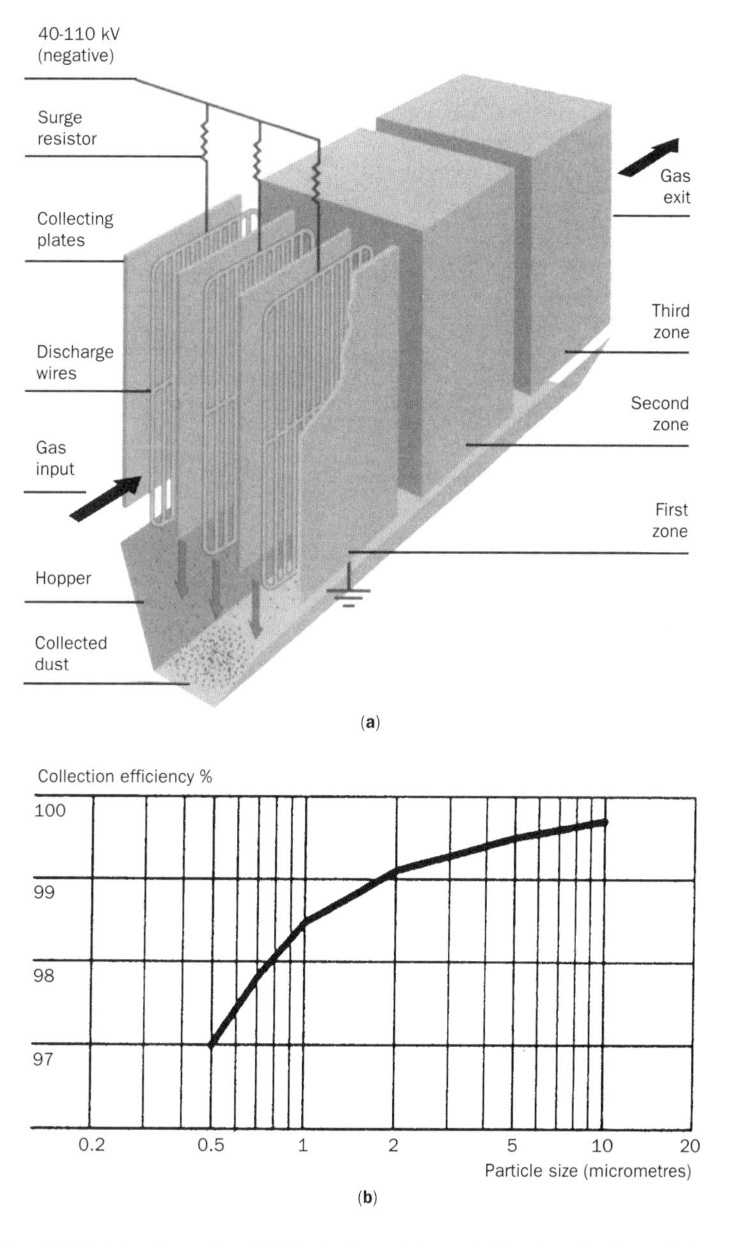

Fig. 4.6 Electrostatic precipitator (ESP): (a) schematic; (b) illustration of the variation in collection efficiency with particle size (*source*: PowerGen plc)

- **Sulphur compounds** – These combine with rainwater to form 'acid rain' that causes widespread destruction of the environment. This problem can be controlled by measures ranging from the use of low sulphur content fuels, to full-scale flue gas desulphurisation (FGD). This latter technique combines the sulphur compounds with limestone to form gypsum which is used for making cements, plasters and plasterboard for the building industry.
- **Ash** – Produced by the burning of coal. PowerGen burnt 20 million tonnes of coal in 1995 and produced 2.8 million tonnes of ash (remember that they are only one of several generating companies in the UK). This is a useful by-product and is used for making lightweight building blocks. Unfortunately the building industry cannot absorb all the ash produced and much of it finds its way to land-fill sites. It is useful since it produces no methane gas and helps to stabilise the site. It is also used in the manufacture of paints, chemicals, filtration equipment and fire-resistant products. The metals vanadium and nickel are also refined from emulsified fuel residues.
- **Water** – Used in the generation of steam and the cooling of the condensers that covert waste steam back to water to be recycled through the boilers. Some water is lost to the atmosphere as water vapour through the cooling towers. This loss is made good from the mains supply or adjacent rivers after treatment. Any water (effluent) returned to rivers, sea or sewers must be at a controlled level of purity, contain sufficient oxygen to maintain life and have a suitable pH value. It must also be at the correct temperature. All this is necessary to avoid upsetting the balance of the plant, fish and insect life in the rivers, estuaries and sea into which the effluent is discharged. It is also necessary to avoid upsetting the bacterial balance of any sewage plants into which the effluent is discharged.

Many of the above problems are reduced by the burning of natural gas or oil fuels. However supplies of these fuels are limited compared with coal. They are also more expensive. Combined cycle gas turbine stations also use natural gas. The gas is burnt in gas turbines, similar to the engines in turbo-prop aircraft, and the gas turbines are coupled to generators to produce electricity directly. The exhaust gases are then passed through boilers. The steam produced in these boilers is then used to power conventional steam turbine generators.

Nuclear power stations also operate without causing any pollution, except in the rare event of a leakage of radioactive waste. In France 75 per cent of all electricity is produced in this way. The remainder being produced by hydro-electric schemes. They no longer use fossil fuels for generating electricity. The main danger is the possibility that the nuclear plant is badly managed and a disaster occurs, as happened at Chernobyl in Russia. The other pollution hazard lies in the problems associated with the safe disposal of spent fuel rods and the decommissioning of plant that is worn out.

The rivers in the UK are not suitable for hydro-electric schemes, except in parts of Scotland. On a smaller scale, wind power is also used and experiments are taking place into the use of tidal power.

Self-assessment tasks

1. State *two* advantages and *two* disadvantages of distributing electricity by overhead cables.
2. State *two* advantages and *two* disadvantages of using nuclear power stations to generate electricity.

Natural gas exploitation and distribution

Any one who is old enough to have lived next to an old-fashioned gas works, where gas was produced from coal, will well remember the dust, dirt and unpleasant smells associated with the process. One hundred and fifty years of such gas production has left the sites of the old gas works heavily polluted with dangerous and toxic wastes. To clean up these sites will be a slow, difficult and costly process. Unfortunately, at the time that the pollution took place, we knew no better.

By comparison, the extraction and use of **natural gas** is much more environmentally friendly. It is the cleanest of the fossil fuels, little or no pollution is caused by its extraction, and it can be piped ashore so that there is no risks such as those associated with oil tankers. However, like all fossil fuels it burns to produce carbon dioxide which is the principal 'greenhouse' gas. Therefore, British Gas plc is constantly helping its consumers to become more efficient in their use of gas. This is done through research into the design of more efficient appliances, and by providing a consultancy service for its customers. Methane (natural gas) is also a 'greenhouse' gas, but the amount escaping to the environment by gas leaks is small compared with other sources and is being constantly reduced. The sources of methane gas in the atmosphere are shown in Fig. 4.7.

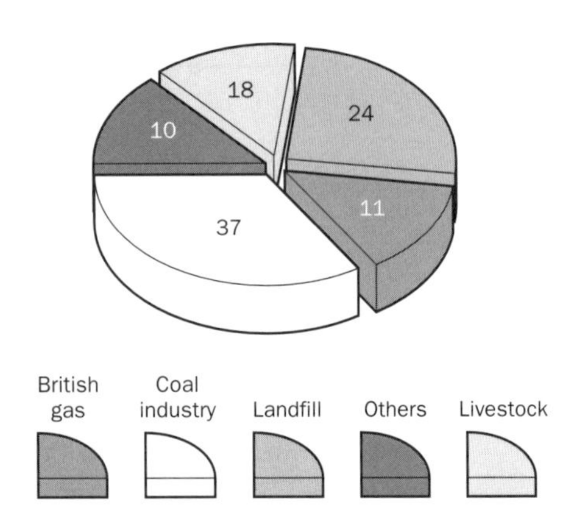

Fig. 4.7 Chart showing the percentage of methane emission in the UK in 1992

Unlike the electricity grid, it is technically possible to bury the gas grid of pipes underground. This has been done with little or no environmental damage, the land being restored and replanted after trenching and pipelaying is complete, as shown in Fig. 4.8. The pumping stations are discreetly positioned and screened by trees and other vegetation to reduce visual impact. Low frequency vibrations from such stations are also being investigated and eliminated as far as possible.

Self-assessment tasks

1. Explain briefly why natural gas is a more environmentally friendly fuel to burn than coal or oil.
2. Name *three* industrial uses for natural gas.
3. Name *three* domestic uses for natural gas.

Fig. 4.8 Gas pipe laying and subsequent restoration of the site (*source*: British Gas plc)

Manufacturing industry

The extraction and power industries described above rely for their equipment on the major engineering companies of the world. Such heavy engineering companies have to be especially careful not to cause undue pollution. However, even light industries can cause pollution and must take care not to adversely affect the environment. Let's look at some causes of pollution and environmental damage.

- The manufacturing processes used. Are any toxic or noxious fumes given off? Is dust produced in large quantities? For example, polishing and grinding equipment must be fitted with dust extractors and filters. This is not only to protect the operators but also people living and working near to the plant.
- Metal machining processes use cutting lubricants that often contain chemical additives. From time to time the tanks on the machines must be cleaned out and replenished. The waste oil must be disposed of by a licensed disposal firm; it must never be poured down the drain.
- Metal finishing (electroplating) plants use some highly toxic chemicals. Great care must be taken to protect the operatives in such plants and, on no account, must the effluent from such plants be poured down the public sewers. Spray painting is another source of pollution. Spray painting booths must offer protection to the operator so that paint particles are not inhaled, nor must they be allowed to escape into the atmosphere. Further, the operatives in metal finishing plants must be trained to observe all the safety procedures and to wear the safety clothing provided. The sites of many old factories have become so polluted by chemical spillages that they can no longer be used for new factories until they have been decontaminated. This is costly and time consuming.
- The sources of materials used. Manufacturers must ask themselves the question: Does the extraction of a particular metal or the chemical processes involved in the manufacture of a particular plastic cause pollution? If it does, they must then search for an alternative material that is less damaging to the environment. Again, hardwoods from the rain forests must not be used if fast growing softwoods from renewable sources can be used instead.
- It is no use manufacturing a product safely if the product itself is unsafe. For example, many hundreds of thousands of aerosol sprays were made perfectly safely but contained a CFC propellant that, when used, damaged the ozone layer. Again, it is no use safely producing an insecticide or a weed killer that will make people ill when it is used.

Responsible manufacturers carry out expensive research to make their products more environmentally friendly and efficient. One such company is GEC Alsthom. This company combines the activities of the UK-based GEC with the French company Alsthom. In addition to the environmental management of their manufacturing plants to comply with or improve upon national and international legislation, they are involved in the manufacture of the following products, all of which are designed to cause as little environmental impact as possible.

- Power station boilers throughout the world. These include the circulating fluidised-bed boilers that enable the pulverised coal to be burnt efficiently at lower temperatures so as to reduce the production of harmful NO_x emissions.
- Firing equipment for pulverised coal, oil and gas including low-NO_x burners: grates, spreader burners and ash removal systems.
- Industrial boilers for all types of industrial and district heating requirements. These include waste-to-energy plants for municipal solid waste incineration, chemical recovery boilers for the pulp and paper industry, waste heat recovery systems and electrically heated boilers.
- Environmental systems, including the electrostatic dust collection and flue gas desulphurisation plant mentioned earlier, rotary kilns for industrial refuse incineration and sewage sludge drying, and the cleaning up of contaminated sites.
- Nuclear power station equipment.
- Advanced treatment plants for producing fresh water.
- Customer services to ensure that the products of the company are maintained and operated at the maximum efficiency and minimum environmental impact.

Self-assessment tasks

Using the resource material provided by your teacher, list *five* ways in which a large engineering company of your choice has reduced the environmental impact of its processes and products since 1990.

Chemical engineering industry

The chemical engineering industry faces a number of problems, not only of a practical and technical nature, but also those arising from media attention following accidents resulting in the loss of life and severe pollution. In addition:

- The plants are large, unsightly and consume large quantities of energy.
- The processes are often potentially hazardous and employ highly dangerous and toxic substances.
- Some of the products from such plants are also potentially dangerous and toxic; great care must be taken in their transportation and use or, in the case of waste materials, their disposal.

Let's look at some of the published figures of ICI, our biggest chemical engineering company.

New plant

ICI requires all its new plants to be built to standards that will meet the regulations it can reasonably anticipate in the most environmentally demanding country in which it operates the process. This will normally require the use of the best environmental practice within the industry.

Waste disposal

Figure 4.9 shows how the levels of waste disposal of this company have been reduced since 1990. These wastes are divided into two categories: hazardous wastes and non-hazardous wastes. Some of these wastes are emitted to air, some to water (rivers and sea) and some to landfill sites. The figures quoted show the position at the end of 1995 (improvements are continuing to be made all the time).

- Overall, hazardous wastes have been reduced by 69% (19% better than the target).
- Hazardous wastes now represent only 3% of the company's total waste.
- Non-hazardous wastes are down by 21% against a target of 50%. This shortfall is due to the fact that hazardous acid waste from the titanium oxide plant is now converted to non-hazardous gypsum which is not only environmentally benign but can be advantageous in the stabilisation of landfill sites. It is also used in the construction industries in the manufacture of cements, plasters and plasterboard. The spent brine from chlorine manufacture is simply salt water and can be disposed of safely into the sea. There is no economically viable way of reducing the volumes of these waste products.

Energy efficiency

Figure 4.10 shows the extent by which ICI reduced its energy consumption by operating its plants more efficiently. The saving in 1995 was 18% of the energy used in 1990. This resulted in a corresponding reduction of 'greenhouse' gas emission.

Recycling

An increasing number of products are now reused, recovered and recycled. These include plastics such as PET and some acrylics. ICI is developing new, recyclable products which have minimum impact on the environment throughout their life.

Product stewardship

The company recognises its responsibilities in all areas of safety, health and the environment for all the products it makes and sells. This responsibility applies to every stage of the products life: from design through production sale, use and final disposal.

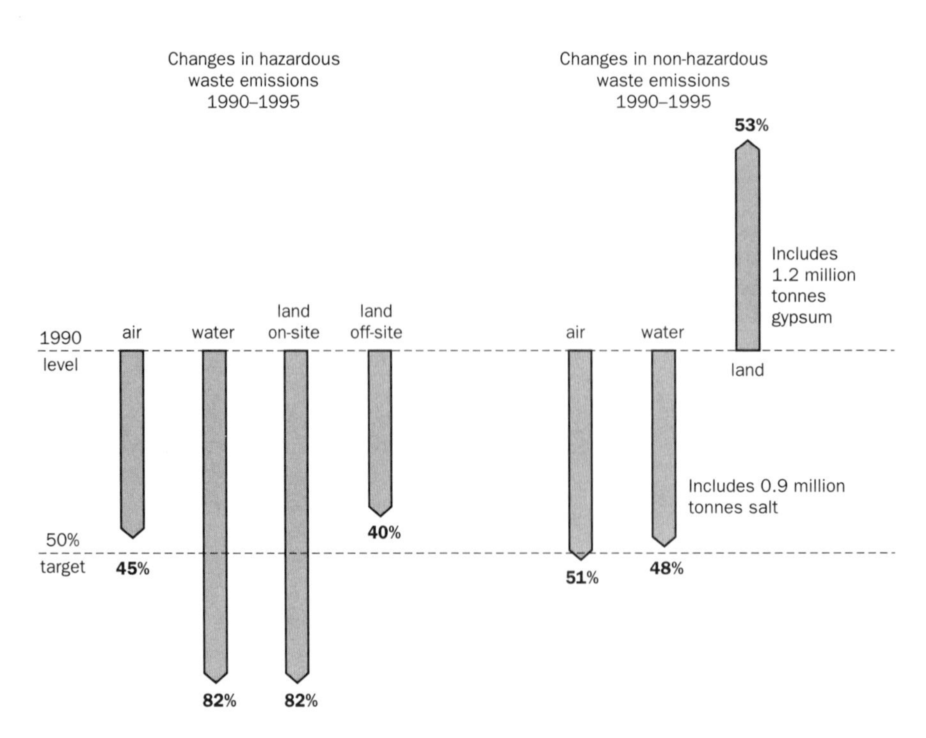

Fig. 4.9 Reduction of waste disposal at ICI (*source*: ICI plc)

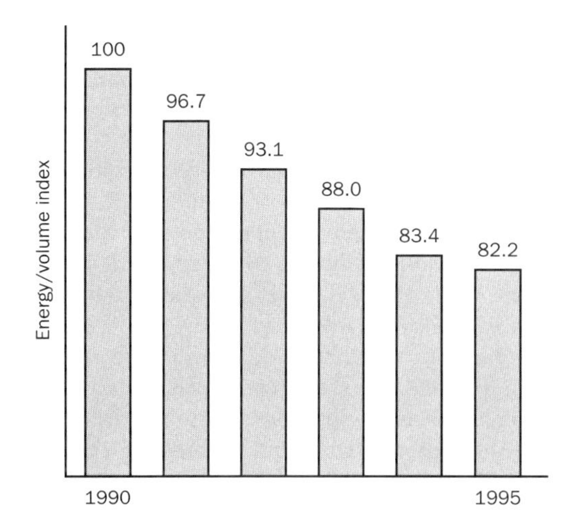

Fig. 4.10 ICI's carbon dioxide emissions (1990–95) (*source*: ICI plc)

Sustainable development

It is the policy of ICI to apply innovative technology and engineering skills to produce new products with lower environmental impact and more product from less raw material, more efficiently and, where possible, from recycled or renewable sources. Global society expects the future to be sustainable.

Bio-diversity

ICI has its own nature conservation programme called 'Nature Link' to protect plants and animals on its property around the world. Information gathered from its sites has been stored on a computerised database and will help in the management of bio-diversity throughout the world.

> **Self-assessment tasks**
>
> Find out how the paints used to decorate our houses have been:
>
> 1. Made more safe to use.
> 2. Made better at protecting the woodwork from damage and weathering.

Nuclear engineering

Because of the adverse publicity surrounding this industry resulting from the after effects of the atomic weapons used in the second world war and, more recently, the Chernobyl disaster, the industry has to be seen to be 'whiter than white'. There is no denying the potential hazards of the industry and, for this reason, it is so closely controlled and rigorously managed that, in practice, it is far safer and more environmentally friendly than many conventional industries.

In nuclear power stations the heat from controlled atomic reactions is used to produce steam to drive conventional turbo-alternator sets. Individually the radioactive 'fuel' rods have too low a mass to start a reaction. When several are brought together in a reactor, however, they achieve 'critical mass' and a reaction starts. Control rods that absorb the radiation can be raised or lowered between the fuel rods and these control the progress of the reaction. If the control rods are all lowered to separate the fuel rods, the reaction stops. The intensity of the reaction is constantly monitored and automatically controlled to keep it at the correct and safe level. In the case of Chernobyl, the automatic safety control systems were manually over-ridden in an unauthorised and unsupervised experiment.

Figure 4.11 shows the main sources of radiation. The radiation emitted by discharges from the nuclear industry in the UK is negligible compared with the background radiation occurring naturally.

Volumes of radioactive waste are small compared with hazardous wastes arising from other sources in the UK. The waste is stored under safe conditions either in liquid or solid form within the industry's own controlled sites. Non-hazardous, non-radioactive wastes discharged from BNFL sites are also controlled by independent Government organisations.

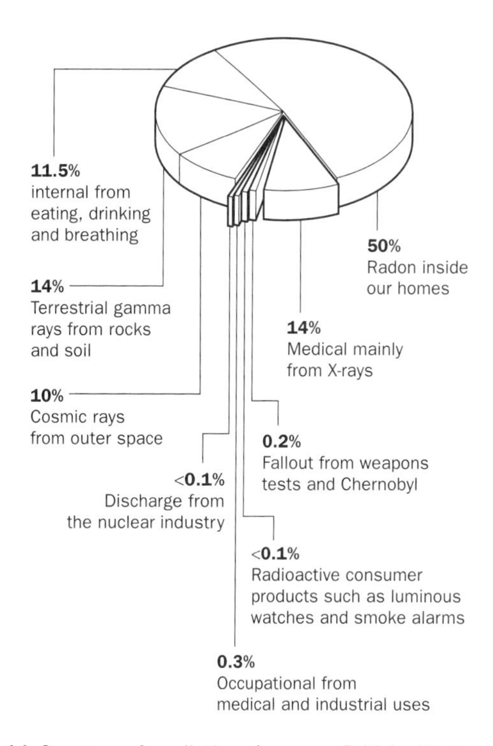

Fig. 4.11 Sources of radiation (*source*: British Nuclear Fuels Limited)

Radioactive materials (radioactive isotopes) are used for a variety of purposes. For example, X-ray testing of welds in pressure vessels and pipe lines. They are also used for the sterilisation of medical equipment and for the sterilisation of cereals and other food stuffs.

> **Self-assessment tasks**
>
> 1. List *three* non-military uses for radioactive products.
> 2. List *two* ways in which radioactive waste can be safely stored until it becomes inactive.

Waste management and disposal

The need for the proper management of waste disposal is summed up clearly by Mr Martin Bettington, the Managing Director of Biffa, in his introduction to a report by his organisation entitled 'Waste: Somebody Else's Problem'. He also draws attention to the fact that we, as consumers, are going to have to pay increasingly for environmental protection and waste disposal within the purchase price of the goods we buy. He states:

> *Society has always produced waste because it consumes – a fact that now provides archaeologists with the detail of history, but one which gives modern society a problem of disposal. As society has developed so the problem of waste has increased but, paradoxically, it is a problem which until recently has been either ignored or sensationalised.*
>
> *Society, industry and consumers, have been embarrassed to confront the issues of waste. Its disposal has simply been 'Somebody else's problem'. However, growing environmental concern, growing consumerism, and the relentless growth in the costs of safe waste disposal are obliging society to face the issues.*
>
> *Industry's attitude to waste has always been a reflection of society's attitude. Until now, waste has been a pariah issue. For all industries, save those capable of spectacular pollution, managing waste has been irrelevant to business performance.*
>
> *Suddenly this is no longer the case. Just as society has been obliged to confront its production of waste, so several factors – finite natural resources, environmental consciousness and consumer pressures – have given rise to increased regulation, spiralling disposal costs and shareholder concern. As a result, every aspect of the commercial environment has been affected.*

Self-assessment task

Explain briefly why we are going to have to pay more for the disposal of waste in the future and why this cost will be related increasingly in the cost of the goods we buy.

Let's now look at how the disposal of waste is managed and regulated. Currently the regulations are based upon the *Control of Pollution Act of 1974: Section 17*, and the *Special Waste Regulations* derived in 1980 from the powers given in Section 17. However, new directives from the European community are imminent.

Planning applications

The establishment of new industrial and commercial plant and activities, or changes of activities and purpose is subject to planning controls. The Local Authority planning officers will request a detailed assessment of the processes to be undertaken and their effect upon the environment.

Waste management (within Local Authority control)

Local Authorities have a three-fold interest in waste management:

- **Waste collection** – Mainly domestic and some small commercial concerns, including a duty for recycling wherever possible (e.g. waste paper, empty glass bottles and garden refuse (compost)). Industrial waste is not included.
- **Waste disposal** – The disposal of all the waste the collection team hands to them in regulated landfill sites and incinerators. It should be noted that waste disposal is due to be privatised in the near future.
- **Emissions to the atmosphere** – This comes within the *Environmental Pollution Act of 1990: Part 1*:
  - *Section A* is concerned with large scale plant such as power stations. It is also concerned with all chemical plant. Such plant falls outside Local Authority control and will be considered later.
  - *Section B* falls within Local Authority control, and is administered by the environmental health officers through a licensing process 'Authorisation for Emission to Atmosphere'. The main method of control is by input to the process and the scale of the process. For example, incinerators must have a through-put not exceeding one tonne per hour. Control is also by the type of plant to be used and the type of particulate and gaseous substances discharged by any means.

Waste management (outside Local Authority control)

From Monday, 1 April 1996 – vesting day for the new Environment Agency – the officers of Her Majesty's Inspectorate of Pollution have been absorbed into the new Agency and are responsible for authorisation for 'Emissions to Atmosphere' and Section A processes under the *Environmental Pollution Act* mentioned above. That is, all incinerators over one tonne through-put per hour, all chemical plant and all other 'specified' processes. The authorisations issued by the Agency:

- Cover inputs to the process and the nature of the process itself.
- Cover lists of polluting substances that may be *emitted direct to the atmosphere* and limits the volume per hour of such pollutants.
- List emissions to sewers and any effects on surface water drainage.
- List the nature of solid residues, the need for further treatment before disposal and the suitability of such residues for direct disposal in landfill sites.

Other industrial processes not covered above, may yet fall within the jurisdiction of the Industrial Pollution Control (IPC) officers of the former National Rivers Authority (NRA) – now incorporated into the Environment Agency – who provide 'control through consent to discharge'. This lists: temperature, rate of discharge at a set volume per hour, pH value, sediment, toxicity and heavy metal content in particular and also the bio-oxygen demand.

The Waste Regulatory Authority (WRA) is now incorporated in the Waste Agency and control is exercised:

- By the issue of waste management licences covering the keeping (storing), treating, transport and disposal of controlled waste.
- Through regulation by (special) waste management licences and associated paperwork. The paperwork giving notice to transport waste must precede every movement and copies must accompany every movement of the waste. This not only applies to persons actively handling and transporting the waste but also to brokers and intermediaries who may not physically handle the waste.

- By the registration of carriers of waste. Only the carrier's head office needs to be registered, and this registration applies to all branch offices throughout the UK. Section 54 places a *general duty of care* on the carrier of controlled wastes. This includes a requirement for the carrier to take such wastes only to authorised sites. The main controlling aspect of the duty of care is the adequate description of the waste and any special hazards. This description must accompany any transfer of the waste both physically and in respect of its ownership.

Self-assessment tasks

1. State, in your own words, the main differences between Section A and Section B emissions as defined in the Control of Pollution Act of 1990.
2. What are the basic legal requirements for the storage, treating, transporting and disposal of 'special' wastes?

Waste disposal

Having looked at the management of waste, let's now look at the most important method of waste disposal in the UK, namely **landfill**. The scale of the problem is enormous. If the annual solid waste disposal to landfill sites in the UK was loaded into heavy lorries, it has been estimated that they would fill *a six lane motorway stretching from London to Tokyo*. Figure 4.12 puts some facts and figures to the problem of waste disposal.

Landfill presently accounts for 90 per cent of controlled waste in the UK. It is, however, not just a question of finding a big hole and filling it! Before a site can be used it must be properly engineered and prepared as shown in Fig. 4.13. The following notes refer to a professionally managed site incorporating the requirements of the Health and Safety Executive (HSE) and the Waste Regulatory Authority (WRA).

The officers of the HSE are responsible for ensuring that owners of the site do not put at risk the health and safety of:

- their employees working at the site
- visitors to the site (e.g. delivery drivers)
- persons living and working adjacent to the site

The officers of the WRA are responsible for ensuring that the well-being of the environment as a whole is not put at risk by operations at the site. For example, no toxic waste must be dumped, no noxious or explosive gases must be allowed to accumulate, and the ground surrounding the site and the water table beneath the site must not be contaminated.

To comply with these requirements the site must be engineered as follows:

- An engineered pit lining is constructed to seal the waste from the surrounding rock, soil strata and water table. Water entering the site must be contained within the site. Capping systems and small working faces restricts the ingress of water.
- Rubbish is deposited in consistent, even layers to strict engineering procedures. This ensures safe decomposition and a stable body of refuse.
- Decomposing waste can generate landfill gas (LFG) and noxious liquids (leachate). The site must be regularly checked for gas migration and ground water quality.
- Currently, 70 per cent of LFG (mainly methane) escapes to the atmosphere and the rest is either flared off or collected for electricity generation. Of the 66 000 MW total of electricity generated in the UK only 32 MW is produced from landfill waste. However, this figure is rising.

Landfill operators must not only provide reassurance of minimal impact on local communities during the site's productive life, but for many years after it has been filled.

Filled sites offer opportunities for landscaping and development of public open spaces in areas of former industrial and mining dereliction. Restoration is now a key part of landfill management since it returns sites to agricultural and recreational use. Thousands of trees are often planted around the perimeter of large modern sites.

Self-assessment task

Describe what you think would be the consequence of filling a hole with rubbish in an uncontrolled manner and then building a housing estate over it. It has happened in the past!

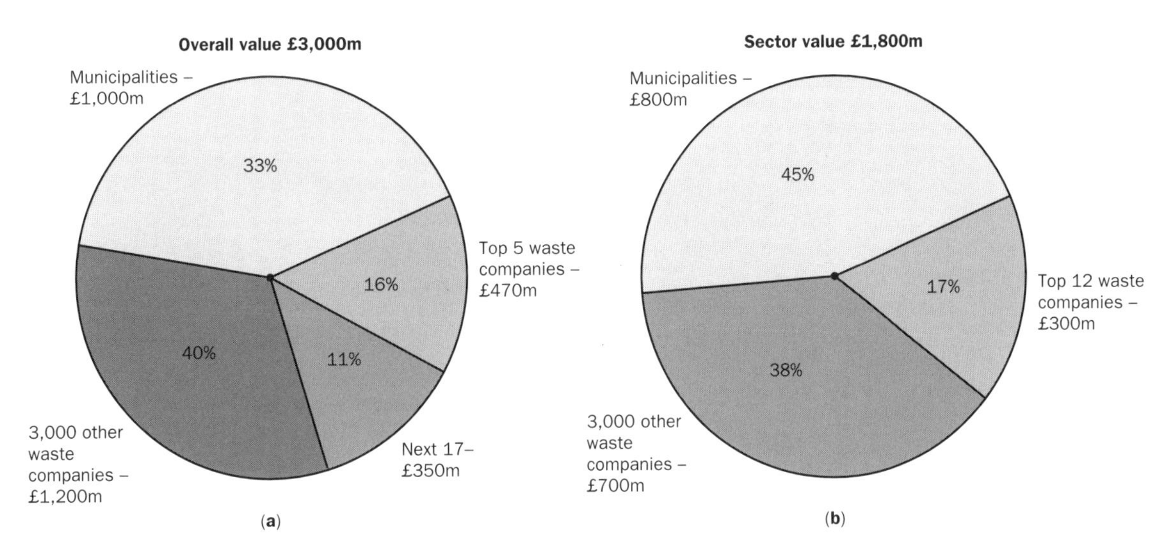

Fig. 4.12 UK waste industry sector: (a) market size and share; (b) non-hazardous collection (*source*: Biffa Waste Services Limited)

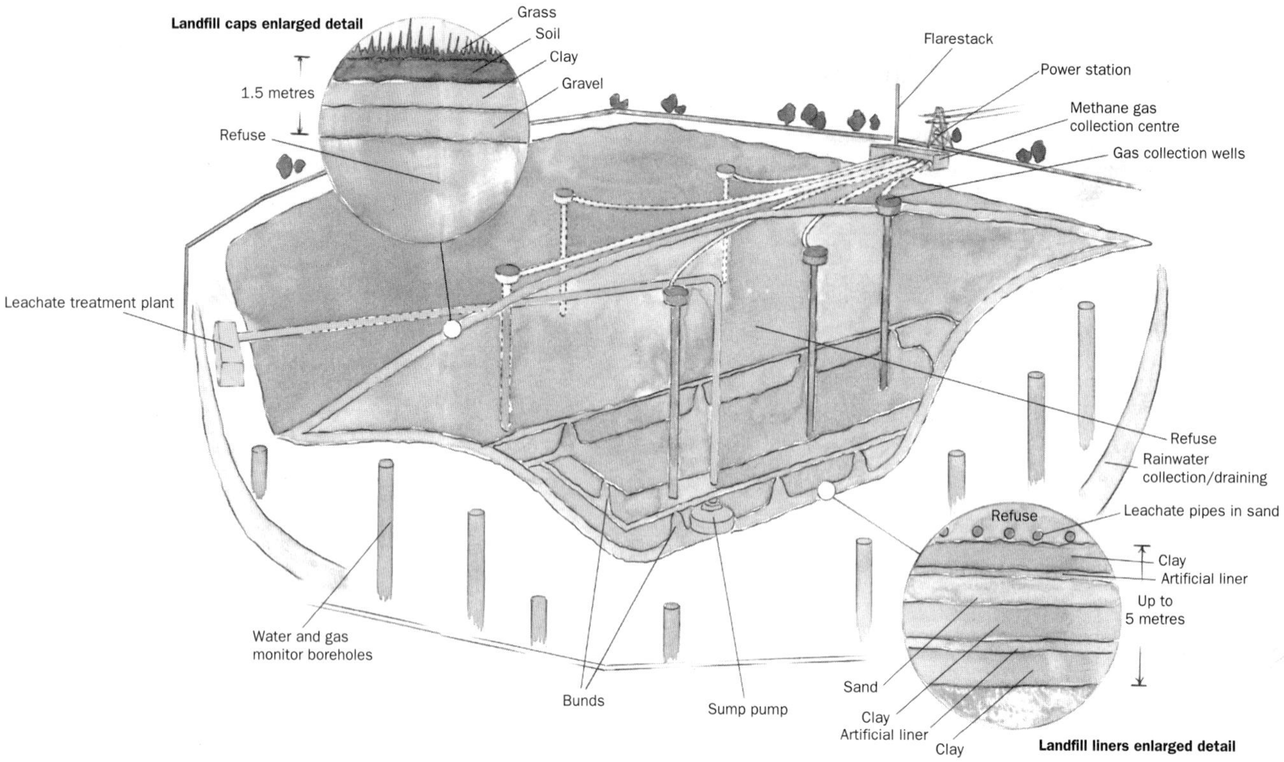

Fig. 4.13 Construction of a modern, fully engineered, landfill site (*source*: Biffa Waste Services Limited)

4.2 Career options and paths in engineering

After reading this section you will be able to:

- Identify and give examples of employers in each of the main sectors of industry.
- Identify suitable career options and pathways using main sources of information.
- Identify and describe qualifications necessary for your own intended career progression and how to acquire them.
- Describe different ways of presenting personal information to prospective employers (job applications).
- Produce suitable curriculum vitae for submission to prospective employers.

We have already seen, in the first part of this chapter, that the days of the 'dark satanic mills' of the industrial revolution have largely passed. Modern engineering skill and ingenuity is aimed at helping us to become 'greener'. For example, the development of aerosols that do not use the CFCs which destroy the ozone layer, finding ways to recycle waste, and designing machinery that is more energy efficient to reduce the risks associated with global warming. But who are engineers, and what is engineering?

The newspapers, radio and television often use the terms **engineering** and **engineer** incorrectly and this can cause confusion. They frequently imply that everyone who works in the engineering industry is an 'engineer'. This is as incorrect as implying that everyone who works in an accounts office is a chartered accountant and that everyone who works in a law court is a barrister.

In the engineering industry, the accounts office and the courts of law, there are people with various levels of qualification doing a range of essential jobs. This section examines various aspects of the engineering industry, the jobs available and the qualifications required.

Self-assessment task

Write down how you would describe an 'engineer' and what he or she does. I will ask you to do this again later in order to see if your views have changed.

The engineering industry

Figure 4.14 shows how the engineering industry is divided into several main sectors. We will now examine what is done by these various sectors of the industry.

Mechanical engineering

This is concerned with all the parts of machinery that move. Mechanical engineers must use its principles to design, develop, install, and maintain all machinery. That is, everything from a dentist's drill to a robot or a giant crane. Design and development has to take into account the stresses and strains that occur between the moving parts and it has to find ways of overcoming problems of vibration, lubrication and weight support. Mechanical engineers involved in this sort of work must also understand the sources of energy that makes the machinery move, whether it is gas, steam, electricity, petrol, diesel, hydraulics or compressed air. They must also have a wide knowledge of engineering materials and metal finishing.

Electrical engineering

This deals with all electrical power generation, power distribution and power use. It is crucial in providing the electrical energy that lights our homes, runs our television sets and video recorders and also keeps the wheels of industry turning. It is concerned with the design and manufacture of the electric motors that are found in domestic appliances such as vacuum cleaners, food mixers, washing machines, power drills and hair dryers. Electrical engineers are concerned with the design and manufacture of the electric motors used for driving the machines used by industry. They are also concerned with the design and manufacture of switch gear for controlling such motors. The electrical engineering industry is concerned with domestic and industrial heating applications. It is concerned with the electrical systems for motor vehicles – the generators, the starter motors and batteries. It also produces all the batteries used in portable radio equipment, digital watches and

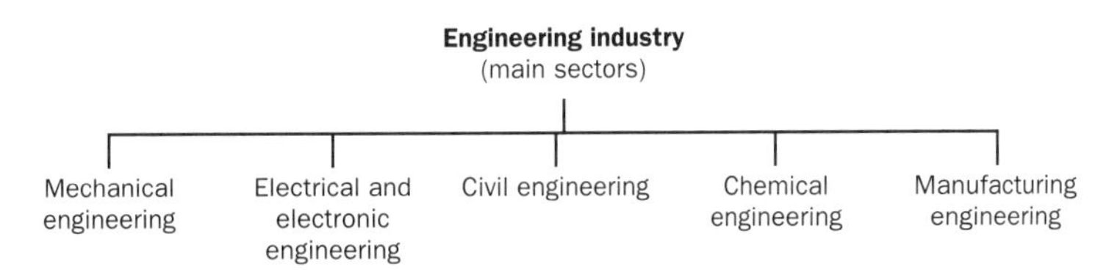

Note that other sectors of the engineering industry combine one or more of the above. For example the automobile engineering combines:

- Mechanical engineering
- Electrical & electronic engineering
- Manufacturing engineering
- Chemical engineering (paints and lubricants)

All the above go to make up a *car*. Then you require:

- Civil engineering to provide the roads to run it on

Fig. 4.14 Main sectors of the engineering industry

clocks, calculators, laptop computers, smoke detectors, alarm systems and portable lighting equipment. As well as the design and manufacture of such equipment, electrical engineers are responsible for its safe installation and the maintenance necessary for its efficient and safe use.

Electronic engineering

In general, electronic engineering is concerned with small current devices. It covers: telecommunications (telephones, radio, television) and computers in all their guises. It is responsible for the signalling systems that makes our journeys safer, for air traffic control using radar equipment, and for satellite navigation, tracking and weather forecasting systems. The electronics industry provides the equipment for all aspects of computer-aided manufacture (e.g. computer-aided design, computer numerical control for machine tools, robotics and programmable logic controllers). At the other end of the scale it provides the programmed controllers for domestic appliances such as washing machines. The engineers employed in this industry are responsible for the design and construction of all forms of electronic equipment including computers, systems installation and writing software, and the maintenance of all such equipment.

Civil engineering

Civil engineers are responsible for the infrastructure of our towns and cities. They are concerned with designing and building structures such as bridges, roads, tunnels, dams, power stations and airports. They are concerned with the water supply and the drains that provide the safe removal of storm water and sewage. Civil engineering made the Channel Tunnel possible. It was their skill that allowed the two ends of the tunnel to meet under the middle of the channel. It is civil engineers that will be co-operating with the mechanical and electrical engineers to bring us the transportation systems of the future.

Chemical engineering

Chemical engineering takes the experimental processes and chemical reactions of the laboratory and reproduces them on a large scale in the chemical manufacturing and processing industry. It takes raw materials and changes them chemically on an industrial scale. For example, changing sand into glass and refining crude oil into useful fuels, lubricants and the feed stocks for the plastics industry. Feed stocks are chemical by-products from the oil refining industry that can be further processed to make materials for plastic moulding and synthetic adhesives.

The chemical engineering industry is wide ranging and is concerned with the manufacture of explosives and fertilisers (which can be chemically similar), with pesticides and fungicides for the farmer, with paints and dyestuffs, and with the brewing, food processing and pharmaceutical industries. Because many of the chemicals used are explosive, flammable and toxic, and because so many of them are produced and stored on a large scale, chemical engineers and their industry have a very considerable responsibility for safety. Even small accidents can have a devastating effect on the local environment and a major disaster, such as the one at Bhopal in India a few years ago, caused widespread environmental damage and massive loss of life.

Manufacturing engineering

This is concerned with the changing of raw materials into manufactured items such as cars and aeroplanes, washing machines and bicycles. Manufacturing engineers must:

- Sort out the best processes and the best machines to convert the specified materials into the finished product at a competitive yet profitable cost.
- Control the rate of production so that the right goods are available in the right quantities and at the right time.
- Monitor the processes so that the required quality is maintained and, if possible, improved.
- Be familiar with all aspects of computer-aided manufacture.

Manufacturing techniques vary with the size of the product and the quantities involved. Large products such as ships and oil platforms are built as 'one-offs'. It is not commercially possible to build prototypes and test them to destruction to prove the design; the design and development has to be right first time. This puts a huge responsibility on the project engineer and the engineering team.

The automobile industry is concerned with the mass production of highly technical products for the mass market. It calls on the skills of chemical engineers for paints, tyres, fuels and lubricants, mechanical engineers for the body shell and machinery, electrical and electronic engineers for all the systems and instruments, and textile designers for the interior trim. Industrial designers (who are artists and sculptors rather than engineers) are called upon to make the shape of the car attractive to the public yet practical in use. The manufacturing engineers have to allow the vehicles to be made at a competitive, profitable price while constantly improving the already high quality and reliability. The manufacturing engineer has to be involved from the start since it is no good producing an ideal design that is impossible to make or to make at a competitive price. Because of the volume of production the manufacturing engineers are at the forefront of their profession and use the very latest, state-of-the-art, computer-aided automated equipment and production management techniques.

Aeronautical engineering produces very complex and large aircraft on a batch production basis. The research and development costs are enormous. Although, like shipbuilding, you cannot produce prototypes to test to destruction, many of the individual components are tested in this way. The first aircraft produced is usually used as a flying laboratory or 'test-bed' to prove the design and obtain approval by the Department of Civil Aviation before production commences. Again, many different engineering skills are combined together by the project director and, although the volume of production is small compared with vehicle engineering, the manufacturing engineers have many problems because of the accuracy, size and relative fragility of many of the airframe elements. Quality control is of major importance in aeronautical engineering; nothing can be left to chance.

In addition to the main sectors mentioned above, there is agricultural engineering, biochemical and biomedical engineering, marine engineering, mining engineering, structural engineering, building services engineering and nuclear engineering.

Self-assessment task

State the engineering sector in which you would most like to work, giving reasons for your choice.

Career options and qualifications

The jobs in engineering are as varied as the industry itself. You may find yourself working in a laboratory or in a workshop, in a design office using a computer or on a building site installing air-conditioning and heating equipment.

There are, however, certain things that people working within engineering have in common. They are likely to be using state-of-the-art technology, often computer driven, for all manner of tasks (designing, testing, welding, machining and so on). You may be working in a factory in your home town or on site anywhere in the world. You will most likely work as part of a team, all of you working to the same end.

Jobs in engineering also depend upon your personal talents and ability, your aims and ambitions, and your aptitude for, and desire to, study. The following job titles are listed in ascending order of skill, ability and qualification:

- **Operator/assembler** (on-the-job training)
- **Craftsperson** (high levels of skill training plus a City & Guilds or an RSA qualification)
- **Engineering technician** (specialised training plus a City & Guilds or an RSA and/or a BTEC qualification)
- **Incorporated technician engineer** or just **incorporated engineer** (specialised training plus a higher BTEC qualification or a degree)
- **Professional (chartered) engineer** (honours degree plus post-graduate training plus a number of years' experience in a responsible post in the industry including management of a technical/research department). Many such persons will also have higher degrees and management qualifications as well.

The vocational qualifications available for persons entering the engineering industry have undergone major changes over the past few years and are unlikely to settle down until the turn of the century. The leading examining bodies for these qualifications are The City & Guilds of London Institute (CGLI), The Royal Society of Arts (RSA) and the Business and Technician Council (BTEC). The examinations taken by undergraduates in obtaining their degrees are governed in a different manner.

The qualifications offered by BTEC are First Certificates, National Certificates and Diplomas, and Higher National Certificates and Diplomas. The certificates are intended for technicians undergoing training in industry and attending college on a part-time basis. The diplomas are for full-time students. Suitable levels of passes by certificate and diploma students are acceptable by most universities as an alternative entry qualification to A-level passes. At a lower level, craft and operative qualifications have been offered for many years by the CGLI.

More recently, an additional vocational qualification has been introduced. This is the General National Vocational Qualification (GNVQ) and it is offered by all the major examining boards given above. This qualification is offered at three levels as follows:

- **Foundation level** – This is equivalent to GCSE grades D, E, F, G.
- **Intermediate level** – This is equivalent to GCSE grades A, B, C or a BTEC First Certificate.
- **Advanced level** – This is equivalent to an A-level or a BTEC National Certificate/Diploma.

In all instances the GNVQ is more vocationally biased compared with its academic counterpart. The current route for BTEC First and National awards and GNVQ qualifications is via further education colleges. However, it is inevitable that many of these qualifications will become available through the sixth form departments of secondary schools and in sixth form colleges and technology colleges. To provide the essential vocational background in equipment and teaching, and to avoid duplication of resources, schools will often work in conjunction with the established technical and commercial departments of their local further education college.

For the advancement of persons already in employment there are the National Vocational Qualifications (NVQs). These must not be confused with the GNVQs already mentioned. The lead body for engineering NVQs is the Engineering Training Authority (EnTra).

Pathways to becoming an engineer

Figure 4.15 shows an outline of the routes by which you can become an engineer in one of the categories described above. It also gives you an indication of the GCSE levels and subjects you should be aiming for if you have not yet left full-time education. Remember that the professional engineering institutions do not recognise all university engineering degrees, only those from universities on their approved list. This should be checked when applying for your course.

The **Modern Engineering Apprenticeship** is gaining increasing importance. It is relevant to companies of all sizes in all sectors of the engineering/manufacturing industry. It has been developed by the Training and Enterprise Councils (TECs) together with EnTra. The former is a Government agency responsible for the local organisation and funding of training and vocational education. The latter is the Government-approved organisation set up to develop the content of training schemes and to set training standards in the engineering industry. It is also the lead body for NVQ qualifications.

Modern apprenticeships are extremely flexible and can be tailored to suit the needs of the company and the trainee (you). The structure of the apprenticeship will draw upon the experience of EnTra and make use of their NVQs to ensure that the training given will be of a consistently high quality. The structure of the modern apprenticeship is shown in Fig. 4.16.

The foundation stage normally takes place 'off-the-job' and provides a broad range of basic skills. It also provides the knowledge and understanding of engineering in general and prepares the apprentice for further training towards their likely occupation. The EnTra level 2 NVQ may form the basis for this. The development of core skills and vocational education start at this stage and continue throughout the apprenticeship.

The post foundation stage takes place largely 'on-the-job' and is devoted to developing the skills, knowledge and understanding needed to achieve the selected NVQ.

Associated vocational education must be provided by any centre approved to deliver engineering apprenticeship programmes. It could be a full qualification (BTEC, City & Guilds, GNVQ) or a selection of units or modules to suit individual needs. This education will most likely be provided by day release to a local further education college.

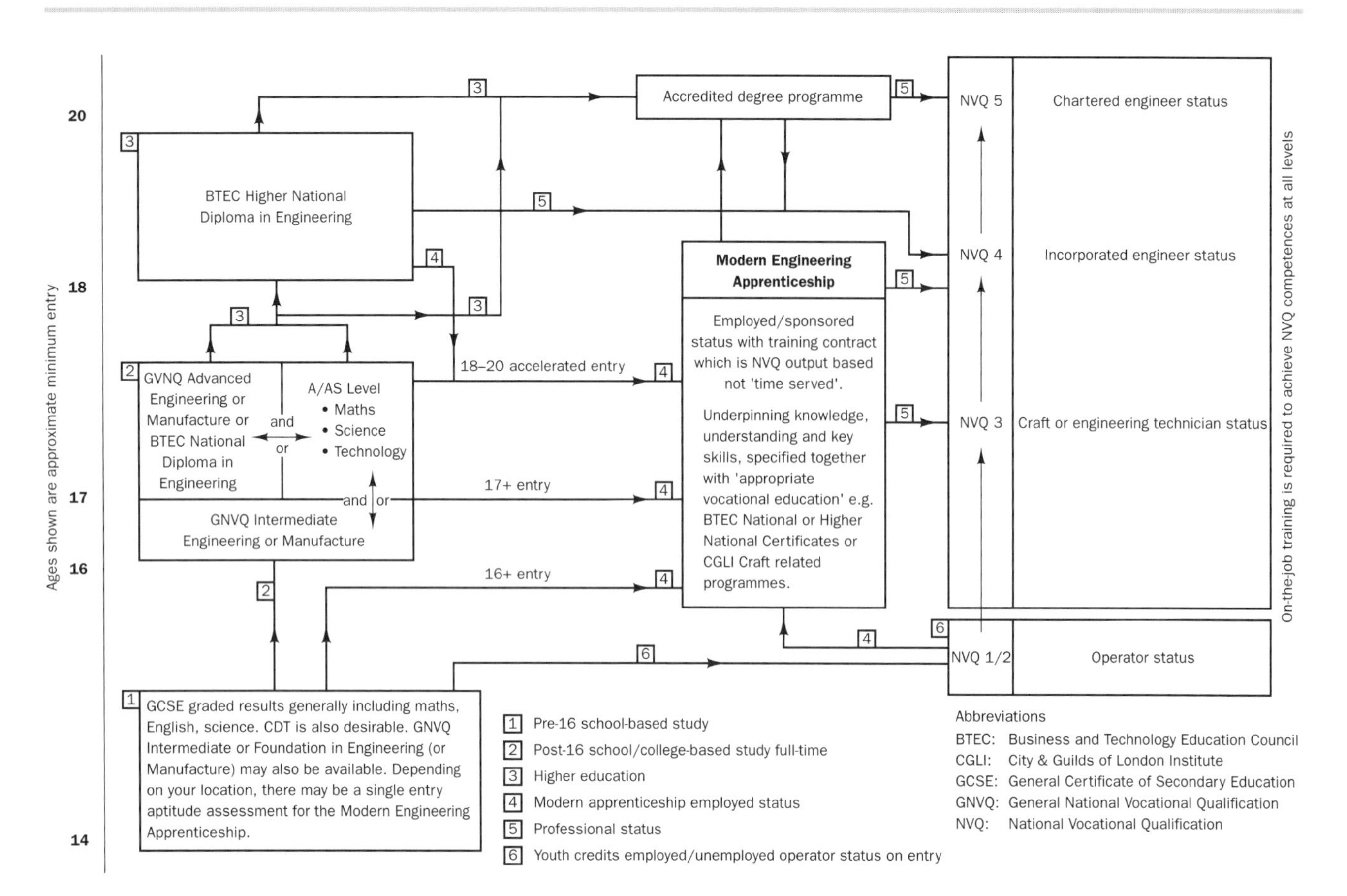

Fig. 4.15 Pathways to becoming an engineering (*source*: West Midlands Industrial Club)

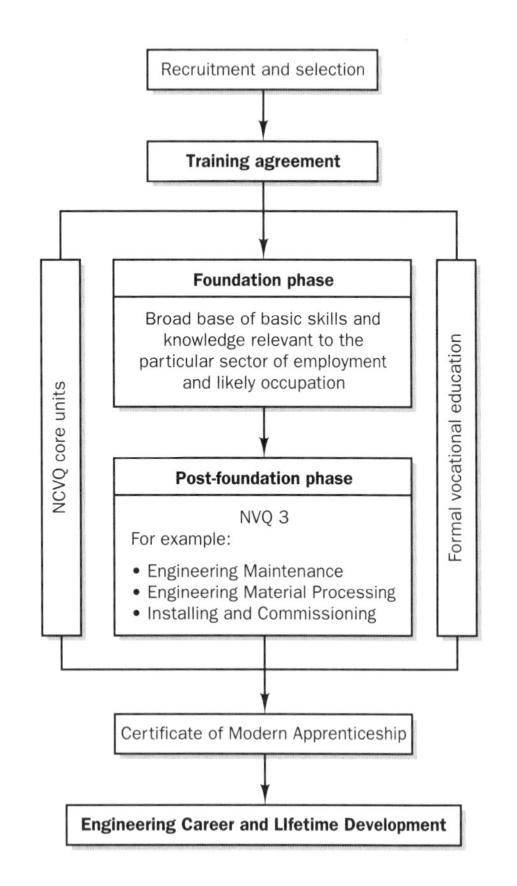

Fig. 4.16 Modern Apprenticeship Framework (*source*: EnTra)

Operators/assemblers

Operators work machines in an engineering factory; assemblers put things together. For this type of work school-leaving qualifications are not essential. However most employers expect to see some evidence of a reasonable level of literacy and numeracy. This is required in order to understand and benefit from the training given and, most importantly, have an understanding of the hazards involved and the safety procedures applicable to the process.

Training is usually carried out in the factory and is often brief, a few days to a few weeks. It is usually limited to a 'need-to-know' basis. Some employers train operators and assemblers in more than one job to promote flexibility and also encourage them to achieve NVQ qualifications at the lower levels to provide greater interest.

Craftspersons

Craftspersons are expected to develop high levels of skill and be capable of working with general purpose machine tools and hand (bench) tools. Employers will expect GCSEs in subjects like maths, science, English and perhaps CDT. Nowadays, a GNVQ would also be appropriate. In view of the type of training to be undertaken, many employers also use their own aptitude tests for manual dexterity. Training will usually follow or be similar to the structure of the modern apprenticeship. Craft trainees who show particular aptitude for the theoretical side of training and the associated further education will often be encouraged to transfer to technician training.

Engineering technicians

Technicians cover a variety of jobs. They may work in drawing and design offices, produce and develop prototypes, work in test laboratories, be responsible for quality control and production control, or they may supervise the maintenance of highly sophisticated computer-controlled machine tools and robots. They usually work under the direct supervision of a professional engineer and are often part of a team.

Employers will usually require a minimum of four GCSEs at grade C or better. These should include maths, science and, preferably, English. Technician apprentices will study for a BTEC National Certificate (which can lead to an NVQ at level 3) when linked to a satisfactory assessment of job skills.

It is then possible to progress to a BTEC Higher National Certificate (which can lead to an NVQ at level 4). It is possible to use a BTEC National Certificate to move into full-time higher education and obtain a BTEC Higher National Diploma or a degree.

Alternatively the school-leaving qualifications mentioned above can also be used to gain entry to a full-time college course leading to an advanced GNVQ or a BTEC National Diploma. Following this you would normally continue in full-time education or on a sandwich course to obtain a BTEC Higher National Diploma or a degree.

Incorporated engineers

Because of the confusion between the engineering technician just described and the higher level, previously known as a technician engineer, the career title of incorporated engineer has been introduced. Incorporated engineers will have studied to at least the Higher National Certificate or Diploma level. Increasingly employers are requiring an engineering degree. Training is usually on the job under the mentorship of a professional engineer, combined with study for professional qualifications.

Incorporated engineers are often team leaders, supervising technicians and craftspersons. They use their knowledge and training to solve all sorts of engineering problems, both in production and design.

Chartered engineers

Chartered engineers have reached the highest engineering career status. They are involved in many aspects of design, research, development and production management. Many also work in consultancy practices or hold professorships in major universities. The minimum qualification is a good honours degree in engineering. Many chartered engineers also have higher (masters) degrees and/or doctorates. In most branches of engineering chartered engineer status is offered through the professional body concerned, and this can be very important for career prospects. In addition to an honours degree, a period of approved work experience lasting several years in a position of responsibility and, usually, the presentation of a project or a thesis is also required.

At this level you do not need to have manual skills, as you will be concerned with research, planning, design, and the organisation of major engineering projects. The problems to be worked out are, however, practical ones which are tackled through a mixture of theory, practical know-how and many years of experience.

As most professional engineers' careers progress, they find themselves more involved in project management and leading teams of engineers and technicians. Here management, communication and interpersonal skills become more important. Many engineers eventually move across into general management where salaries are higher than in engineering posts.

Self-assessment tasks

1. Once again define an 'engineer' and what he or she does.
2. Has your concept of an engineer and his or her work changed since you first responded to this exercise, and in what way?
3. Comparing your own skills and aptitudes for study with the above criteria, at which of the above categories should you aim in order to achieve job satisfaction and success.

Applying for jobs

No matter which level of job you are applying for and no matter how well qualified you are, or whether it is your first job or a change of job to further your career, the following facts are true:

- Many people go after every job.
- Just a few get an interview.
- Only one gets the job.
- First impressions count.

When you hear of a job, or see one advertised in the paper or at an employment office, you usually have to write or telephone for further details and an application form. Small firms don't often have forms, the employer will want to find out about you from your letter or may even ask you some questions on the phone. It is only if you handle these initial stages correctly that you will get a chance of going for an interview. Remember that writing letters, completing forms or making telephone calls, is the first chance you have of impressing the employer.

Letters

Here are a few general points about letter writing:

- Always write neatly and clearly – it's a good idea to make a practice copy first.
- Use black or blue ink.
- Use a fountain pen or a fine roller-ball pen rather than a plain biro.
- Use good quality plain white or pale blue paper. If you can't write straight use bold lines drawn on a sheet placed under your writing paper as a guide.
- Envelopes should either match the paper or be brown 'office' ones.
- If you start your letter using 'Dear Sir' or 'Dear Madam', you should end it 'Yours faithfully'.

- If you start your letter with somebody's name ('Dear Mr Brown'), you should end it 'Yours sincerely'.
- Make your signature clear or print your name underneath your signature.
- Apply for a job as soon as you can after hearing about it.
- If you name people who will give you a reference, make sure you obtain their permission first.
- Always say what job you are writing about and where you saw the advert – one firm may have several vacancies at the same time.

Writing the letter

If the job advert says 'write for further details' or 'write for an application form' do just that. Don't write an essay about the story of your life. Make your letter short, neat and clear: even this sort of letter can affect the impression you make on an employer. Read it through and make sure there are no spelling or grammatical errors. Don't make corrections, write it out again.

If, however, the advert says 'apply in writing' you should then give full details about yourself and your reasons for applying for the job.

Figure 4.17 shows a typical advert and the application letter. It also highlights some of the more important points.

Writing 'on the off-chance'

Not all letters of application need be sent in response to a specific job advert. You may want to write to a firm to see if they have any intention of recruiting staff in the near future. It could be useful to include a 'CV' (see below) as well. Always enclose a stamped and addressed envelope for their reply, and ask them if they would note your name and address so that they can contact you at some future date should a vacancy arise.

Sending a 'CV'

Sometimes you are asked to send a CV – (curriculum vitae) which is Latin for an outline of your education, qualifications and career to date. This is more often asked for in the case of jobs for older people, rather than for school-leavers. An example is shown in Fig. 4.18.

- A CV should be presented on a separate sheet of paper, and preferably typed or word-processed.
- It is worthwhile getting your CV photocopied, so that you can use it over and over again.
- Always write a covering letter with a CV, explaining your reasons for applying for that particular job.

> **Self-assessment task**
>
> Write out your own CV now. You will require it in one of the extended exercises at the end of this chapter.

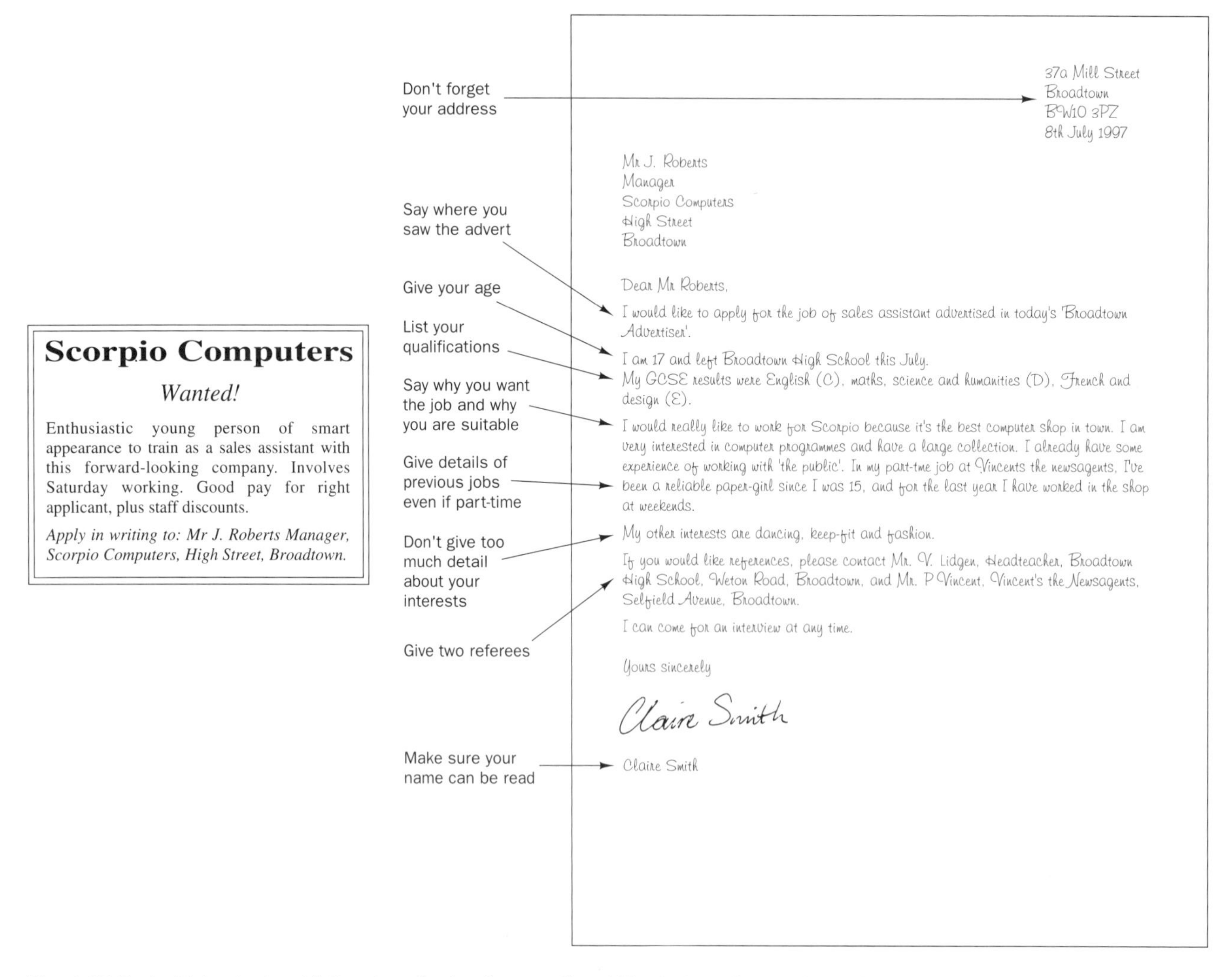

Scorpio Computers

Wanted!

Enthusiastic young person of smart appearance to train as a sales assistant with this forward-looking company. Involves Saturday working. Good pay for right applicant, plus staff discounts.

Apply in writing to: Mr J. Roberts Manager, Scorpio Computers, High Street, Broadtown.

37a Mill Street
Broadtown
BW10 3PZ
8th July 1997

Mr J. Roberts
Manager
Scorpio Computers
High Street
Broadtown

Dear Mr Roberts,

I would like to apply for the job of sales assistant advertised in today's 'Broadtown Advertiser'.

I am 17 and left Broadtown High School this July.
My GCSE results were English (C), maths, science and humanities (D), French and design (E).

I would really like to work for Scorpio because it's the best computer shop in town. I am very interested in computer programmes and have a large collection. I already have some experience of working with 'the public'. In my part-tme job at Vincents the newsagents, I've been a reliable paper-girl since I was 15, and for the last year I have worked in the shop at weekends.

My other interests are dancing, keep-fit and fashion.

If you would like references, please contact Mr. V. Lidgen, Headteacher, Broadtown High School, Weton Road, Broadtown, and Mr. P Vincent, Vincent's the Newsagents, Selfield Avenue, Broadtown.

I can come for an interview at any time.

Yours sincerely

Claire Smith

Claire Smith

Fig. 4.17 Typical job advert and letter of application (*source*: City of Birmingham Careers Service)

Curriculum Vitae

Personal details

Name: **Matthew Wilkinson**
Title: Mr
Address: 64 Preston Street Nextham, Near Broadtown BW19 4AL
Date of birth: 10th February 1975

Education

1986–91 *Nextham Comprehensive School*
GCSEs in English Language (A), Geography (B), Spanish (B), French (C), Mathematics (D), History (D), Biology (E), Chemistry (E)

1994–96 *Broadtown Technical College*
Part-time BTEC National Certificate in Mechanical Engineering

Employment

1992–1994 A–Z Engineering Ltd, Broadtown
Trainee service engineer assisting in the maintenance of computer-controlled machine tools

1994–Present A.N. Other Engineering, Nextham
Operator/setter for CNC machine tools for two years. I am now a CNC machine tool programming assistant and am familiar with both turning and centres and machining centres fitted with FANUC controllers

Other achievements

Clean driving licence
St John's Ambulance First Aid Certificate

Interests

Canoeing, basketball (member of Broadtown Superstars 1st Team – have represented Broadtown in twin-town matches in France and Germany), photography, rally driving.

Referees

Mr M. Linton,
Personnel Manager,
A–Z Engineering Limited,
Broad Street,
Broadtown

Mr F. Davidson,
Head of Engineering & Science,
Broadtown College of Further Education,
College Road,
Broadtown

Fig. 4.18 Typical curriculum vitae (*source*: City of Birmingham Careers Service)

Figure 4.19 shows a typical advert and covering letter. The CV shown in Fig. 4.18 above would also be enclosed in this example.

Application forms

Here are some general points about completing application forms:

- Read through the whole form before you start to fill it in, so that you put all the information in the right places.
- You may wish to complete it lightly in pencil before using ink.
- When you are ready, use black ink – the firm may wish to photocopy it.
- Write clearly and neatly, and get the spelling correct. It looks especially bad if you can't spell the name of your last school or the street you live in!
- Try to answer all the questions, even if you think they are irrelevant.
- Write a short covering letter to send with the completed form. If the form does not give you space to explain your reasons for applying for the job, and you feel you are particularly suitable, then make the covering letter longer to include these points. Some employers would not interview people who merely send in a 'bald facts' form.
- Again, keep a photocopy of the form and your covering letter for future reference.

Broadtown Borough Council

Industrial Museum

Post of Assistant Curator

This interesting and varied post involves helping the Curator to establish an industrial museum to reflect the past skills and industries of Broadtown. We are seeking an experienced engineer who is friendly and outgoing and who can communicate with the general public.

The ideal applicant will be over 25 years of age with a BTEC National Certificate as a minimum requirement plus industrial experience. Minimum starting salary will be around £15000 p.a.

Send CV, with letter of application, to the The Curator, Broadtown Industrial Museum, The Old Station, Broadtown BW138LP.

64 Preston Street
Nextham
Near Broadtown
BW19 4AL
Tel: 668237

21st April 1997

The Curator,
Broadtown Industrial Museum
The Old Station
Broadtown
BW13 8LP

Dear Sir or Madam,

With reference to your advertisement in the 'Broadtown Advertiser of 20 May, for an Assistant Curator, I enclose my Curriculum Vitae and hope this will be of interest to you and that you will grant me an interview

I am currently employed by A.N. Other Engineering Co., Nextown as a CNC machine tool programmer. Since I frequently have to visit customers of my company to install the programmes in their machines and also train their staff, I am used to dealing with people on a personal basis. I also advise the Governors of my old school on the purchase of any equipment.

My family have long been associated with the skills and industries of this area and I am well versed in its industrial history.

I look forward to hearing from you.

Yours faithfully,

Matthew Wilkinson

Matthew Wilkinson

Fig. 4.19 Typical job advert and covering letter (*source*: City of Birmingham Careers Service)

Figure 4.20 shows a typical application form and Fig. 4.21 shows a typical covering letter.

Addressing envelopes

- The envelope containing your form or letter is the very first chance you have to impress a prospective employer.
- Choose sensible-looking envelopes. They should match your writing paper or they should be brown office envelopes. The latter are cheap and business like. Don't use scented envelopes with pretty flowers, or ones coloured pink or lilac; these will not impress employers. Make sure the envelopes are big enough so that you don't have to fold the form or letter more than twice.
- Don't start the address too high up, or the post mark may cover part of it.
- Write clearly. Use the person's name and their department if you have this information – it can speed up delivery within a large firm – and use the postcode at the end of the address.
- Use a first class stamp.

An example of a correctly addressed envelope is shown in Fig. 4.22.

Telephoning about jobs

When you ring up about a job vacancy, always have the advert beside you so that you know:

- The **number** you want to ring.
- The **extension** number, if one is given – this is the number you give when the firm's switchboard operator answers.
- The **name** of the person you need to speak to, or the **department** (usually Personnel). If no name is given, say 'I'm ringing about the job advertised in...' and let the person at the other end put you through to the correct extension.

If you are ringing from a call box have:

- Plenty of coins handy so that you do not get cut off before you have finished your call.
- A pen and paper or your diary, so that you can write down details (time, place, name of person to see) if you are offered an interview.

The interview

Most people are nervous at their first interview. Give it your best shot but don't be too upset if you fail to get the job. Look upon it as a learning experience; you will be better prepared for the next time. Your chances of success will be better if you:

Busby Engineering Co. Application Form

Application for the post of: ....

Please use block letters

Surname .... Forenames in full ....

Full postal address ....

....

Daytime telephone number ....

Date of birth ....../....../...... Age .... Nationality ....

Schools attended from age 11 ....

....

Examinations passed (Please indicate if you are waiting for any results)

Subject *Grade* *Level (O level, GCSE, CSE, A level, GNVQ)*

Details of previous employment

Employer's name and address *Dates of employment* *Duties*

Qualifications gained through employment

Activities and interests
(e.g. sports, hobbies, social; please mention special achievements or official positions, held, etc.

Please state here your particular reasons for applying for this job

References
Please give the names and addresses of two people (not relatives) to whom reference may be made. These should include your last employer, or headteacher if you are a school leaver.

1. .... 2. ....

....

Fig. 4.20 Typical application form (*source*: City of Birmingham Careers Service)

- Arrange to arrive slightly before the time you are due. This gives you time to settle down and calm your nerves. You must never rush in at the last moment all flustered. On the other hand, if you are too early you may get in the way and, if you are too late, you may lose the interview.
- Make sure you know the name of the person who is to interview you and the name of the department in which he or she works. If the name is unusual check with the switchboard operator how to pronounce the name.
- Dress simply but neatly in a business-like way for the interview.
- Clean your shoes and your fingernails and have a neat hair style.
- Take copies of your application and CV so that you have something to refer to if your nerves affect your memory – a common problem.
- Make a list of questions about the job: rate of pay, training, further education, hours of employment, prospects for advancement and anything else that you may think is relevant. This will help to keep the interview going during the awkward pauses that always seem to occur.
- Do some research into the company to which you are applying. What does it make and who are its customers?

48b Johns Road
Monkton
near Downtown
Bartonshire
BD8 4HS

The Personnel Manager,
Busby Engineering plc.
Park Lane Estate
Downtown

20th April 1997

Dear Sir or Madam,

Please find enclosed my completed application form for the post of trainee engineering operative..

Yours faithfully

Marcia Howard.

Marcia Howard

Fig. 4.21 Typical covering letter (*source*: City of Birmingham Careers Service)

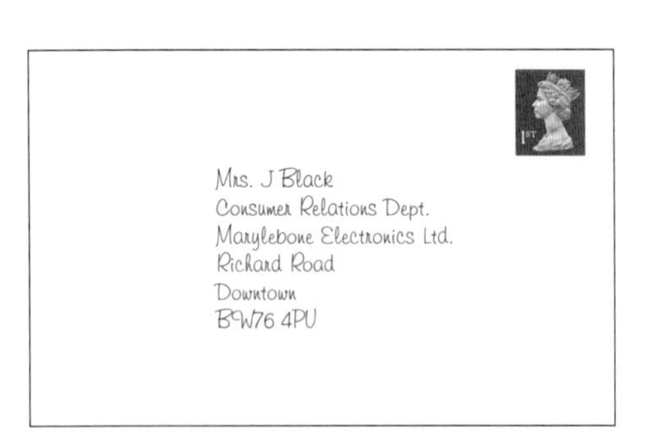

Fig. 4.22 An addressed and stamped envelope (*source*: City of Birmingham Careers Service)

What is the structure of the company? Is it part of a famous national group or is it a family business? Think hard about those inevitable awkward questions: 'Why do you want the job?' and 'What makes you think you can do the job?' Be positive in your answers, show an interest in the company and not just in the job as a means of earning some money.

- Show an interest in engineering (e.g. maintaining and repairing your cycle, motorcycle, etc., and/or having a constructional hobby such as building model aeroplanes or assembling electronics kits). This will also help to strike up a rapport with the interviewer. A weekend job to earn money to support your hobby also shows initiative and provides proven experience of working with other people and meeting the public.
- Show common courtesy. Providing you are given sufficient time, always acknowledge invitations for interviews in writing. You should also reply even if you decide against accepting the invitation. If you have not been given time to reply in writing you should still phone in with your decision.

Hopefully your school or college will give you the experience of several mock interviews using different interviewing techniques before you set out for your first real interview. Good luck!

Self-assessment tasks

1. You see an advert for a job that interests you and which includes a name and address to whom you should apply for an application form.
2. Assuming you are successful, list the sequence of events from seeing the advertisement to presenting yourself for interview.

Brief 1
Technology in society

1. Select any major domestic appliance (electric or gas) in your home and:
   (a) Draw a schematic block diagram, showing the major elements of the device and how they are inter-related.
   (b) Describe how the device functions and suggest how it could be improved to make it:
      (i) easier to operate
      (ii) more energy efficient
2. Approach the general practice with which you are registered and report on how modern technology is used:
   (a) in the management of the practice
   (b) clinically
3. Approach a large local motor dealer and report on how modern technology is used in the diagnosis of engine faults and in the tuning of engines in order that they remain energy efficient and that exhaust gas emissions are kept within legal limits.

Brief 2
Pollution and waste disposal

Most large companies in main sectors of the engineering industry are very aware of the importance of public relations, particularly over environmental issues. Within the confines of this unit it has only been possible give a brief overview of their environmental activities.

1. Approach a large public company for information on what they are doing to make their manufacturing processes and their products more environmentally friendly. Some companies are more co-operative than others and are better geared to providing information. You may have to try more than one company. Using the material you receive, write a detailed report summarising the environmental activities of the company:
   (a) Describe how the company has improved its environmental performance over the last five years in terms of the manufacturing processes they employ, the products they manufacture, and the type of waste produced and how they dispose of it.
   (b) Describe the intentions of the company for the next five years.
   (c) Support your report with relevant facts and figures and the use of graphs and diagrams.
2. Approach the environmental health department of your Local Authority as a first point of contact, in order to obtain the information to write a report on how your Local Authority does the following:
   (a) Organises waste collection, including sorting for recycling.
   (b) Disposes of this waste:
      (i) by landfill sites, including the recovery and use of LFG
      (ii) incineration, including the recovery and use of waste heat
   (c) Deals with any special local issues.
3. Write a report on how your local water company:
   (a) Collects, filters, purifies and distributes domestic water in your area and any special problems involved. Illustrate your report with relevant diagrams.
   (b) (i) Treats sewage before returning the water to the environment.
      (ii) Disposes of the solid residue in a safe manner.
      (iii) Uses any methane gas produced.

Brief 3
Applying for a job

The following advertisement appeared in the Broadtown Advertiser.

Engineering Apprenticeships. Axle Engineering Co. Ltd., is offering 6 engineering apprenticeships commencing in September 1996. These will be organised in conjunction with the Engineering Training Authority (EnTra) and will follow the structure of their 'Modern Apprenticeship' scheme. Day release to an appropriate technician course at a college of further education will be an integral part of the course.

Applicants should have successfully completed their secondary education and should have obtained O level GCSEs in mathematics, physical or general science, English language and CDT at grade C or better. Additional GCSEs at any grade will be taken into account.

Apply in writing, including a CV and the names of two referees, to: Mrs J M Smith (Personnel Manager), Axle Engineering Co. Ltd, Broadtown Trading Estate, Broadtown, BW19 1AA.

1. Using the above advertisement as a starting point:
   (a) Write a letter of application.
   (b) Draw up an appropriate CV.
   (c) Address a suitable envelope.
2. This time assume that you are asked to write in for an application form.
   (a) Write a letter requesting for an application form.
   (b) Draw up an example of the sort of application form you would expect to receive.
   (c) Copy this form and complete it.
3. Assuming you are a successful candidate and are called for an interview:
   (a) Describe how you would dress and present yourself for the interview.
   (b) Draw up a list of the questions that you will want to ask at the interview.

While you can use the examples in this book as a starting point, your response to this brief must be your own original work.

4.3 Unit test

Test yourself on this unit with these sample multiple-choice questions.

1. The control unit of a modern dish washing machine is most likely to contain a:

(a) microchip
(b) mechanical timer
(c) bi-metal strip thermostat
(d) modem

2. Worktop surfaces in a modern kitchen are most likely to be made from:

(a) steel
(b) aluminium
(c) wood
(d) laminated plastic

3. The engine management system of a car is most likely to be controlled by a:

(a) laptop computer
(b) dedicated computer
(c) mainframe computer
(d) personal computer

4. A high quality tennis racket frame will nowadays be made from:

(a) aluminium
(b) stainless steel
(c) carbon-fibre-reinforced plastic
(d) uPVC plastic

5. Up-to-the-minute holiday information is best obtained from:

(a) brochures
(b) teletext
(c) journals
(d) guide books

6. Spreadsheet software is used to produce:

(a) charts, graphs and tables
(b) letters
(c) engineering drawings
(d) restaurant menus

7. Documents containing text and illustrations are best transmitted by a:

(a) telephone
(b) teleprinter
(c) fax machine
(d) photocopier

8. Automated chemical processing equipment is most likely to be controlled by a:

(a) modem
(b) CNC controller
(c) scanner
(d) programmable logic controller

9. Long distance telephone cables are most likely to be made from:

(a) copper wires
(b) aluminium wires
(c) expanded polystyrene
(d) glass fibres

10. A modern factory is most likely to carry out the batch machining of small engineering components using a:

(a) CNC lathe
(b) capstan lathe
(c) centre lathe
(d) turret lathe

11. The computer software package most suitable for storing patients' records in a hospital is:

(a) a database
(b) word processing
(c) spreadsheet
(d) file server

12. An ultra-sound scanner is used for:

(a) checking a patient's blood pressure
(b) listening to a patient's breathing
(c) visually examining a patient's internal organs
(d) testing a patient's speech

13. A microchip is likely to be found in:

(a) a heart pacemaker
(b) an artificial leg
(c) a stethoscope
(d) a first-aid kit

14. Hospital consultants are likely to consult a national database to:

(a) obtain their patients' addresses
(b) match donor organs with their patients
(c) organise staff appointments
(d) control their finances

15. Surgical instruments used for cutting (e.g. scalpels) are made from:

(a) alloy steel
(b) tungsten carbide
(c) Teflon
(d) titanium nitride

16. Fuel for road vehicles is extracted by:

(a) mining
(b) quarrying
(c) desalination
(d) drilling

17. A renewable energy source is:

(a) natural gas
(b) wind power
(c) fuel oil
(d) coal

18. The 'greenhouse effect' resulting from burning fossil fuels in power stations and road vehicles is caused by:

(a) water vapour
(b) nitrogen
(c) carbon dioxide
(d) ash

19. The waste from an electroplating plant contains cadmium. This waste is classified as:

(a) toxic
(b) non-toxic
(c) radioactive
(d) biodegradable

20. Waste paper is recycled in order to:

(a) reduce the greenhouse emissions created by burning it
(b) preserve the trees from which it is made
(c) reduce acid rain
(d) raise money for the local councils

21. Landfill sites are used for the disposal of:

(a) radioactive waste
(b) liquid waste
(c) toxic waste
(d) non-hazardous waste

22. The methane gas given off from landfill sites:

(a) is best allowed to disperse into the atmosphere
(b) can be used as a fuel for generating electricity
(c) is non-flammable
(d) can be used for making fizzy drinks

23. The authority responsible for the disposal of hazardous waste is:

(a) any local council
(b) any regional health authority
(c) the Health and Safety Executive
(d) the Environment Agency

24. The person who services your car is:

(a) a professional engineer
(b) a mechanic
(c) an operative
(d) a craftsperson

25. The person responsible for heading the design team for a new car will most likely be:

(a) a civil engineer
(b) a chemical engineer
(c) a manufacturing engineer
(d) a mechanical engineer

26. A curriculum vitae (CV) is:

(a) a course of technical study
(b) an industrial disease
(c) an account of your education, training and career to date
(d) a medical software system

27. An engineering technician would need:

(a) skill training alone
(b) a GNVQ and skill training
(c) a Higher National Certificate or Diploma and skill training
(d) an engineering degree alone

28. The letters TEC:

(a) are an abbreviation for your local college of further education
(b) stand for: Technical Education Committee
(c) stand for: Training and Enterprise Council
(d) stand for: Technician Examination Consortia

29. If you start a letter of application 'Dear Mr Smith', you should finish the letter:

(a) yours truly
(b) kind regards
(c) yours faithfully
(d) yours sincerely

30. When offered an interview for a job you should:

(a) reply only if you are going to attend
(b) just turn up at the time stated
(c) reply only if you are not going to attend
(d) reply in either case stating whether or not you are interested in attending

Answers to numerical self-assessment tasks

3.1 1. (a) 150×10^3 W; (b) 75×10^{-3} V
2. (a) 350×10^{-6} A; (b) 210×10^9 N

3.3 1. (a) 56.81; (b) 0.003601; (c) 208000
2. (a) 3.142; (b) 10 0002; (c) 0.100
3. (a) 23.693; (b) 1.023; (c) 0.90
4. (a) 4.01704; (b) 0.0346 4; (c) 2.0000

3.4 1. (a) 3161; (b) 104
2. (a) 0.03322; (b) −3.07

3.5 1. (a) 40×10^{12}; (b) 36×10^{-3}
2. (a) 0.9×10^6; (b) 4×10^9

3.6 1. (a) 75×10^3; (b) 36×10^{-3}; (c) 11.3×10^{-6}
(a) 1.89×10^6; (b) 4.51×10^9

3.7 1. (a) $k = F$; (b) $F = \dfrac{pA}{x}$
2. (a) $g = E$; (b) $t = \dfrac{Fs}{mzp}$
3. (a) $t = \dfrac{v - u}{a}$; (b) $T_1 = T_2 - \dfrac{x}{l\alpha}$

3.8 1. 0.44 mmN^{-1}
2. 0.075 $\Omega\ {}^\circ\text{C}^{-1}$
3. 5 ms^{-1}
4. $v = 9 + 1.8t$

3.9 1. 12.6 kN
2. 1.96 kPa
3. 995 kPa
4. 25.2 mm

3.10 1. 11 kN
2. 7 kN
3. 12 kN, 10.4 kN

3.11 1. 308 N
2. 75 N
3. 4 kN

3.12 1. 300 kNm^{-1}, 750 N
2. 40 kNm^{-1}, 20.8 mm
3. 125 kNm^{-1}, 3.92 mm

3.13 1. 37.5 J, 30 W
2. 30 kJ, 3.75 kW
3. 20 J, 16 J, 75%
4. 2.5 kJ, 313 W, 42%

3.14 1. (a) 2.4 ms^{-1}; (b) 30 m
2. (a) 30.6 ms^{-1}; (b) 8.18 s; (c) 917 m
3. (a) 1 min 40 s; (b) 31.5 kJ; (c) 315 W
4. (a) 4.67 s; (b) 1.10 kW; (c) 74%

3.15 1. (a) 220 N; (b) 165 J; (c) 110 W
2. (a) 2.09 ms^{-1}; (b) 1.57 kN; (c) 54.9 kW
3. (a) −0.667 ms^{-1}; (b) 1.2 km
4. (a) 0.3 ms^{-2}; (b) −0.36 ms^{-2}; (c) 22.5 m

Answers to unit tests

Question	Unit			
	1	2	3	4
1	(c)	(b)	(c)	(a)
2	(b)	(c)	(a)	(d)
3	(b)	(d)	(d)	(b)
4	(d)	(a)	(b)	(c)
5	(b)	(c)	(c)	(b)
6	(a)	(c)	(d)	(a)
7	(c)	(a)	(b)	(c)
8	(d)	(c)	(a)	(d)
9	(a)	(c)	(c)	(d)
10	(c)	(a)	(d)	(a)
11	(d)	(d)	(b)	(a)
12	(b)	(b)	(c)	(c)
13	(b)	(b)	(c)	(a)
14	(d)	(d)	(c)	(b)
15	(d)	(a)	(a)	(a)
16	(d)	(c)	(c)	(d)
17	(b)	(d)	(b)	(b)
18	(a)	(b)	(d)	(c)
19	(d)	(c)	(c)	(a)
20	(c)	(a)	(d)	(b)
21	(b)	(c)	(c)	(d)
22	(a)	(d)	(a)	(b)
23	(a)	(c)	(d)	(d)
24	(c)	(a)	(a)	(b)
25	(b)	(b)	(b)	(d)
26				(c)
27				(c)
28				(c)
29				(d)
30				(d)

Index

abbreviations 72
acceleration systems, uniform 101–3
accidents 38
acidified zinc chloride 22
acids 35
adhesives 22–3, 33
airbrush rendering 52
alloys 2, 3
alternating current 110–12
amalgam 21
ammeter 116
ampere 107
amplifier 12
angle plate 24, 26, 28
annealing 33
anode 35
anodising 35
application forms 149–50
apron (lathe) 17
arbors 27
arc welding 22, 32
assembly drawings 76–7
auto-dimensioning 67
AutoCAD 70
auxiliary views 70, 71
availability (of a material) 7

Bakers Fluid 22
balanced moment systems 95–6
balances 113–14
bar charts 56, 57
barrier creams 38
batch production 42
bearings 72
bio-diversity 139
bipolar transistors 12
block diagram 40, 58
boiler suit 39
bolts 8, 20–1
bonding *see* adhesives; powder bonding
borax 22
brazing spelter 22
Brinell test 6
British Association Thread (BA) 7
British Standards 39, 61, 62, 67, 69, 72
  BS 308 67
  BS 4335 58
  BS 4500 75
Business and Technician Council
    (BTEC) 145

cable 10, 13
calorising 37
capacitors 11, 13–14
carburising 34
careers
  job applications 147–52
  options 143–4
  qualifications 145–7
carriage assembly (saddle) 17
cars: components, accessories 129
case hardening 34
catchplate 17, 26
cathode 35
cathode ray oscilloscope (CRO) 117–18
Celsius scale 104
ceramics 5, 11, 13
chartered engineer 145, 147
charts 40, 55, 56-7, 58-9
chemical engineering industry 138, 144
chemical treatment 23, 35
chucks 17, 23, 25-6, 28
circuit diagrams 61, 63
circuits *see* electrical circuits; integrated
    circuits; printed circuit boards
City & Guilds of London Institute
    (CGLI) 145
civil engineering 144
clearance angles 18
clothing *see* protective equipment/clothing
cold-working 33
colour code 13
commercial applications 130–2
communication *see* graphical
    communication
communications equipment 128–9
components
  electrical product 10–13, 13–14
  electromechanical product 14–15
  library files of 67
  mechanical product 7–10
composite materials 5
compound slide 17
computer-aided design (CAD) 50, 52–4,
    67, 74, 77, 130
computer-aided design and draughting
    (CADD) 49
computer-aided manufacturing
    (CAM) 67, 130
computer numerical control (CNC) 130
computers, types of 126, 129

concrete 5
concurrence 94
concurrent engineering 58
conductivity 6
conductors 107, 109
continuous production 42
Control of Pollution Act 1974 140
coolant 27, 28
coplanar forces 94
corrosion 6–7
cost factors 7
coulomb 107
counterboring 73, 74
countersinking 73, 74
craftsperson 145, 146
crating technique 51
'Crocodile Clips' 62
cross-links 4
cross-slide 17
cumulative errors 41
current *see* electric current
cutting 18–19, 19–20
cutting plane 70
cutting speeds 24, 27, 28
CV (curriculum vitae) 148–9

data 57, 118
  charts 55
databases 56, 57
decimals 86
degreasing 35
design *see* computer-aided design
designers 50
diagrams 55, 61, 63, 94
dimensioning 67-8
dimensions 41–2
diodes 11–12, 14
domestic appliances 126–8
'doping' 5
drawings 51, 52, 61
  see also assembly drawings;
    engineering drawing; scale
    drawings
drills and drilling 16–17, 23–7
ductility 6

efficiency
  energy 138
  system 99
elastic systems 96–7
elasticity 6

electric current 107, 110–12
electrical
  circuit 107
  connections 29
  engineering 143–4
  potential difference 107–8
  power 108
    generation and distribution 135–6
  product *see* components
  properties 6
  systems 107
electrical/electronic systems 61, 62
electrolyte 35
electrolytic corrosion 7, 21, 37
electromechanical product *see* components
electronic engineering 144
electroplating 35
elevations 69, 71
emergencies 38–9
emery 5
employee's responsibilities 44
employer's responsibilities 43–4
enamels 36
energy 98–9, 104, 138
engineered products 1
engineering *see* chemical engineering industry; civil engineering; electrical engineering; electronic engineering; mechanical engineering; nuclear engineering
engineering drawing 67–77
  case study 77–9
engineering drawing conventions 72
engineering industry 137, 138, 143–4
engineering materials 132
engineering technician 145, 147
engineering technology
  in health and medicine 132–3
  in industry and commerce 130–2
  in leisure activities 128–30
  in the home 126–8
  see also environmental impact of technology; waste management
Engineering Training Authority (EnTra) 145
engineers 145, 147
engineer's rule 114
Environment Agency 140
environmental impact of technology 133–9
  see also waste management
Environmental Pollution Act 1990 140
equilibrium *see* static equilibrium
etching 35
expansivity 6
extinguishers, types of 38–9

face plate 17, 26
feedshaft 27
filler rod 22
finishing *see* surface finishing
fire 38–9
first aid 39
fits 75
flow chart 40, 58–9
flux 21, 21–2, 30
food processors 128
force
  active 98
  electro-motive 107
  reactive 98
force systems, concurrent coplanar 94
formulae, transposing 88
'forward bias' 12
fusion 105

galvanising 35, 36
gas *see* natural gas
gears 72
General National Vocational Qualification (GNVQ) 145
grain growth 33
graphical communication 49
graphical formats 58
graphs 55–7, 58, 88–90
gravity 91, 92–3
grinding 35–6
guards 27, 38

hardening *see* case hardening; quench hardening
hardness 6
health and medicine applications 132–3
health and safety *see* safety; workplace
Health and Safety at Work etc. Act 1974 43
Health and Safety Commission 43
Health and Safety Executive 43
heat energy 104–6
heat treatment 23, 33
histogram 56
Hooke's law 96
hot dipping 36
hot spraying 37
hydraulic systems *see* pneumatic/hydraulic systems
hygiene 38

ICI 138, 139
illustrations 52
incorporated engineer (technician) 145, 147
inductors 11, 14
industrial applications 130–2, 137–8, 139
industry *see* engineering industry
inertia 98
information technology 132–3
insulators 11, 13, 107
integrated circuits 62
interview techniques (job application) 150–2
ISO metric screw 7
isometric drawing 51, 68
Issigonis, Alec 50

Jacobs chuck 17, 23, 25
job applications 147–52
joining methods 20, 29
joints 21
joule 98
junction diodes 11

Kelvin scale 104
kilogram 84

laminae 92
landfill 141
laquers 36
lathe 17, 24
legislation 43, 140
leisure activities applications 128–9
letters and letter writing 147–8, 150
library files (components) 67
light-emitting diodes 12, 14
limits 75
linear expansivity 106
lines 67
Local Authority control 140
locking devices 8

machining symbols 72–4
maintenance 44–5
maintenance procedures 44–5
malleability 6
manufacturing industry 137, 144
material removal 16, 23
materials
  properties of 6–7
  selection of 6–7
  types of 2–5
  see also engineering materials; metals
mathematical techniques 80–90
measurement systems 65, 84–5
measuring devices 113–18
mechanical engineering 143
mechanical product *see* components
mechanical properties 6–7
medical hardware 133
metallic arc welding *see* arc welding
metals
  ferrous 2–3
  non-ferrous 3–4
metre 84
micrometer 114–15
Microsoft Excel 57
milling 19–20, 27–9
millscale 7, 35
mining 134–5
Modern Engineering Apprenticeship 145
morse taper 23
motion 98
multimeter 117
multiples 87
music equipment (Hi-fi, CD) 129

National Vocational Qualification (NVQ) 145
natural gas 136–7
neutral flame 31
Newton's Laws of Motion 98
normalising 34
nuclear engineering 139
nuts 8, 9

oblique drawing 51, 68
ohmic conductors 109
Ohm's law 107–8
operator/assembler 145, 146
orthogonal cutting 18
orthographic projection 68–9
oscilloscope 117–18
overarm steady 27
oxidation 7
oxide colour films 34
oxy-acetylene welding 22, 31–2

paints and painting 36
parallel bars 24, 26, 28
personal conduct 38
photodiodes 12, 14
pickling 35
pictograms 56
pictorial drawing 51, 68
pie charts 56, 57
planning applications 140
plant, new, standards for 138
plastics 4, 5, 11, 128
plating 35, 36–7
pneumatic/hydraulic systems 61–2
polishing 36–7
polymers, polymer materials 4
powder bonding 37
power 98, 108
  diodes 14
PP307 (BS) 58, 61, 67, 72
PP308 (BS) 67
precedence network 58
prefixes 84
presentation techniques 52
pressure 91–2
  regulator 22
principle of moments 95
printed circuit boards 62
process planning sheet 42
processes
  identification 16–23
  production 40–5
  safety procedures and equipment 37–9
  selection 37
  techniques and procedures 23
product stewardship 138
production
  batch 42
  continuous 42
  see also processes
production schedule 43
programmable logic controllers (PLCs) 131
projection *see* orthographic projection
properties *see* materials
protective equipment/clothing 38–9
pyramid test, Vickers 6

qualifications 145–7
quantities, scalar and vector 85
quarrying 134–5
quench hardening, quenching 34

radian 84
rake and clearance angles 18
reamers, machine 17
recrystallisation 33
recycling 138
refrigerants 128
refrigerators and freezers 128
relationship
  linear 89
  proportional 89
re-scaling 67
resistance
  to chemical attack 6–7
  temperature coefficient of 6
resistance strain gauges 113
resistivity 6
resistors 11, 13
  in parallel 110
  in series 109
responsibilities 38, 43–4
reverse bias 12
riveting 21, 30
rivets 9–10
robots, industrial 131
root mean square value 111
Royal Society of Arts (RSA) 145
rubbers 4–5, 11
rust 7

'sacrificial' protection 37
safety
  equipment 38–9, 43
  legislation 43
  procedures 37–8, 43
scale drawings 74
schematic drawings 61
schematic wiring diagram 63, 65
screw threads 7–8, 72
screwed fastening 20–1, 29–30
screws 7–8
  lead 27
  self-tapping 9
  see also setscrews
sections, sectional views 70, 71
Seebeck effect 116
sensible heat 105
setscrews 8–9
Sherardising 37
SI system of units 84, 107
signal diodes 14
significant figures 80
sketches and sketching 50–2
sodium borate 22
software 53, 57
soldering, hard 22, 31
soldering, soft 21–2, 30–1
solids, three dimensional 92–3
soluble oils 27, 28
solvent attack 7
solvents 35
Special Waste Regulations 140
specific heat capacity 104–5
specific latent heat 105–6
spotfacing 73, 74
spreadsheets 56, 57

stainless steel 128
static equilibrium 91, 93, 93–6
stopwatches 116
studs 9
sub-multiples 87
'super-glues' 22–3, 33
surface finishing 23, 35–7, 42, 73
sustainable development 139
swarf 24
symbols 61, 69, 72, 107
  machining 72–4

tables 57
tailstock 17, 25
technology *see* engineering technology; information technology
telecommunications 128–9
telephone technique (job application) 150
temperature
  coefficient of resistance 6
  measurement 104, 115–16
  recrystallisation 33
tempering 34
tensile strength 6
terminations 29–30
tests: hardness 6
text (drawings) 67
texture 42
thermal
  properties 6
  systems 104
thermocouples 116
thermometers 115
thermoplastics 4, 11
thermo-setting plastics 4, 11
time measurement 116
tinplate 37
title block (drawings) 76
toasters, electric 127
tolerance 13
tolerances 41–2, 74
tool post 17, 24–5
torque wrench 29
toughness 6
training 145–7
transistors 14
  bipolar 12
transport
  personal 129
  public 129
triangle of forces 94
turning 17–19, 24–5
turning moments 93

units 84–5

vacuum cleaners 127
vaporisation 105
varnishes 36
V-blocks 24, 28
velocity systems, uniform 99–101
Vernier calliper 115
vice, machine 24, 28

Vickers pyramid test 6
visual images 52
volt 107
voltmeter 117

washing machines 126–7
Waste Agency 140–1
waste disposal 138, 139, 140, 141
waste management 139–42
Waste Regulatory Authority (WRA) 140–1
watt 98
weight 91
welded joints 72
welding 22, 31–2
wire, connecting 10, 13
wiring *see* schematic wiring diagram
wood composites 5
work 98
work holding 23–4, 28
workplace
  maintenance 44–5
  safety, comfort, health 38–9
wrench, torque 29